Communications
in Computer and Information Science **2006**

Rationale

The CCIS series is devoted to the publication of proceedings of computer science conferences. Its aim is to efficiently disseminate original research results in informatics in printed and electronic form. While the focus is on publication of peer-reviewed full papers presenting mature work, inclusion of reviewed short papers reporting on work in progress is welcome, too. Besides globally relevant meetings with internationally representative program committees guaranteeing a strict peer-reviewing and paper selection process, conferences run by societies or of high regional or national relevance are also considered for publication.

Topics

The topical scope of CCIS spans the entire spectrum of informatics ranging from foundational topics in the theory of computing to information and communications science and technology and a broad variety of interdisciplinary application fields.

Information for Volume Editors and Authors

Publication in CCIS is free of charge. No royalties are paid, however, we offer registered conference participants temporary free access to the online version of the conference proceedings on SpringerLink (http://link.springer.com) by means of an http referrer from the conference website and/or a number of complimentary printed copies, as specified in the official acceptance email of the event.

CCIS proceedings can be published in time for distribution at conferences or as post-proceedings, and delivered in the form of printed books and/or electronically as USBs and/or e-content licenses for accessing proceedings at SpringerLink. Furthermore, CCIS proceedings are included in the CCIS electronic book series hosted in the SpringerLink digital library at http://link.springer.com/bookseries/7899. Conferences publishing in CCIS are allowed to use Online Conference Service (OCS) for managing the whole proceedings lifecycle (from submission and reviewing to preparing for publication) free of charge.

Publication process

The language of publication is exclusively English. Authors publishing in CCIS have to sign the Springer CCIS copyright transfer form, however, they are free to use their material published in CCIS for substantially changed, more elaborate subsequent publications elsewhere. For the preparation of the camera-ready papers/files, authors have to strictly adhere to the Springer CCIS Authors' Instructions and are strongly encouraged to use the CCIS LaTeX style files or templates.

Abstracting/Indexing

CCIS is abstracted/indexed in DBLP, Google Scholar, EI-Compendex, Mathematical Reviews, SCImago, Scopus. CCIS volumes are also submitted for the inclusion in ISI Proceedings.

How to start

To start the evaluation of your proposal for inclusion in the CCIS series, please send an e-mail to ccis@springer.com.

Jia Jia · Zhenhua Ling · Xie Chen · Ya Li ·
Zixing Zhang
Editors

Man-Machine Speech Communication

18th National Conference, NCMMSC 2023
Suzhou, China, December 8–10, 2023
Proceedings

 Springer

Editors
Jia Jia
Tsinghua University
Beijing, China

Xie Chen
Shanghai Jiao Tong University
Shanghai, China

Zixing Zhang
Hunan University
Hunan, China

Zhenhua Ling
University of Science and Technology
of China
Anhui, China

Ya Li
Beijing University of Posts
and Telecommunications
Beijing, China

ISSN 1865-0929 ISSN 1865-0937 (electronic)
Communications in Computer and Information Science
ISBN 978-981-97-0600-6 ISBN 978-981-97-0601-3 (eBook)
https://doi.org/10.1007/978-981-97-0601-3

This Springer imprint is published by the registered company Springer Nature Singapore Pte Ltd.
The registered company address is: 152 Beach Road, #21-01/04 Gateway East, Singapore 189721, Singapore

Paper in this product is recyclable.

Preface

This volume contains the papers from the 18th National Conference on Man–Machine Speech Communication (NCMMSC 2023), the largest and most influential event on speech signal processing in China, which was hosted in Suzhou, China, December 8–10, 2023 by the Chinese Information Processing Society of China and China Computer Federation, co-organized by AISPEECH Co., Ltd. and Shanghai Jiao Tong University. It also received strong support from various leading domestic and international enterprises, universities, and institutions engaged in speech technology.

NCMMSC 2023 is also the annual academic meeting of the technical committee of Speech Dialogue and Auditory Processing of China Computer Federation (CCF TFS-DAP). This conference is centered on intelligent speech and language processing and other topics, attracting the active participation of experts and scholars engaged in speech technology-related industries at home and abroad, and jointly promoting the continuous innovation and development of speech technology in China.

Papers published in these proceedings of the "National Conference on Man–Machine Speech Communication (NCMMSC 2023)" are focused on the topics of speech recognition, synthesis, enhancement and coding, audio/music/singing synthesis, avatar, speaker recognition and verification, human–computer dialogue systems, large language models as well as phonetic and linguistic topics such as speech prosody analysis, pathological speech analysis, experimental phonetics, and acoustic scene classification. Each submission underwent a rigorous peer-review process, with each paper being single-blind evaluated by three or more reviewers, and each reviewer handled no more than six papers. This year, we received a record-breaking 207 submissions, including 119 in English. The number of full papers accepted by NCMMSC 2023 was 165, of which 106 were in English. Finally, only 31 papers were meticulously selected for inclusion in these proceedings.

The proceedings editors wish to thank the dedicated Scientific Committee members and all the other reviewers for their contributions. We also thank Springer for their trust and for publishing the proceedings of NCMMSC 2023.

December 2023

Jia Jia
Zhenhua Ling
Xie Chen
Ya Li
Zixing Zhang

Organization

Honorary Chairs

Jianwu Dang   Tianjin University, China
Fang Zheng   Tsinghua University, China

Conference Chairs

Changchun Bao   Beijing University of Technology, China
Jianhua Tao   Tsinghua University, China
Kai Yu   Shanghai Jiao Tong University, China

Program Chairs

Jia Jia   Tsinghua University, China
Zhenhua Ling   University of Science and Technology of China, China
Xie Chen   Shanghai Jiao Tong University, China

Organizing Chairs

Ming Li   Duke Kunshan University, China
Zhiyong Wu   Tsinghua Shenzhen International Graduate School, China
Shengrong Gong   Changshu Institute of Technology, CCF Suzhou Branch, China
Shuai Fan   AI Speech Co., Ltd., China

Industry Liaison Chairs

Lei Xie   Northwestern Polytechnical University, China
Chao Zhang   Tsinghua University, China
Rui Liu   Inner Mongolia University, China

Publicity Chairs

Yu Wang   Shanghai Jiao Tong University, China
Yuexian Zou   Peking University Shenzhen Graduate School,
                        China
Qingyang Hong   Xiamen University, China

Publication Chairs

Ya Li   Beijing University of Posts and
                        Telecommunications, China
Zixing Zhang   Hunan University, China

Tutorial Session Chairs

Yanmin Qian   Shanghai Jiao Tong University, China
Zhizheng Wu   Chinese University of Hong Kong, Shenzhen,
                        China

Special Session Chairs

Dong Wang   Tsinghua University, China
Xuan Zhu   Samsung R&D Institute, China

Youth Forum Chairs

Xiaolei Zhang   Northwestern Polytechnical University, China
Jun Du   University of Science and Technology of China,
                        China
Shengchen Li   Xi'an Jiaotong-Liverpool University, China

Student Forum Chairs

Longbiao Wang   Tianjin University, China
Jiangyan Yi   Institute of Automation, Chinese Academy of
                        Sciences, China
Shaofei Xue   AI Speech Co., Ltd., China

Finance Chair

Mengyue Wu · Shanghai Jiao Tong University, China

Contents

Ultra-Low Complexity Residue Echo and Noise Suppression Based on Recurrent Neural Network

Jianquan Zhou [ID], Yi Gao[⊠] [ID], and Siyu Zhang [ID]

Wecom, WXG, Tencent, Shenzhen, China
{damonjqzhou,jackyyigao,serriezhang}@tencent.com

Abstract. Deep learning residue echo suppression (RES) exhibits superior performance compared with traditional methods in recent years. However, a low-resource system or preemptive multi-tasking system requires low-complexity model that should be very computationally efficient to reduce race-condition issues which could cause system delay jitter and echo delay changes. In this paper we do an extensive study on low-complexity recurrent network models with different topologies, feed in with different combinations of the far-end signal, microphone signal, predicted linear echo and residue error signal. The proposed RES models can achieve comparable echo cancellation and noise reduction capabilities to the AEC Challenge 2022 baseline model at a complexity lower than 5% of the baseline model.

Keywords: AEC · residue echo suppression · recurrent neural network · GRU

1 Introduction

DNN-based RES exhibits superior performance in many systems. The benefits include: a) The RES model is able to leverage the information provided by the linear filter, then achieve better echo suppression. b) Noise suppression(NS) task could also be completed in this single model to save computation, c) An easily overlooked benefit is that for a commercial VOIP system, the hybrid system is more robust when any one of the Linear AEC(LAEC) or RES model is malfunctioning, this could reduce the chance of serious echo leakage.

In the latest AEC Challenges [1, 2], works that utilizing the LAEC-DNN RES hybrid architecture claim the model sizes of more than 1 million parameters [3–14]. The parameter size in [5] is relatively small, but it adopts a U-Net-based network backbone, with 1.6 GFLOPs computation complexity. [3] tried different configurations with sizes ranging from 800k to 8M. Even with the vectorized code, the computation of the optimal model is still about five times that of its linear AEC. Most of the RTF values listed in the literature show that the complexity of the models is high. We also deployed a transformer-based dual-path model in our product which is deeply optimized [15], but the computation is still high for mobile device applications.

© The Author(s), under exclusive license to Springer Nature Singapore Pte Ltd. 2024
J. Jia et al. (Eds.): NCMMSC 2023, CCIS 2006, pp. 1–8, 2024.
https://doi.org/10.1007/978-981-97-0601-3_1

For the low complexity AEC, the LSTM-based system in [16] has a computation of 31-758k MACs, which is comparable with the LAEC and is a good candidate for a commercial system.

In this paper, motivated by RNNoise [17] model devised for speech enhancement, which with non-vectorized C implementation claims a small parameter size (88k) and 17.5M FLOPs or 1.3% CPU consumption of a single x86 core (Haswell i7-4800MQ), we propose a set of RNN-based RES topologies and compare their performances. We do an extensive study on the RNN-based models with different configurations of the inputs, e.g., different combinations of the far-end signal, mic signal, prediction linear echo and residue error, and different network topologies. Performance of different models are evaluated.

2 Problem Formulation

The problem is formulated by

$$d(n) = x(n) * h + s(n) + v(n) \tag{1}$$

where $d(n)$ denotes the microphone input signal that includes near-end signal $s(n)$ and echo signal $y(n)$ and noise signal $v(n)$. h is the impulse response from the delay compensated far-end signal to the microphone input, $*$ denotes convolution.

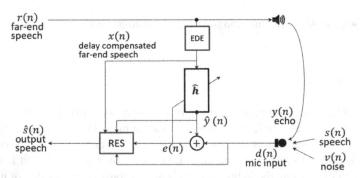

Fig. 1. Overview of the LAEC-RES system.

As shown in Fig. 1, adaptive filter $\hat{h}$ is an estimate of h.. Echo delay is estimated by the echo delay estimator (EDE) through far-end speech $r(n)$, and $d(n)$. Echo compensated far-end speech $x(n)$ is filtered by $\hat{h}$ to get the predicted echo $\hat{y}(n)$, which is then subtracted from $d(n)$ to get residue error $e(n)$. Because of the limited performance of the linear filter $\hat{h}$, residue echo may exist in $e(n)$ as well as near-end speech and noise signal. For this reason, we use an NN-based RES for further residue echo and noise suppression such that the output $\hat{s}(n)$ is as close as possible to the input speech $s(n)$.

3 Proposed Method

3.1 Network Topologies

The proposed RES models are listed in Fig. 2. Motivated by the RNNoise model, we adapt it to handle the residue echo and noise suppression tasks.

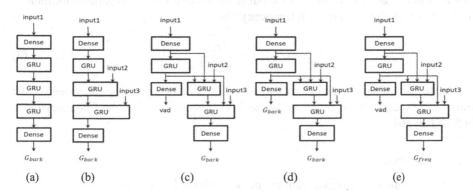

Fig. 2. The proposed RES model topologies

Figure 2(a) is of the simplest form where the input and output layers are dense and three hidden layers are GRU network. Intput1 represents the feature set by concatenating the features calculated from different streams, e.g., $d(n)$, $e(n)$, $x(n)$, and $\hat{y}(n)$. For a single stream, the feature calculation is similar with that of RNNoise but only 18 Bark-frequency cepstral coefficients(BFCC) are included to cover 8 kHz bandwidth, e.g., the audio is sampled at 16 kHz rate. It also includes the temporal derivative and the second temporal derivative of the first 6 BFCCs. It also includes the first 6 coefficients of the DCT of the pitch correlation across frequency bands. Finally, we include the pitch period as well as a spectral non-stationarity metric that can help in differentiating speech from noise and residue echo. In total, 38 input features are used for each stream, e.g., $\mathcal{F}_d$ is to indicate the 38-dim features calculated from one frame of $d(n)$, and the same for $\mathcal{F}_e$, $\mathcal{F}_x$, and $\mathcal{F}_{\hat{y}}$. The input1 could be the concatenation of the features from all the four streams, e.g., $[[\mathcal{F}_d\ \mathcal{F}_e, \mathcal{F}_x\ \mathcal{F}_{\hat{y}}]]$, or just from some of them.

The output gains are for 18 Bark bands, $\hat{G}_{bark}(b)$, represented as the square root of the ideal ratio mask (IRM) for each band b:

$$\hat{G}_{bark}(b) = \sqrt{\frac{E_s(b)}{E_e(b)}} \qquad (2)$$

where $E_s(b)$ is the energy of the near-end speech frame and $E_e(b)$ is the energy of the residue error frame. We tried to use $E_d(b)$ as the denominator but seems it causes some over-suppression on near-end speech.

In Fig. 2(b), input2 is concatenated and fed into the second GRU layer, and input3 is concatenated and fed into the third GRU layer. Intput2 or input3 could be $\mathcal{F}_e$ or $\mathcal{F}_{\hat{y}}$.

In Fig. 2(c), the topology is the similar with that of RNNoise model, the network includes a VAD output, that may improve training by ensuring that the corresponding GRU indeed learns to discriminate speech from noise and residue echo.

In Fig. 2(d), the network includes a branch of gains output instead of VAD output to see if it's better to help the corresponding GRU to detect near-end speech.

In Fig. 2(e), in order to increase the resolution of the output gains, based on the topology of Fig. 2(d), we replace the output gains, which are subband gains as formulated by Eq. (2), with the gains of each frequency bin:

$$G_{freq}(k) = \sqrt{\frac{E_s(k)}{E_e(k)}} \tag{3}$$

At the same time, the input2 and input3 are replaced with the log-magnitude of the frame of $\hat{y}(n)$ and $e(n)$, denoted as $\mathcal{M}_{\hat{y}}$ and $\mathcal{M}_e$ respectively. k is the frequency index in STFT domain.

3.2   Loss Functions

For Fig. 2(a) and Fig. 2(b), we train the model with the loss function:

$$L_{GB} = \sum_{b=1}^{K} \left(G_{band}(b)^{0.5} - \hat{G}_{band}(b)^{0.5} \right)^2 \tag{4}$$

here, $\hat{G}_{band}(b)$ is the gain estimate for each bark band, e.g., $\hat{G}_{bark}(b)$, and $G_{band}(b)$ is the corresponding groundtruth. K is the number of BFCC.

For Fig. 2(c) we use the loss function:

$$L_{final} = \alpha \times L_{GL} + \beta \times L_{GB} \tag{5}$$

where L_{GL} is the VAD loss by binary cross-entropy function. α and β are the weights, larger β is to emphasis the gain loss.

Similarly, for Fig. 2(d), L_{GL} is the gain loss at the left branch. For Fig. 2(e), $G_{band}(b)$ is the gain for each frequency, K is the number of frequency bins.

4   Experiments

To illustrate the performance of the different topologies with different input-output combinations, the training date include an 700 h audio database recorded in real VOIP scenarios, including recordings from PC, Macbook, iPhone, and Android mobile devices, as well as external audio equipment such as iLoud Stereo Bluetooth speakers, AdamX5 loudspeaker, Genelek high-fidelity loudspeaker, and Edifier desktop speakers. In addition, we also use the training data from the AEC Challenge 2022 single-talk dataset, which includes recordings from both PC and mobile devices. All data were converted to 16 kHz sampling rate data. Then echo data is mixed with the clean near-end speech data at a random signal-to-echo ratio from −20 dB to 40 dB. A commercial 160-h noise database is used for training, which is mixed with the microphone speech at SNR ranging from −5 dB to 40 dB. By doing so, the proposed RES models have the ability to suppress both reside echo and noise.

A simple linear post-processing is applied to the output of the model $\hat{G}(k)$ to better remove the echo and noise:

$$\tilde{G}(k) = \begin{cases} 0 & \hat{G}(k) < 0.20 \\ \frac{\hat{G}(k)-0.20}{0.25} & 0.20 \le \hat{G}(k) < 0.45 \\ 1 & \hat{G}(k) \ge 0.45 \end{cases} \qquad (6)$$

AEC Challenge 2022 blind test samples are used for evaluation. The objective measures include AECMOS, fwsegSNR and ERLE. To measure the STOI, test data are synthesized by mixing AEC Challenge 2022 near-end clean speech data with the echo data from AEC Challenge 2022 blind test far-end single talk folder. AEC Challenge far-end singletalk samples and doubletalk samples are used for evaluation, but a few samples with non-causal echo delay is excluded.

EDE is a fine-tuned version of algorithm based on counting of the co-occurrences of "signal present" events in the $d(n)$ and $r(n)$ signals [18]. Signals are aligned at a granularity of 10 ms block. The LAEC is based on the SpeexDSP LAEC [19] where the default filter length is 1920 with frame size of 160.

The Adam optimizer with a learning rate of 0.0001 is used to train the models. α is set as 0.05 and β is 0.95 in Eq. (5).

As the LSTM_DY [15] model used for comparison is lack of noise reduction capability, its output was further processed by RNNoise model to suppress noise. The first 5 s of each output audio is deleted before calculating the objective measures to allow sufficient convergence of the LAEC filter.

4.1 Complexity

The default configuration of the network layers from top to bottom is Dense(18), GRU(18), GRU(36), GRU(72), and Dense(18). The number of parameters for different topology are listed in Table 1.

Table 1. Parameter Size of different models

Algorithm	Params
Figure 2(a)/(b)/(c)/(d)/(e)	36K/48K/54K/54K/104K
RNNoise [17]	88K
LSTM_DY [15]	528K
Baseline [21]	1.3M

As a comparison, we listed the size of the RNNoise model, as well as that of the commercially used system, LSTM_DY [15], which adopts a Dense-LSTM-Dense structure and uses $d(n)$ and $\hat{y}(n)$ as model input, its best model has about 528k parameters, and the objective measures in next section are tested with this model. They are also compared with the Baseline model [21], which is provided by the AEC Challenge 2022 organizers based on a 3-layer GRU network.

The non-vectorized C/C++ code of the topology in Fig. 2(c) is executed on Intel(R) Core(TM) i5-10210U CPU@1.60 GHz. The Real-Time Factor is 0.014, including FFT/IFFTs and network inference. The vectorized (X86 SSE2 Intrinsics) code achieves a lower RTF as 0.008, which makes it to be capable of running smoothly on a PC or mobile device from low-tier to high-tier configuration.

4.2 Results

As Table 2 shows, with the topology shown in Fig. 2(c), where Input1 uses $[[\mathcal{F}_d, \mathcal{F}_e, \mathcal{F}_x, \mathcal{F}_\varphi]]$, Input2 uses $\mathcal{F}_{\hat{y}}$, and Input3 uses $\mathcal{F}_e$, the AECMOS score outperforms the Baseline model in both single talk and double talk scenarios. At the same time, its standard deviation (σ) is also the smallest, that means the performance is more stable. So we use this topology as the default model for the following comparison.

In the far-end single-talk scenario, the ERLE is 60.07dB, which is significantly higher than that of the baseline and the standard deviations are very close.

In double talk scenario, we are more concerned with the intelligibility score. For STOI [20] measurement, we use randomly selected utterances from the AEC Challenge 2022 synthetic near-end speech corpus as clean speech $s(n)$, then mixed with the microphone signal, e.g., above AEC Challenge 2022 blind test farend single talk data, at the signal-to-echo-ratio (SER) of 0 dB, 8 dB, 16 dB and 30 dB to simulate the double talk scenario. The results show that the proposed RES models have STOI values which are close to that of the Baseline model.

Table 2. RES performance. DT: double talk, ST: single-talk, NE: near-end, FE: far-end. AECMOS used for ST-FE and DT scenarios, fwsegSNR used for ST-NE scenario, ERLE used for ST-FE scenario and STOI used for DT scenario in ICASSP 2022 blind test set.

Model	Input1	Input2	Input3	AECMOS ST FE		AECMOS DT		ERLE ST FE		STOI DT	
				Avg	σ	Avg	σ	Avg	σ	Avg	σ
Figure 2(a)	$[[\mathcal{F}_d, \mathcal{F}_e, \mathcal{F}_x, \mathcal{F}_\varphi]]$	–	–	3.98	0.53	4.45	0.29	54.13	**17.3**	0.73	0.16
	$[[\mathcal{F}_d, \mathcal{F}_e, \mathcal{F}_\varphi]]$	–	–	4.15	0.46	4.61	0.27	58.7	19.36	0.72	0.17
	$[[\mathcal{F}_e, \mathcal{F}_x]]$	–	–	4.12	0.55	4.55	0.31	**60.27**	20.36	0.74	0.17
Figure 2(b)	$[[\mathcal{F}_d, \mathcal{F}_e, \mathcal{F}_x, \mathcal{F}_\varphi]]$	$\mathcal{F}_{\hat{y}}$	$\mathcal{F}_e$	4.23	**0.44**	**4.64**	**0.23**	58.21	19.84	**0.75**	0.16
Figure 2(c)	$[[\mathcal{F}_d, \mathcal{F}_e, \mathcal{F}_x, \mathcal{F}_\varphi]]$	$\mathcal{F}_e$	$\mathcal{F}_{\hat{y}}$	4.14	0.45	4.56	0.26	59.41	18.41	0.72	0.17
	$[[\mathcal{F}_d, \mathcal{F}_e, \mathcal{F}_x, \mathcal{F}_\varphi]]$	$\mathcal{F}_{\hat{y}}$	$\mathcal{F}_e$	**4.29**	**0.41**	**4.66**	**0.21**	**60.07**	18.83	0.74	0.17
	$[[\mathcal{F}_d, \mathcal{F}_e, \mathcal{F}_\varphi]]$	$\mathcal{F}_{\hat{y}}$	$\mathcal{F}_e$	4.21	0.46	4.61	0.24	59.7	19.47	**0.75**	0.16
	$[[\mathcal{F}_e, \mathcal{F}_x]]$	$\mathcal{F}_{\hat{y}}$	$\mathcal{F}_e$	4.16	0.45	4.58	0.27	57.88	18.55	0.74	0.17
Figure 2(d)	$[[\mathcal{F}_d, \mathcal{F}_e, \mathcal{F}_x, \mathcal{F}_\varphi]]$	$\mathcal{F}_{\hat{y}}$	$\mathcal{F}_e$	4.02	0.46	4.51	0.27	57.68	19.37	0.74	0.18
Figure 2(e)	$[[\mathcal{F}_d, \mathcal{F}_e, \mathcal{F}_x, \mathcal{F}_\varphi]]$	$\mathcal{M}_{\hat{y}}$	$\mathcal{M}_e$	3.60	0.64	4.26	0.47	52.55	18.89	0.74	0.18
LSTM_DY[15]				3.17	0.82	4.05	0.68	30.95	**10.69**	**0.75**	**0.08**
Baseline[21]				**4.22**	0.61	4.58	0.33	41.98	20.22	**0.76**	**0.1**

We use frequency-weighted segmental signal-to-noise ratio, fwsegSNR [20], for noise suppression capability evaluation. As shown in Table 3, the proposed RES model list in Fig. 2(c) have the highest average fwsegSNR improvement, that means the noise suppression is better than the Baseline and the LSTM_DY model, even slightly better than that of RNNoise model, at the same time, the standard deviation is comparable with that of the RNNoise model.

The PESQ score of the proposed RES models is slightly worse than that of RNNoise, but better than the Baseline and LSTM_DY models. That means the proposed RES model could achieve a good noise-suppression capability.

Table 3. fwsegSNR Improvement and PESQ for different models

	fwsegSNR Improv.(dB)		PESQ	
	Avg	σ	Avg	σ
Figure 2(c)	**3.02**	3.39	**2.92**	**0.55**
LSTM_DY [15]	0.62	**0.25**	2.76	0.62
Baseline [21]	0.21	**0.44**	2.89	0.60
RNNoise [17]	**2.93**	3.48	**2.99**	**0.54**

5   Conclusions

In this paper, we aimed to design an ultra-low complexity RES network for an efficiency demanding AEC system. To achieve this, we extend the RNNoise model and proposed a set of RNN-based RES topologies. These models use different form of input, including different combination of the far-end signal, mic signal, prediction linear echo and residue error. The test results show that the proposed RES models shown in Fig. 2(c) can achieve good echo cancellation and noise suppression performance comparing with the AEC Challenge 2022 Baseline model and the smaller LSTM_DY model. The complexity of our best model is only around 5% of the Baseline model, also much smaller than the LSTM_DY model. It can run smoothly on mid/low-tier mobile devices.

References

1. Sridhar, K., et al.: ICASSP 2021 acoustic echo cancellation challenge: datasets, testing framework, and results. In: ICASSP 2022, Singapore, Singapore, pp. 9107–9111 (2022)
2. https://www.microsoft.com/en-us/research/academic-program/acoustic-echo-cancellation-challenge-icassp-2023/
3. Valin, J.-M., et al.: Low-complexity, real-time joint neural echo control and speech enhancement based on percepnet. In: ICASSP 2021–2021 IEEE International Conference on Acoustics, Speech and Signal Processing (ICASSP). IEEE (2021)
4. Peng, R., et al.: ICASSP 2021 acoustic echo cancellation challenge: integrated adaptive echo cancellation with time alignment and deep learning-based residual echo plus noise suppression. In: ICASSP 2021, Toronto, Canada, pp. 146–150 (2021)

5. Halimeh, M.M., et al.: Combining adaptive filtering and complex-valued deep postfiltering for acoustic echo cancellation. In: ICASSP 2021–2021 IEEE International Conference on Acoustics, Speech and Signal Processing (ICASSP). IEEE (2021)
6. Zhang, G., et al.: Multi-scale temporal frequency convolutional network with axial attention for speech enhancement. In: ICASSP 2022–2022 IEEE International Conference on Acoustics, Speech and Signal Processing (ICASSP). IEEE (2022)
7. Zhao, H., et al.: A deep hierarchical fusion network for fullband acoustic echo cancellation. In: ICASSP 2022–2022 IEEE International Conference on Acoustics, Speech and Signal Processing (ICASSP). IEEE (2022)
8. Zhang, S., et al.: Multi-task deep residual echo suppression with echo-aware loss. In: ICASSP 2022–2022 IEEE International Conference on Acoustics, Speech and Signal Processing (ICASSP). IEEE (2022)
9. Sun, X., et al.: Explore relative and context information with transformer for joint acoustic echo cancellation and speech enhancement. In: ICASSP 2022–2022 IEEE International Conference on Acoustics, Speech and Signal Processing (ICASSP). IEEE (2022)
10. Zhao, H., et al.: A low-latency deep hierarchical fusion network for fullband acoustic echo cancellation. In: ICASSP 2023–2023 IEEE International Conference on Acoustics, Speech and Signal Processing (ICASSP) (2023)
11. Chen, Z., et al.: A progressive neural network for acoustic echo cancellation. In: ICASSP 2023–2023 IEEE International Conference on Acoustics, Speech and Signal Processing (ICASSP) (2023)
12. Sun, J., et al.: Multi-task sub-band network for deep residual echo suppression. In: ICASSP 2023–2023 IEEE International Conference on Acoustics, Speech and Signal Processing (ICASSP) (2023)
13. Xu, W., Guo, Z.: Tayloraecnet: a taylor style neural network for full-band echo cancellation. In: ICASSP 2023–2023 IEEE International Conference on Acoustics, Speech and Signal Processing (ICASSP) (2023)
14. Zhang, Z., et al.: Two-step band-split neural network approach for full-band residual echo suppression. In: ICASSP 2023–2023 IEEE International Conference on Acoustics, Speech and Signal Processing (ICASSP) (2023)
15. Chen, H., et al.: Ultra dual-path compression for joint echo cancellation and noise suppression. arXiv preprint arXiv:2308.11053 (2023)
16. Pfeifenberger, L., Pernkopf, F.: Nonlinear residual echo suppression using a recurrent neural network. In: Interspeech, pp. 3950–3954 (2020)
17. Valin, J.-M.: A hybrid DSP/deep learning approach to real-time full-band speech enhancement. In: 2018 IEEE 20th International Workshop on Multimedia Signal Processing (MMSP), Vancouver, BC, Canada, pp. 1–5 (2018)
18. Voelcker, B., Kleijn, W.B.: Robust and low complexity delay estimation. In: IWAENC 2012; International Workshop on Acoustic Signal Enhancement, Aachen, Germany, pp. 1–4 (2012)
19. Speex-dsp. Website. Accessed 12 Feb 2023. https://github.com/xiph/speexdsp
20. Loizou, P.C.: Speech Enhancement: Theory and Practice, pp. 1–394. CRC Press, Taylor Francis Group, Boca Raton (2007)
21. Sridhar, K., et al.: ICASSP 2021 acoustic echo cancellation challenge: Datasets, testing framework, and results. In: ICASSP 2021–2021 IEEE International Conference on Acoustics, Speech and Signal Processing (ICASSP). IEEE (2021)

Semi-End-to-End Nested Named Entity Recognition from Speech

Min Zhang[✉], XiaoSong Qiao, Yanqing Zhao, Chang Su, Yuang Li, Yinglu Li, Mengyao Piao, Song Peng, Shimin Tao, and Hao Yang

Huawei Translation Services Center, Beijing, China
{zhangmin186,qiaoxiaosong,zhaoyanqing,suchang8,liyuang3,
liyinglu,piaomengyao1,pengsong2,taoshimin,yanghao30}@huawei.com

Abstract. There are two approaches for Named Entity Recognition (NER) from speech: two-step pipeline and End-to-End (E2E). In the pipeline approach, cascading errors are inevitable. In the E2E approach, its annotation method poses a challenge to Automatic Speech Recognition (ASR) when Named Entities (NEs) are nested. This is because multiple special tokens without audio signals between words will exist, which may even cause ambiguity problems for NER. In this paper, we propose a new paradigm and name it semi-E2E, as it completes parts of NER in ASR. Specifically, we introduce a novel annotation method for nested NEs, where only two special tokens are used to annotate the heads (The head of an NE is its first word, for examples, "western" is the head of the NE "western Canadian" in Fig. 1. If an NE has only one word, its head is itself.) of NEs, regardless of the number of NE categories. Also, we use a span classifier to classify only the spans that start with the predicted heads in transcriptions. From the experimental results on the nested NER dataset of Chinese speech CNERTA, our semi-E2E approach gets the best $F1$ score (1.84% and 0.53% absolute points higher than E2E and pipeline respectively).

Keywords: semi-E2E · nested NER · NE head annotation · span classification · speech recognition

1 Introduction

As a fundamental subtask of information extraction, Named Entity Recognition (NER) aims to locate and classify NEs appearing in unstructured texts into predefined semantic categories, such as *person* (PER), *location* (LOC), *organization* (ORG), and *geo-political entity* (GPE). NER plays a crucial role in Natural Language Processing (NLP) applications (information retrieval, question and answering, etc.) [15] and has been well studied in the context of written languages [23]. It is pointed out that nested NEs are very common in the field of NER [6], i.e., an NE can contain or embed other NEs. As illustrated in Fig. 1, there are four nested NEs.

© The Author(s), under exclusive license to Springer Nature Singapore Pte Ltd. 2024
J. Jia et al. (Eds.): NCMMSC 2023, CCIS 2006, pp. 9–22, 2024.
https://doi.org/10.1007/978-981-97-0601-3_2

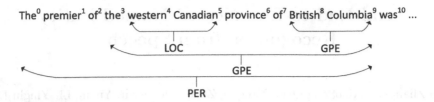

Fig. 1. A sentence from ACE2005 [4] with 4 nested NEs. The superscript of each word indicates its index in the sentence.

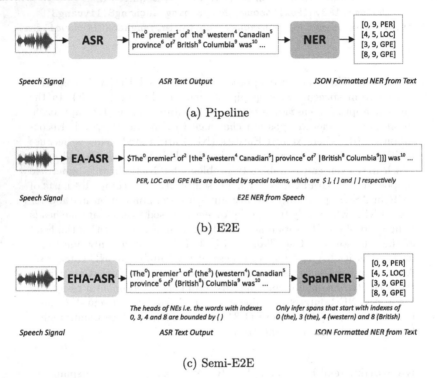

Fig. 2. (a) Pipeline, (b) E2E and (c) Semi-E2E for nested NER from speech. The NEs in (a) and (c) are denoted by triples (start indexes, end indexes and NE types).

NER from speech has been actively studied in the Spoken Language Understanding field [2], and is usually implemented through a two-step pipeline (as shown in part (a) of Fig. 2) that consists of (1) processing the audio using an ASR system and (2) applying an NER tagger to the ASR output. However, there are cascading errors in this pipeline approach, i.e., NER results will be greatly influenced by the quality of ASR outputs.

Recently, the End-to-End (E2E) approach for NER from French [8], English [22] and Chinese [3,24] speech has been proposed, which adopts Entity-aware ASR (EA-ASR) to predict NEs during the decoding process. In EA-ASR, special tokens are added to the ASR vocabulary list to annotate NEs in the transcriptions. However, these studies only consider the case of flat NEs, where there are no more than 2 special tokens between two words. When the NEs are nested, this annotation method poses a challenge to EA-ASR because there are more special tokens without any audio signals between two words. For example, the sentence in part (b) of Fig. 2 shows that there are 3 special tokens (|||) between the words "Columbia^9" and "was^{10}". Furthermore, this annotation method even causes ambiguity problems for NER. For example, according to the annotation in part (b) of Fig. 2, both "the^3 western^4 Canadian^5" and "western^4 Canadian^5" could be recognized as NEs. However, only "western^4 Canadian^5" is an NE actually.

In this paper, we propose a semi-E2E approach to alleviate this challenge, which consists of Entity Head-aware ASR (EHA-ASR, developed by us) and a span classifier for NER (SpanNER), as illustrated in part (c) of Fig. 2. EHA-ASR is used to predict the heads of NEs during the ASR decoding process, where only two special tokens are needed to annotate the heads of NEs in transcriptions. And the span classifier is used to classify the spans that start with the predicted heads of NEs. Since the heads of NEs are predicted in ASR, i.e., parts of NER are completed by EHA-ASR, we name this approach semi-E2E. We compare this approach with the pipeline and E2E approaches on the nested NER dataset CNERTA [19] of Chinese speech. Experimental results show that our semi-E2E approach gets the best $F1$ score (1.84% and 0.53% absolute points higher than E2E and pipeline respectively).

The primary contributions of this work can be summarized as follows:

- To the best of our knowledge, we are the first to propose a semi-E2E approach for nested NER from speech, which is a new paradigm different from the pipeline and E2E approaches.
- The annotation for nested NEs is novel, scalable and robust: It only needs two special tokens regardless of the number of NE types, and does not have ambiguity problems.
- The experimental results on the dataset CNERTA demonstrate that our semi-E2E approach outperforms the pipeline and E2E approaches on $F1$ scores.

2 Related Work

2.1 NER

There has been a long history of research involving NER [16]. Traditional approaches are based on Hidden Markov Model (HMM) [31] or Conditional Random Field (CRF) [14]. With the rapid development of deep learning technology [11], sequence labeling methods such as LSTM-CRF [12] and BERT-LSTM-CRF [5] achieve very promising results in the field of NER. However, these methods cannot directly handle the nested structure because they can only assign one label to each token.

As NEs are often nested [6], various approaches for nested NER have been proposed in recent years [21]. One of the most representative directions is Span-NER that recognizes nested entities by classifying sub-sequences of a sentence [7,27,28]. SpanNER methods are naturally suitable for the nested structure because nested entities can be easily detected in different sub-sequences, and the strengths and weaknesses of these methods are systematically investigated in [7].

It should be pointed out that the above methods are designed in the context of written languages.

2.2   NER from Speech

There are two categories of methods for NER from speech: two-step pipeline [26] and E2E [3,8,22].

The two-step pipeline approach consists of processing the given audio on an ASR system and then feeding the transcribed ASR output to the NER tagger. Although ASR and NER are well studied in their respective fields [7,9,20,30], there is little research on how to better combine ASR and NER for better results, and the pipeline approach suffers from cascading errors.

Recently, the E2E approach is proposed for NER from French [8], English [22] and Chinese [3,24] speech. This method adds special tokens to the ASR vocabulary list to annotate NEs in transcriptions, and adopts EA-ASR to predict NEs during the decoding process. However, this method only discusses the case of flat NEs, and does not consider the case of nested NEs. And the annotation methods [3,8,22] may result in ambiguity problems when NEs are nested. In this paper, we propose a semi-E2E approach for nested NER from speech, where a novel robust annotation method for nested NEs is developed.

Character sequence: An[0] IL-2[1] promoter[2] bearing[3] a[4] defective[5] NF-[6] chi[7] B[8] site[9] was[10] completely[11] inactive[12] in[13] EBV-transformed[14] B[15] cells[16] ,[17] while[18] it[19] still[20] had[21] activity[22] in[23] Jurkat[24] T[25] cells[26] .[27]

NEs: [1, 1, protein], [1, 2, DNA], [7, 8, DNA], [6, 8, protein], [6, 9, DNA], [15, 16, cell_type], [14, 16, cell_line], [24, 26, cell_line]

Our annotation: An[0] (IL-2[1]) promoter[2] bearing[3] a[4] defective[5] (NF-[6]) (chi[7]) B[8] site[9] was[10] completely[11] inactive[12] in[13] (EBV-transformed[14]) (B[15]) cells[16] ,[17] while[18] it[19] still[20] had[21] activity[22] in[23] (Jurkat[24] T[25] cells[26] .[27]

Annotation in [7]: An[0] ($IL-2[1]) promoter[2]) bearing[3] a[4] defective[5] ($NF-[6] $chi[7] B[8]]] site[9]) was[10] completely[11] inactive[12] in[13] EBV-<transformed[14] {B[15] cells[16]]] ,[17] while[18] it[19] still[20] had[21] activity[22] in[23] {Jurkat[24] T[25] cells[26]] .[27]

Fig. 3. Our annotation and the annotation in [22] for a sentence from GENIA [13] with 7 nested NEs out of 8. Our annotation has less special tokens than the annotation in [22].

3 Methodology

The semi-E2E approach for nested NER from speech is illustrated in part (c) of Fig. 2, which consists of two components, EHA-ASR and SpanNER. EHA-ASR aims to predict the heads of NEs while recognizing the speech, and SpanNER aims to classify the spans that start with the predicted NE heads.

3.1 NE Head Annotation

In this section, we design a novel annotation method for heads of nested NEs in transcriptions, which is very simple and only needs two special tokens to annotate the heads of NEs. Figure 3 provides an example for this annotation, where the sequence has 28 words, with 8 NEs of 4 types identified by triples (start indexes, end indexes and NE types, for example, in the row 2 of this figure, $[1, 1, \text{protein}]$ denotes NE "IL-2^1" of type protein).

In the row 3 of Fig. 3, the heads of the 8 NEs are annotated by two special tokens, i.e., ( ). Since there are nested NEs that have the same heads ($[1, 1, \text{protein}]$ and $[1, 2, \text{DNA}]$, $[6, 8, \text{protein}]$ and $[6, 9, \text{DNA}]$), only 6 heads need to be annotated.

For comparison, in the row 4 of Fig. 3, we show the results of the annotation in [22] for this example, where 5 special tokens are used to annotate the 4 types of NEs, i.e., (] for protein, $] for DNA, <] for cell_type and {] for cell_line. It could be clearly seen that special tokens used in our annotation are less than that in the compared annotation (12 vs. 16), and the number of cases where more than 1 special token is contained between two words is less in our annotation than that in the compared annotation (2 vs. 4). Since these special tokens do not have any audio signals, our annotation is more concise and reduces the challenge to ASR.

More importantly, when NEs are nested, the compared annotation may result in ambiguity problems, but our annotation will not. For the sentence in Fig. 3, the sub-sequence "($NF-^{6} chi7 B^8]] site^9]" in the compared annotation can be parsed as NEs "chi B", "NF- chi B" and "NF- chi B site", or as NEs "chi B", "NF-chi B" and "chi B site". And these problems also exist in the annotation methods proposed in [8, 22]. Since only the heads of NEs are annotated, our annotation does not have these problems and is robust for nested NEs.

It should be noted that our annotation does not annotate NE types in NE heads because the same NE heads can belong to different types of NEs. For example, in Fig. 3, the word "IL-2^1" is the head of NE "IL-2^1" of type protein and the head of NE "IL-2^1 promoter^2" of type DNA. This is why SpanNER is needed in our approach.

3.2 EHA-ASR

E2E ASR systems based on neural networks have seen large improvements in recent years. Since Transformer has been shown to be more effective than

RNN/LSTM for ASR on various corpora [10,29], we build the ASR components with Transformer-based encoder-decoder architecture, where the encoder is enhanced with Convolution Neural Network (CNN), namely, Conformer [9]. The Conformer encoder maps the input sequence of audio features to a sequence of hidden states by a convolution subsampling layer and a number of conformer blocks. A conformer block is composed of four modules stacked together, i.e., a feed-forward module, a self-attention module, a convolution module, and a second feed-forward module at the end. The decoding process is enhanced with the joint CTC/attention approach and the Transformer-based Language Model (LM).

Our method EHA-ASR does not require ASR framework modification and just adds two special tokens to the ASR vocabulary list to identify the heads of NEs. EHA-ASR learns an alignment between the speech and the annotated transcription. During the decoding, as illustrated in part (c) of Fig. 2, the heads of NEs in the transcription are bounded by the two special tokens, which are used for SpanNER to predict NEs.

3.3 SpanNER

In this paper, we adopt the framework of SpanNER in [7], which consists of three major modules: token representation layer, span representation layer and span prediction layer.

Token Representation Layer. Considering a sentence $X = \{x_1, \cdots, x_n\}$ with n tokens, we convert the tokens to their contextual embeddings with BERT [5] encoder. We generate the input sequence by concatenating a CLS token, $\{w_i\}_{i=1}^n$ and a SEP token, and use a series of L stacked Transformer blocks (TBs) to project the input to a sequence of contextual vectors, i.e.,

$$h_0, \cdots, h_n = TB_L(\text{[CLS]}, w_1, \cdots, w_n, \text{[SEP]}). \tag{1}$$

Span Representation Layer. First, we enumerate all the possible m spans $S = \{s_1, \cdots, s_i, \cdots, s_m\}$ for sentence $X = \{x_1, \cdots, x_n\}$ and then re-assign a semantic tag for each span s_i. For example, in the sentence "This^0 is^1 London^2", the possible span's (start, end) indexes are $\{(0,0), (1,1), (2,2), (0,1), (1,2), (0,2)\}$, and the tags of these spans are all "O" except that $(2,2)$ (London) is "LOC". Then we use b_i and e_i to denote the start and end indexes of the span s_i, and represent the span with boundary embeddings and span length embedding as:

$$s_i = [h_{b_i}; h_{e_i}; z_{e_i - b_i + 1}], \tag{2}$$

where $z_{e_i - b_i + 1}$ is the span length embedding, which could be obtained by a learnable look-up table.

Table 1. Statistics of the dataset CNERTA, where "Avg" is for average, "Sent" for sentence, "Len" for length, "Prop" for proportion and "#" for number

	Train	Dev	Test
Audio Duration	56.68 h	7.50 h	7.59 h
Avg Sent Len	19.69	19.77	19.75
Max Sent Len	39	44	39
Prop Nested NEs	31.25%	29.50%	28.35%
# Instance	34,102	4,440	4,445
# Entity	23,805	5,889	7,263
# ORG	7,066	2,187	2,794
# PER	5,846	1,116	1,072
# LOC	10,893	2,586	3,397

Span Prediction Layer. We feed the span representation s_i into a Multi-Layer Perceptron (MLP) classifier, and apply a softmax layer to obtain the probabilities p on all semantic tags:

$$p = \mathrm{softmax}(\mathrm{MLP}(s_i)). \tag{3}$$

During training, we minimize the following cross-entropy loss function:

$$\mathcal{L} = -\sum_{i=1}^{k} y_t \log(p_t), \tag{4}$$

where k is the number of semantic tags, and y_t denotes a label indicating whether the span s_i is in tag t.

During inferring, only the spans that start with the NE heads predicted by EHA-ASR are used for NE classification.

4 Experiments

4.1 Experimental Setup

In this paper, we compare the semi-E2E approach with the pipline and E2E approaches on the dataset CNERTA [19], which is a large-scale human-annotated Chinese multimodal nested NER dataset (42,987 annotated sentences accompanying by 71 h of speech data). Table 1 shows the high level statistics of data splits for CNERTA. To the best of our knowledge, CNERTA is currently the only dataset for nested NER from speech.

CNERTA[1] is based on the AISHELL-1 [1] dataset, which is an open-source speech corpus containing over 170 h of Chinese speech data and has been commonly used for evaluating the performances of ASR systems on Chinese. The

---

[1] https://github.com/DianboWork/CNERTA.

Table 2. # Special Tokens and # Multiple in the ground-truth transcriptions of CNERTA for EA-ASR and EHA-ASR

Approach	# Special Tokens			# Multiple (>1)		
	Train	Dev	Test	Train	Dev	Test
EA-ASR	47,610	11,778	14,526	4,324	1,727	2,376
EHA-ASR	42,648	9,286	11,162	253	116	180
	↓**10.4%**	↓**21.2%**	↓**23.2%**	↓**94.2%**	↓**93.3%**	↓**92.4%**

corpus covers five domains: "finance", "science and technology", "sports", "entertainments" and "news".

The Conformer-based ASR models for the pipeline, E2E and semi-E2E approaches are built with WeNet [25] and make use of both the joint CTC/attention decoding and LM. The encoders have 4 attention heads with 2048 linear units. They stack 12 layers of encoder blocks, and their output size is 256. The decoders use the same configuration, except that they have only 6 layers. SpecAugment is used to randomly apply time and frequency masking blocks to log mel-filterbank features. Speed perturbation with factors of 0.9, 1.0 and 1.1 is also applied to Conformer, resulting in 3-fold data augmentation. The CNN module of the Conformer encoder uses a kernel size of 15. These ASR models are trained with the Adam optimizer, with the learning rate set to 0.0005 and a linear warmup scheduler. During decoding, the width of the beam search is set to 5.

The pretrained model `bert-base-chinese`[2] is used for the encoder of the SpanNER model, where the embedding size of the span width is set to 50, the max span width is set to 20 and the MLP for span classification takes two layers. The SpanNER model is trained on the transcriptions of CNERTA, which is optimized by AdamW with the learning rate set to 0.00001 and a linear warmup scheduler. The training epoch number is set to 10, and the batch size is set to 16. For fair comparison, the SpanNER model is also chosen for the NER tagger in the pipeline approach. During inference, all possible spans are used in the pipeline approach, while only the spans starting with the heads predicted by NEA-ASR are used in the semi-E2E approach.

The commonly used metrics Character Error Rate (CER) [17] and $F1$ score [18] are chosen to evaluate ASR and NER respectively.

4.2 Annotation Results

In EA-ARS, we adopt the annotation method in [3] for Chinese with 6 special tokens, i.e., [] for PER, ( ) for LOC, and < > ORG. In EHA-ASR, we use two special tokens ( ) for heads of NEs. Table 2 reports the number of special tokens (denoted as # *Special Tokens*) and the number of consecutive occurrences of more than 1 special token (denoted as # *Multiple*) in EA-ASR and EHA-ASR on the ground-truth transcriptions of CNERTA.

---

[2] https://huggingface.co/bert-base-chinese.

Table 3. CER(%) of ASR, EA-ASR and EHA-ASR on CNERTA

Model Name	Approach	CER
ASR	Pipeline	8.22
EA-ASR	E2E	8.35
EHA-ASR	Semi-E2E	8.30

Table 4. NER results (precision, recall and $F1$ score) of the pipeline, E2E and semi-E2E approaches on CNERTA

Approach	Precision	Recall	$F1$
Ground-Truth	**80.69**	**81.30**	**81.00**
Pipeline	61.11	**58.15**	59.59
E2E	73.67	48.21	58.28
Semi-E2E	**73.79**	50.73	**60.12**

From Table 2, it could be clearly seen that our annotation reduces not only "# Special Tokens", but also "# Multiple" significantly (more than 90% reduction). Since the special tokens have no audio signals, our annotation obviously reduces the challenge to ASR.

4.3   ASR Results

We compare the CER of ASR, EA-ASR and EHA-ASR models in the pipeline, E2E and semi-E2E approaches. EA-ASR and EHA-ASR are trained on CNERTA with special tokens indicating NEs or heads of NEs. So when CER is computed for EA-ASR and EHA-ASR on the test set, these special tokens are ignored. Table 3 shows CER for the three models on the test set of CNERTA. It could be seen that both EA-ASR and EHA-ASR are faced with a small loss (0.13% and 0.08% CER) in performance when being compared with ASR, which is consistent with the results in [3]. And EHA-ASR has a small advantage (0.05%) over EA-ASR in CER since less special tokens are annotated in the transcriptions.

4.4   NER Results

In this section, we compare the NER results of the pipeline, E2E and semi-E2E approaches on the test set of CNERTA. For the pipeline approach, the ASR model outputs the speech transcriptions, and then the SpanNER model predicts the NEs on all the possible spans. For the E2E approach, NEs can be directly identified by special tokens based on nearest greedy matching. However, the ambiguity problems are inevitable and affect the NER results. For the semi-E2E approach, the EHA-ASR model outputs the speech transcriptions which annotate the heads of NEs, and then the SpanNER model predicts the NEs on the spans that start with these heads.

Table 5. Effect of NE heads in the semi-E2E approach

SpanNER	Precision	Recall	$F1$
All Spans	61.15	**57.10**	59.06
Selected Spans	**73.79**	50.73	**60.12**

Table 6. Two cases for the pipeline, E2E and semi-E2E approaches on CNERTA, where the sentence

Case ID	Ground-Truth Labels	Method	ASR Text	NEs
1	链家地产市场研究部张旭向南都记者分析称	Pipeline	链家地产市场研究部张旭向南都记者分析称	链家地产:ORG, 链家地产市场研究部:ORG, 张旭:PER, 南都:ORG
	链家地产:ORG, 链家地产市场研究部:ORG, 张旭:PER	E2E	<<链家地产>市场研究部>张旭向<南都>记者分析称	链家地产:ORG, 链家地产市场研究部:ORG, 南都:ORG
	English: Zhang Xu from Lianjia Real Estate Market Research Department analyzed to Nandu reporter that (Lianjia Real Estate:ORG, Lianjia Real Estate Market Research Department:ORG, Zhang Xu:PER)	Semi-E2E	(链)家地产市场研究部(张)旭向(南)都记者分析称	链家地产:ORG, 链家地产市场研究部:ORG, 张旭:PER, 南都:ORG
2	六名驴友在广东英德峡谷中露营遇山洪失联	Pipeline	六名驴友在广东英德峡股中陆营玉山红失联	广东:LOC, 英德:LOC, 峡股:LOC, 玉山:LOC, 中陆营:LOC, 玉山红:LOC
	广东:LOC, 英德:LOC, 英德峡谷:LOC	E2E	六名驴友在(广东)(英德)峡股中[陆营玉山红失联	广东:LOC, 英德:LOC
	English: Six donkey friends lost contact in a flash flood while camping in Yingde Canyon, Guangdong (Guangdong:LOC, Yingde:LOC, Yingde Canyon:LOC)	Semi-E2E	六名女友在(广)东(英)德峡股中陆营遇山红失联	广东:LOC, 英德:LOC

Table 4 reports the precision, recall and $F1$ scores of the three approaches on CNERTA, as well as the results of the SpanNER model on the ground-truth speech transcriptions (denoted as "Ground-Truth" in Table 4). It can be clearly seen that results of "Ground-Truth" are much better than the three approaches. This is because the errors in ASR output affect the performance of NER.

As shown in Table 4, the semi-E2E approach gets the best $F1$ score among the three approaches (0.53% absolute points higher than pipeline, and 1.84% absolute points higher than E2E, which are significant improvements in the field of NER). Since the semi-E2E approach outperforms the E2E approach in precision, recall and $F1$ score (+0.12%, +2.52% and +1.84% respectively), our annotation of NE heads is more effective for nested NER. However, both the semi-E2E and E2E approaches have a severe drop in recall compared to the pipeline approach (-7.42% and -9.96%), despite a more significant improvement in precision (+12.68% and +12.56%). The high recall in the pipeline approach may be because it has the global text context for entire NER, which is not available for the semi-E2E and E2E approaches. This suggests that a more robust approach should be developed.

Table 7. Results in different NE categories of the pipeline, E2E and semi-E2E approaches on CNERTA

Method	Category	Precision	Recall	$F1$
Ground-Truth	PER	84.74	82.60	83.66
	LOC	83.52	82.99	83.26
	ORG	75.92	78.74	77.30
Pipeline	PER	43.63	42.00	42.80
	LOC	68.57	63.39	65.88
	ORG	59.21	58.05	**58.63**
E2E	PER	57.28	37.89	45.61
	LOC	79.25	55.35	65.18
	ORG	72.78	43.55	54.49
Semi-E2E	PER	55.63	38.82	**45.73**
	LOC	82.51	56.13	**66.81**
	ORG	70.48	48.78	57.66

In order to further analyze the effect of the annotation of NE heads in the semi-E2E approach, we report the NER results of the semi-E2E approach on the test set of CNERTA in Table 5 when the SpanNER model classifies all possible spans (denoted as "All Spans"). "Selected Spans" means the SpanNER model classifies the spans that start with the annotated NE heads by EHA-ASR. From Table 5, it could be seen that the annotated NE heads improve the $F1$ score by 1.06% absolute points, showing the effectiveness of this annotation. Meanwhile, as the CER of ASR outputs in the semi-E2E approach is sightly worse than that in the pipeline approach, the $F1$ score of "All Spans" is sightly worse than that of the pipeline approach (−0.53%).

4.5 Case Study

In this section, we show two cases from the test dataset of CNERTA for the pipeline, E2E and semi-E2E approaches in Table 6, where the ground-truth sentences and NEs are provided, and the error characters in ASR text and the error NEs recognized are in red for each approach.

From Table 6, it could be seen that the E2E approach fails to recognize the PER NE 张旭 in Case 1, whereas the pipeline and semi-E2E approaches can. This indicates that our semi-E2E approach is better than the E2E approach in NER recall. In Case 2, the pipeline approach outputs four error NEs since the error characters in its ASR text affect the NER model. However, because only the spans than start with the predicated NE heads (广 and 英) are used for the NER model, the semi-E2E approach does not output those error NEs. This shows that our semi-E2E approach is significantly better than the pipeline approach in NER precision.

4.6   Results of Different NE Categories

Table 4 demonstrates that there is a large gap between NER results from the speech signal and the ground-truth text. Like [3], we report the NER results of different NE categories in Table 7.

As shown in Table 7, the major loss occurs in PER, which is consistent with the results in [3]. This is because there are a lot of homophones and polyphones in Chinese, especially for person names. In addition, the semi-E2E method outperforms the E2E method on all NE categories (PER, LOC and ORG) and outperforms the pipeline method in PER and LOC in terms of $F1$ scores.

5   Conclusion

In this paper, a semi-E2E approach consisting of EHA-ASR and SpanNER is proposed for nested NER from speech, which is a new paradigm. For EHA-ASR, only two special tokens need to be added to the ASR vocabulary list to identify the heads of NEs, regardless of the number of NE categories. For SpanNER, only the spans that start with the heads predicted by EHA-ASR are used to predict NEs. The semi-E2E approach is compared with the pipeline and E2E approaches on the dataset CNERTA. Experimental results show that the semi-E2E approach gets the best $F1$ score and outperforms the E2E approach in all aspects.

References

1. Bu, H., Du, J., Na, X., Wu, B., Zheng, H.: Aishell-1: an open-source mandarin speech corpus and a speech recognition baseline. In: 2017 20th Conference of the Oriental Chapter of the International Coordinating Committee on Speech Databases and Speech I/O Systems and Assessment (O-COCOSDA), pp. 1–5 (2017)
2. Caubrière, A., Rosset, S., Estève, Y., Laurent, A., Morin, E.: Where are we in named entity recognition from speech? In: Proceedings of the Twelfth Language Resources and Evaluation Conference. pp. 4514–4520. European Language Resources Association, Marseille (2020). https://aclanthology.org/2020.lrec-1.556
3. Chen, B., Xu, G., Wang, X., Xie, P., Zhang, M., Huang, F.: AISHELL-NER: named entity recognition from Chinese speech. In: IEEE International Conference on Acoustics, Speech and Signal Processing, ICASSP 2022, Virtual and Singapore, 23–27 May 2022, pp. 8352–8356. IEEE (2022). https://doi.org/10.1109/ICASSP43922.2022.9746955
4. Walker, C., Stephanie Strassel, J.M., Maeda, K.: ACE 2005 Multilingual Training Corpus. Linguistic Data Consortium, Philadelphia 57 (2006)
5. Devlin, J., Chang, M.W., Lee, K., Toutanova, K.: BERT: pre-training of deep bidirectional transformers for language understanding. In: Proceedings of the 2019 Conference of the North American Chapter of the Association for Computational Linguistics: Human Language Technologies, vol. 1 (Long and Short Papers), pp. 4171–4186. Association for Computational Linguistics, Minneapolis (2019). https://doi.org/10.18653/v1/N19-1423. https://aclanthology.org/N19-1423

6. Finkel, J.R., Manning, C.D.: Nested named entity recognition. In: Proceedings of the 2009 Conference on Empirical Methods in Natural Language Processing, pp. 141–150. Association for Computational Linguistics, Singapore (2009). https://aclanthology.org/D09-1015

7. Fu, J., Huang, X., Liu, P.: SpanNER: named entity re-/recognition as span prediction. In: Proceedings of the 59th Annual Meeting of the Association for Computational Linguistics and the 11th International Joint Conference on Natural Language Processing, vol. 1: Long Papers), pp. 7183–7195. Association for Computational Linguistics, Online (2021). https://doi.org/10.18653/v1/2021.acl-long.558. https://aclanthology.org/2021.acl-long.558

8. Ghannay, S., et al.: End-to-end named entity and semantic concept extraction from speech. In: 2018 IEEE Spoken Language Technology Workshop (SLT), pp. 692–699 (2018). https://doi.org/10.1109/SLT.2018.8639513

9. Gulati, A., et al.: Conformer: convolution-augmented transformer for speech recognition. In: Proceedings of Interspeech 2020, pp. 5036–5040 (2020)

10. Han, W., et al.: Contextnet: improving convolutional neural networks for automatic speech recognition with global context. In: Meng, H., 0011, B.X., Zheng, T.F. (eds.) Interspeech 2020, 21st Annual Conference of the International Speech Communication Association, Virtual Event, Shanghai, China, 25–29 October 2020, pp. 3610–3614. ISCA (2020). https://doi.org/10.21437/Interspeech.2020-2059

11. Hinton, G.E., Salakhutdinov, R.: Reducing the dimensionality of data with neural networks. Science **313**(5786), 504–507 (2006). https://doi.org/10.1126/science.1127647

12. Huang, Z., Xu, W., Yu, K.: Bidirectional LSTM-CRF models for sequence tagging (2015). http://arxiv.org/abs/1508.01991

13. Kim, J.D., Ohta, T., Tateisi, Y., Tsujii, J.: Genia corpus - a semantically annotated corpus for bio-textmining. Bioinformatics **19**(Suppl 1), i180-2 (2003)

14. Lafferty, J.D., McCallum, A., Pereira, F.C.N.: Conditional random fields: probabilistic models for segmenting and labeling sequence data. In: Proceedings of the Eighteenth International Conference on Machine Learning, ICML 2001, pp. 282–289. Morgan Kaufmann Publishers Inc., San Francisco (2001)

15. Liu, P., Guo, Y., Wang, F., Li, G.: Chinese named entity recognition: the state of the art. Neurocomputing **473**(C), 37–53 (2022). https://doi.org/10.1016/j.neucom.2021.10.101

16. McCallum, A., Li, W.: Early results for named entity recognition with conditional random fields, feature induction and web-enhanced lexicons. In: Proceedings of the Seventh Conference on Natural Language Learning at HLT-NAACL 2003, pp. 188–191 (2003). https://aclanthology.org/W03-0430

17. Morris, A.C., Maier, V., Green, P.D.: From WER and RIL to MER and WIL: improved evaluation measures for connected speech recognition. In: Interspeech (2004)

18. Sasaki, Y.: The truth of the F-measure (2007)

19. Sui, D., Tian, Z., Chen, Y., Liu, K., Zhao, J.: A large-scale Chinese multimodal NER dataset with speech clues. In: Proceedings of the 59th Annual Meeting of the Association for Computational Linguistics and the 11th International Joint Conference on Natural Language Processing, vol. 1: Long Papers), pp. 2807–2818. Association for Computational Linguistics, Online (2021). https://doi.org/10.18653/v1/2021.acl-long.218. https://aclanthology.org/2021.acl-long.218

20. Wang, X., Jiang, Y., Bach, N., Wang, T., Huang, Z., Huang, F., Tu, K.: Automated concatenation of embeddings for structured prediction. In: Proceedings of

the 59th Annual Meeting of the Association for Computational Linguistics and the 11th International Joint Conference on Natural Language Processing, vol. 1: Long Papers, pp. 2643–2660. Association for Computational Linguistics, Online (2021). https://doi.org/10.18653/v1/2021.acl-long.206. https://aclanthology.org/2021.acl-long.206

21. Wang, Y., Tong, H., Zhu, Z., Li, Y.: Nested named entity recognition: a survey. ACM Trans. Knowl. Discov. Data **16**(6) (2022). https://doi.org/10.1145/3522593

22. Yadav, H., Ghosh, S., Yu, Y., Shah, R.R.: End-to-end named entity recognition from English speech. In: INTERSPEECH, pp. 4268–4272 (2020)

23. Yadav, V., Bethard, S.: A survey on recent advances in named entity recognition from deep learning models. In: Proceedings of the 27th International Conference on Computational Linguistics, pp. 2145–2158. Association for Computational Linguistics, Santa Fe (2018). https://aclanthology.org/C18-1182

24. Yang, H., Zhang, M., Tao, S., Ma, M., Qin, Y.: Chinese ASR and NER improvement based on whisper fine-tuning. In: 2023 25th International Conference on Advanced Communication Technology (ICACT), pp. 213–217 (2023). https://doi.org/10.23919/ICACT56868.2023.10079686

25. Yao, Z., et al.: Wenet: production oriented streaming and non-streaming end-to-end speech recognition toolkit. In: Interspeech, pp. 4054–4058 (2021)

26. Zhang, M., et al.: Incorporating pinyin into pipeline named entity recognition from Chinese speech. In: 2023 Asia Pacific Signal and Information Processing Association Annual Summit and Conference (APSIPA ASC), pp. 947–953 (2023). https://doi.org/10.1109/APSIPAASC58517.2023.10317287

27. Zhang, M., Qiao, X., Zhao, Y., Tao, S., Yang, H.: Smartspanner: making spanner robust in low resource scenarios. In: Findings of the Association for Computational Linguistics: EMNLP 2023 (2023)

28. Zhang, M., et al.: Fastspanner: speeding up spanner by named entity head prediction. In: 2023 5th International Conference on Natural Language Processing (ICNLP), pp. 198–202 (2023). https://doi.org/10.1109/ICNLP58431.2023.00042

29. Zhang, Q., et al.: Transformer transducer: a streamable speech recognition model with transformer encoders and rnn-t loss. In: ICASSP 2020–2020 IEEE International Conference on Acoustics, Speech and Signal Processing (ICASSP), pp. 7829–7833 (2020)

30. Zhao, Y., Li, J., Wang, X., Li, Y.: The speech transformer for large-scale mandarin Chinese speech recognition. In: ICASSP 2019–2019 IEEE International Conference on Acoustics, Speech and Signal Processing (ICASSP), pp. 7095–7099 (2019). https://doi.org/10.1109/ICASSP.2019.8682586

31. Zhou, G., Su, J.: Named entity recognition using an HMM-based chunk tagger. In: Proceedings of the 40th Annual Meeting of the Association for Computational Linguistics, pp. 473–480. Association for Computational Linguistics, Philadelphia (2002). https://doi.org/10.3115/1073083.1073163. https://aclanthology.org/P02-1060

A Lightweight Music Source Separation Model with Graph Convolution Network

Mengying Zhu[1,2] , Liusong Wang[1,2] , and Ying Hu[1,2(✉)]

[1] School of Computer Science and Technology, Xinjiang University, Urumqi, China
{zmy,wls}@stu.xju.edu.cn, huying@xju.edu.cn
[2] Key Laboratory of Signal Detection and Processing in Xinjiang, Urumqi, China

Abstract. With the rapid advancement of deep neural networks, there has been a significant improvement in the performance of music source separation methods. However, most of them primarily focus on improving their separation performance, while ignoring the issue of model size in the real-world environments. For the application in the real-world environments, in this paper, we propose a lightweight network combined with the Graph convolutional network Attention (GCN_A) module for Music Source Separation (G-MSS), which includes an *Encoder* and four *Decoders*, each of them outputs a target music source. The G-MSS network adopts both time-domain and frequency-domain L1 losses. The ablation study verifies the effectiveness of our designed GCN Attention (GCN_A) module and multiple *Decoders*, and also make a visualization analysis of the main components in the G-MSS network. Comparing with the other 13 methods on the MUSDB18 dataset, our proposed G-MSS achieves comparable separation performance while maintaining the lower amount of parameters.

Keywords: Music source separation · Graph convolutional network · Dual-path transformer

1 Introduction

Music Source Separation (MSS) has a wide range of applications in the field of music information retrieval (MIR), such as automatic lyrics recognition [1], music genres classification [2], and music education [3,4]. Furthermore, the research on music source separation also contributes to the investigation of speech separation [5–7] and speech enhancement [8–10]. The current music source separation methods mainly focus on separating four well-defined instruments: *vocals*, *bass*, *drums*, and *other* accompaniment. This task has been challenging due to its simultaneous isolation of multiple sources and other interfering factors.

Recently, with the rapid development of deep learning, the deep neural network (DNN)-based methods achieve excellent performances in the MSS field, which can be divided into two categories: one is based on the time domain and the other time-frequency domain. The former directly processes the song and

J. Jia et al. (Eds.): NCMMSC 2023, CCIS 2006, pp. 23–36, 2024.
https://doi.org/10.1007/978-981-97-0601-3_3

outputs the waveforms of individual sources [5, 11–13]. The latter first subjects the input waveforms to a Short-Time Fourier Transform (STFT) operation to obtain a time-frequency (T-F) spectrogram, which is then fed into a MSS network [14–17].

Stoller et al. firstly proposed a singing voice separation model based on time-domain, an adaptation of the U-Net to the one-dimensional, namely Wave-U-Net [11]. Luo et al. proposed ConvTasNet for speech separation [5], which consists of stacked one-dimensional dilated convolutional blocks. It has been developed for MSS tasks [12]. Défossez et al. proposed Demucs, a MSS model with a U-Net structure and bidirectional LSTM [12]. Kim et al. recently proposed a Wave-U-Net with the contrastive learning technique (Wave-U-Net-CL), achieving a comparable performance on MSS task [13].

Recently, there exists a number of MSS methods based on T-F domain. Stöter et al. proposed UMX, a typical MSS architecture based on the T-F domain, which consists of three consecutive bidirectional LSTM (BLSTM) layers [18]. Based on their work, Sawata et al. proposed X-UMX, which utilized two new loss functions, multi-domain loss and combination loss, as well as a bridging operation, leading to further improvement in separation performance [19]. Koyama et al. further designed a DEQ-UMX model, which replaces the BLSTM layers in the original UMX with a DEQ-based BLSTM layers, resulting in a reduction in parameters and achieving an improvement in MSS performance [14]. D3Net [15] is another model that combined the multi-dilated convolution with DenseNet [20]. Choi et al. proposed LaSAFT, a U-Net architecture that combined the Gated Point-wise Convolutional Modulation module (GPoCM) for MSS task [16]. In addition, Hu et al. proposed CDE-HTCN, a cross-domain encoder and the hierarchic temporal convolutional network, achieving a comparable performance on the MUSDB18 dataset [21]. Luo et al. proposed band-split RNN (BSRNN), a frequency-domain model that explictly splits the spectrogram of the mixture into subbands and perform interleaved band-level and sequence-level modeling [22].

Many previous studies about music source separation based on deep learning have primarily focused on improving performance without considering the issue of model size in practical applications. Additionally, previous methods often only employed the time-domain L1 loss, which neglects the relationships among the frequency bins. In this paper, we propose a lightweight Music Source Separation network combined with a Graph convolutional network (G-MSS), which maintains a small model size while simultaneously improving the separation performance.

Our contributions mainly include the following aspects:

(1) We propose a MSS network that adopts dual-path transformer backbone and utilizes Graph Convolutional Network (GCN) as attention module, which keeps a comparable performance with a smaller amount of parameters.
(2) We explore two L1 losses of the time domain and frequency domain simultaneously.
(3) We also provide the visualization analysis of the main components in the G-MSS network.

2   Related Work

Recently, Transformer-based and GCN-based methods have been widely proposed and obtained splendid performance in the field of speech signal processing.

2.1   Transformer-Based Methods

Transformer was introduced by Vaswani A et al. [23] in 2017, which introduced the self-attention mechanism, and abandoning the traditional architectures of recurrent neural networks (RNNs) and convolutional neural networks (CNNs). This enables the model to capture the long-term dependencies and parallelly process the sequential data, achieving excellent performance. Lately, it has been widely applied in the fields of natural language processing [24, 25], image processing [26, 27] and speech processing, such as speech separation [28] and music source separation [29]. Luo et al. proposed a Dual-Path RNN network for time-domain single-channel speech separation [30]. Inspired by their work, Chen et al. proposed a Dual-Path Transformer network for modeling speech sequences with direct context-awareness [31], which combines two transformers, an intra-transformer and inter-transformer, to learn the order information of the speech sequences efficiently. Wang et al. proposed a denoising and dereverberation network based on a dual-path transformer [10], achieving the best performance on the task of simultaneous denoising and dereverberation.

2.2   Graph Convolutional Network

Kipf T.N et al. proposed Graph Convolutional Network (GCN) for semi-supervised classification tasks on graph-structured data [32]. Let $G = (\boldsymbol{\nu}, \boldsymbol{\varepsilon})$ denote a graph with N nodes, $\boldsymbol{\nu}_i \in \boldsymbol{\nu}, i \in [1, N]$, and a set of edge links $\boldsymbol{\varepsilon}$. The adjacency matrix, $A \in \mathbb{R}^{N \times N}$, represents graph data, describing the connectivity between nodes. $A(\boldsymbol{\nu}_i, \boldsymbol{\nu}_j)$ is a non-zero element if and only if $(\boldsymbol{\nu}_i, \boldsymbol{\nu}_j) \in \boldsymbol{\varepsilon}$. The value of the $A(\boldsymbol{\nu}_i, \boldsymbol{\nu}_j)$ denotes the relationship of edge links between $\boldsymbol{\nu}_i$ and $\boldsymbol{\nu}_j$. For each node, GCN considers its own feature vector as well as the feature vectors of its neighboring nodes. The representation of each node and its neighbors are obtained by weighted averaging the feature vectors of neighboring nodes according with adjacency matrix A. When stacking multiple GCN layers, each layer further updates the nodes, capturing the information of different scales. The hierarchical rules of a multi-layer GCN are as follows:

$$H^{(l+1)} = \sigma(\tilde{D}^{-1/2}\tilde{A}\tilde{D}^{-1/2}H^{(l)}W^{(l)}) \tag{1}$$

where $H^{(l)}$ is the output of the l-th layer. $\tilde{A} = A + I_N$, I_N is the identity matrix, $\tilde{D}_{ii} = \sum_j \tilde{A}(i, j)$. $\sigma(.)$ denotes an activation function, such as ReLU or Sigmoid. $W^{(l)}$ is a trainable weight matrix for layer l, and H^0 is the input of the first layer of GCN.

In recent years, GCN has gradually made widespread application in the field of speech processing. Wang T et al. proposed to utilize GCN as an auxiliary encoder to encode structural details for speech separation [33]. Tzirakis et al. proposed to utilize GCN as integrated into the embedded space of the U-Net architecture for speech enhancement [34]. Furthermore, GCN has also been applied in the task of speech emotion recognition [35, 36].

3   Proposed Method

The illustration of our proposed G-MSS is shown in Fig. 1. The network consists of an encoder for feature generation, a Graph Convolution Network Attention (GCN_A) for capturing the attention along frequency dimension, a mask estimator composed of multiple consecutive dual-path transformers for focusing on the music-related features and predicting the masks of various target music sources. Each is multiplied with the input of the mask estimator and then fed into four decoders, and each outputs a T-F spectrogram to reconstruct the waveform of a target source. The stereo input song, a 2-D tensor $x \in \mathbb{R}^{2 \times T}$, each channel is firstly transformed by Short-Time Fourier Transform (STFT) to obtain a complex spectrogram containing real and imaginary parts. Then, these complex spectrogram of two channels, a 3-D tensor $X \in \mathbb{R}^{4 \times T \times F}$, are served as the input of the G-MSS network and fed into the encoder, whose output is denoted as $\tilde{X} \in \mathbb{R}^{32 \times T \times F}$. $\tilde{X}$ is then multiplied with the output of GCN-A, and passed through a Conv block, increasing the number of channels to 128 while downsampling on the frequency dimension.

3.1   Encoder/Decoder

The *Encoder* and *Decoder* in Fig. 1 adopt the same structure except that the *Decoder* also contains an upsampling operation as shown in Fig. 2. The *Conv* block consists of a convolutional layer with the kernel size of 1×1, normalization and *PRelu* activation operation. By this *Conv* block, the channel number of feature maps in *Encoder* increases to 32, while that in *Decoder* decreases to 32. There is an extra upsampling operation in *Decoder*, which is implemented along the frequency dimension by a transposed convolution layer. There exist upper and lower branches. Each branch contains four consecutive dilated *RH* convolutional blocks with dilation factors of 0, 1, 2, and 3, which is a combination of two parallel CNN layers with the kernel size of $1 \times k$ and $k \times 1$ followed by a CNN layer with that of $k \times k$. k of the upper and lower branches are 3 and 5, respectively. Different from the *RHConv* proposed in [37], here, we adpot dilated *RH* convolutional blocks. This makes the network gradually expand the receptive field to capture a broader range of contextual information, resulting in extracting more global features. The feature maps from the upper and lower branches are combined using a Feature Selection Fusion block (FSFblock) [21], which merges the features of various scales while preserving important features and discarding unimportant ones.

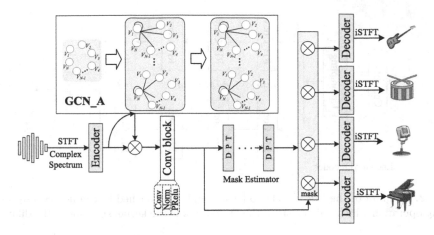

Fig. 1. Illustration of the G-MSS architecture. GCN_A denotes Graph Convolutional Network which is designed to calculate the attention weight of frequency dimension. There are four consecutive Dual-Path Transformer (DPT) in mask estimator. $\otimes$ the element-wise multiplication.

3.2 Graph Convolution Network Attention

We construct a graph representation $G = (\boldsymbol{\nu}, \boldsymbol{\varepsilon})$, for the feature maps generated by the *Encoder*, as shown in the GCN_A module in Fig. 1, where $\boldsymbol{\nu}$ represents a set of nodes, $\boldsymbol{\varepsilon}$ that of edges denoted as $(\boldsymbol{\nu}_i, \boldsymbol{\nu}_j)$, where $\boldsymbol{\nu}_i$ and $\boldsymbol{\nu}_j$ are the i-th and j-th nodes in the graph, respectively.

We adopt frame-to-node transformation for graph construction following the prior works [33, 35]. Each node has four neighboring nodes corresponding to the previous and following two frames. Here, we define the relationship of node neighborhood as a cycle construction where the node of the first time frame is connected with the last time frame. So, the number of frames related to the current frame, n, is 5. The graph adjacency matrix $\boldsymbol{A}_n$ is as follows:

$$
A_5 = \begin{bmatrix} 1 & 1 & 1 & 0 & \dots & 1 & 1 \\ 1 & 1 & 1 & 1 & \dots & 0 & 1 \\ \vdots & \vdots & \vdots & \vdots & \ddots & \vdots & \vdots \\ 1 & 0 & 0 & 0 & \dots & 1 & 1 \\ 1 & 1 & 0 & 0 & \dots & 1 & 1 \end{bmatrix} \tag{2}
$$

Following Eq.(1), the output of GCN_A with two-layer GCN can be represented as follows:

$$
F = \tilde{D}^{-1/2} \tilde{A}_5 \tilde{D}^{-1/2} (\sigma(\tilde{D}^{-1/2} \tilde{A}_5 \tilde{D}^{-1/2} \tilde{X} W^{(0)})) W^{(1)} \tag{3}
$$

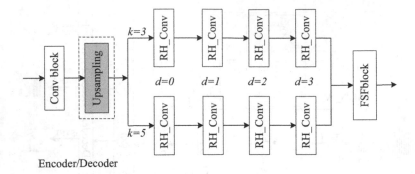

Fig. 2. Diagram of the Encoder/Decoder in Fig. 1. The dashed box indicates a upsampling operation which only exists in the Decoder. k is the kernel size, and d the dilation rate.

where $\tilde{X} \in \mathbb{R}^{32 \times T \times F}$, $\tilde{D} = \sum_j A_5(i,j)$ is a diagonal matrix, W the weight matrix, and σ the *ReLU* activation function.

3.3   Mask Estimator

In order to reduce computational complexity, we first use a convolutional layer with the kernel size of 1×1 to halve the number of channels, which is followed by a *PReLU* operation. The mask estimator consists of four consecutive Dual-Path Transformers (DPTs). Each DPT module consists of right-and-left two transformers, as shown in Fig. 3. The input $Y \in \mathbb{R}^{C \times T \times F}$ is firstly fed into the Multi Head Self Attention (MHSA) block of the first transformer.

$$Y' = Norm(Y + MHSA(Y)) \tag{4}$$

where Y' denotes the output Y after undergoing an MHSA operation, followed by residual connection and normalization. Then, the Y' is passed through a Gated Recurrent Unit (GRU) layer.

$$Y_{out} = Norm(Y' + Linear(ReLU(GRU(Y')))) \tag{5}$$

Y_{out} is reshaped to $C \times F \times T$ and fed into the second transformer, which has the same structure as the first transformer. After four consecutive DPT modules, the feature maps passing through a Gated Linear Unit (GLU) to output the masks of four target music sources. Each mask is multiplied with the input of the mask estimator and fed into a *Decoder*, respectively. Each *Decoder* outputs the spectrogram of a target music source.

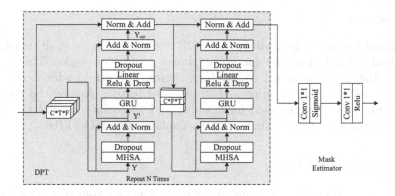

Fig. 3. Diagram of the mask estimator architecture. It includes four consecutive DPT modules, $N = 4$.

3.4 Loss Function

The loss function combines both time-domain loss and frequency domain loss functions. It is calculated as follows:

$$Loss = \alpha * Loss_T + \beta * Loss_F \tag{6}$$

where α and β are hyperparameters, we set them to 0.5 and 0.5, $Loss_T$ and $Loss_F$ both are L_1 loss:

$$Loss_T = \sum_i^S ||y_i - x \odot \tilde{m}_i||_1 \tag{7}$$

$$Loss_F = \sum_i^S ||Y_i - iSTFT(x \odot \tilde{m}_i)||_1 \tag{8}$$

where x denotes the waveform of the input song, y_i the waveform of the i-th target music source, $\tilde{m}_i$ the mask of the i-th target music source, S the number of target music sources, and Y_i the spectrogram of the i-th target music source. Here, S equals 4.

4   Experiments

In this section, we first introduce the experimental dataset, followed by model configurations and experimental settings. Finally, we analyze the experimental results and draw a conclusion.

4.1 Dataset

We utilized the publicly available MUSDB18 dataset [38], a multi-track fully supervised dataset designed for music source separation tasks. This dataset includes four target sources: *vocals*, *bass*, *drums*, and *other*, which consists of 150 songs, 100 songs for training and 50 songs for testing. Among the 100 songs for training, 86 songs are used for training and 14 songs for validation. Each song is provided in stereo format with a sampling rate of 44.1 kHz.

4.2 Experimental Setup

In our experiment, the sample rate of the songs was maintained at 44.1 kHz and split into two-second segments. Each segment of a song was processed through STFT to obtain the complex spectrogram, adopting a *Hanning* window with a length of 2048 points and a hop size of 1024 points. We utilized the Adam optimizer with an initial learning rate of 3e–4 and trained for a total of 300 epochs.

We evaluate our proposed G-MSS network using three objective metrics [39], namely, Source-to-Distortion Ratio (SDR), Signal-to-Interference Ratio (SIR), and Signal-to-Artifact Ratio (SAR). SDR, SIR, and SAR are defined as follows:

$$SDR = 10log_{10}\left(\frac{||s_{target}||^2}{||e_{interf} + e_{noise} + e_{artif}||^2}\right) \tag{9}$$

$$SIR = 10log_{10}\left(\frac{||s_{target}||^2}{||e_{interf}||^2}\right) \tag{10}$$

$$SAR = 10log_{10}\left(\frac{||s_{target} + e_{interf} + e_{noise}||^2}{||e_{artif}||^2}\right) \tag{11}$$

where s_{target} represents the true source, e_{interf}, e_{noise}, e_{artif} respectively denote interference, noise, and artifact error terms. These metrics are used to objectively assess the qualities of predicted music sources and the performance of the MSS network. These evaluation metrics have larger numerical values, and the MSS network has better performance. These evaluation metrics were calculated using the Python package museval [40].

5 Results and Analysis

We conducted an ablation study to verify the effectiveness of main components in the G-MSS network, and make a visualization analysis. We will compare our proposed G-MSS network with other 13 MSS models on SDR metric, and 6 MSS models on SAR and SIR metrics.

5.1 Ablation Study

We designed an ablation experiment to verify the effectiveness of the components in our proposed G-MSS network. (i) denotes a complete G-MSS network, (ii) our proposed G-MSS without the GCN_A module, and (iii) that without GCN_A module and without multi decoders. That is four masks all fed into a decoder to output four spectrograms. (iv) denotes building upon the (iii), the network only utilizes time domain $L1$ loss. The parameters of the above four networks are 1.7M, 1.7M, 0.8M, 0.8M, respectively.

As shown in Table 1. In comparison with Experiment (i), when the network removes the GCN_A module, the average scores of SDR, SAR, and SIR decrease by 0.17 dB, 0.28 dB, and 0.19 dB, respectively. Specially, for the sources of *bass* and *drums*, the average SDR score decreases by 0.2 dB and 0.36 dB, respectively. These verify the effectiveness of our proposed GCN_A module. When the network further removes multi decoders instead of using one decoder (iii), the average scores of SDR, SAR and SIR further decrease by 0.14 dB, 0.28 dB and 0.19 dB, respectively. Finally, when the network further removes frequency domain loss instead of only using time domain loss (iv), the average scores of SDR further decrease by 1.1 dB.

To provide a more intuitive view of the impact of each component in the G-MSS network, we extracted a portion of the spectrograms and visualized them. As shown in Fig. 4, the first column illustrate the spectrograms of reference, the second column denotes the spectrograms output from the network corresponding to (iii) in Table 1, the third column from the network corresponding to (ii) in Table 1, and the fourth column from the network corresponding to (i) in Table 1. Respectively compared with the spectrograms of the first column, the frequency components in the red box of the fourth column are relatively approximate to the reference spectrograms. Especially, for the spectrograms of the first row representing *other* sources, there exists some artificial low-frequency components in the spectrograms of the second and third colunms, while a little one in that of the fourth column. For the spectrograms of the fourth row representing *vocals* source, many frequency components in the red box of the spectrogram output from three networks are lost while a little one in that of the spectrogram from the complete G-MSS network.

Table 1. The Ablation Study on MUSDB18 Dataset

	SDR					SAR					SIR				
	Vocals	Bass	Drums	Other	Avg.	Vocals	Bass	Drums	Other	Avg.	Vocals	Bass	Drums	Other	Avg.
G-MSS(i)	7.90	5.98	7.02	5.56	6.62	7.87	6.23	7.55	5.54	6.80	17.30	12.45	13.53	9.44	13.18
-GCN_A(ii)	7.88	5.78	6.66	5.49	6.45	7.84	5.89	6.95	5.43	6.52	16.98	12.06	13.71	9.22	12.99
-Decoders(iii)	7.30	5.66	7.04	5.25	6.31	7.39	5.51	7.03	4.98	6.23	16.70	11.38	13.34	9.48	12.72
-L1_F(iv)	6.31	5.74	5.81	2.98	5.21	7.76	6.31	7.26	4.73	6.52	16.70	12.23	13.76	8.80	12.87

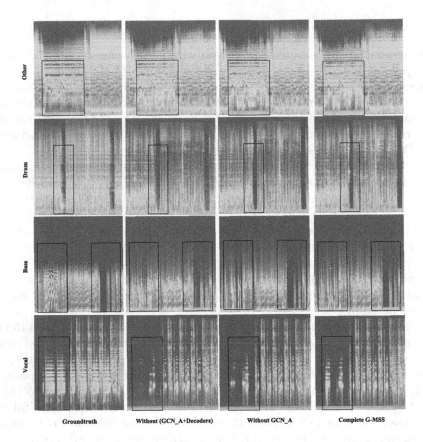

Fig. 4. Visulization comparsion. The horizontal axis represents time (s), and the vertical axis represents frequency (kHz). The second column denotes the spectrograms output from the network corresponding to (iii) in Table 1, the third column from the network corresponding to (ii) in Table 1, and the fourth column from the network corresponding to (i) in Table 1.

Table 2. Comparison between G-MSS and the state-of-the-art methods at the metric of median SDR(DB) on MUSDB18 dataset. ✓ denotes the method modeling directly on the spectrogram, while ✗ that on the waveform.

Models	Vocals	Bass	Drums	Other	Avg.	Parm.	Spec.
IRM oracle	9.43	7.12	8.45	7.85	8.22	N/A	✓
Wave-U-Net [11]	3.25	3.21	4.22	2.25	3.23	10.2M	✗
UMX [18]	6.32	5.23	5.73	4.02	5.33	8.9M	✓
Meta-TasNet [41]	6.40	5.58	5.91	4.19	5.52	45.5M	✗
DEQ-UMX [14]	6.60	5.14	6.17	4.20	5.53	25.06M	✓
Sams-Net [17]	6.61	5.25	6.63	4.09	5.65	3.7M	✓
X-UMX [19]	6.61	5.43	6.47	4.64	5.79	9.5M	✓
Conv-TasNet [5]	6.21	6.53	6.23	4.26	5.81	8.9M	✗
LaSAFT [16]	7.33	5.63	5.68	4.87	5.88	N/A	✓
D3Net [15]	7.24	5.25	7.01	4.53	6.01	7.9M	✓
DEMUCS [12]	6.84	7.01	6.86	4.42	6.28	648M	✗
Wave-U-Net-CL [13]	6.91	7.29	7.05	4.62	6.47	5.7M	✗
CDE-HTCN [21]	7.37	7.92	7.33	4.92	6.89	12.6M	✗
G-MSS	**7.90**	5.98	7.02	**5.56**	6.62	**1.7M**	✓

Table 3. Comparison of median SAR and SIR metrics (DB) with five methods on the MUSDB18 dataset

	SAR					SIR				
Models	Vocals	Bass	Drums	Other	Avg.	Vocals	Bass	Drums	Other	Avg.
IRM oracle	9.51	7.40	8.40	7.93	8.31	16.52	11.78	15.15	14.33	14.45
Wave-U-Net [11]	4.05	4.64	5.29	3.99	4.49	7.45	8.31	6.97	7.06	7.45
Conv-TasNet [5]	6.87	6.60	6.19	4.88	6.13	14.27	12.62	12.94	7.15	11.75
UMX [18]	6.52	6.34	6.02	4.74	5.90	11.93	9.27	10.51	9.31	10.25
DEMUCS [12]	6.54	6.41	6.18	5.18	6.08	14.48	13.46	11.93	6.37	11.93
CDE-HTCN [21]	7.50	7.95	7.43	5.47	7.09	14.65	12.72	13.10	10.35	12.71
G-MSS	**7.87**	6.23	**7.55**	**5.54**	6.80	**17.30**	12.45	**13.53**	9.44	**13.18**

5.2 Comparison with Other MSS Models

We compared our proposed G-MSS network with other 13 MSS models on the SDR metric. To ensure a fair comparison, we only compared G-MSS with the models that did not utilize additional data during the training phase. The results of the compared models listed in Table 2 are all from their reported paper. As shown in Table 2, our proposed G-MSS achieves the best SDR metrics of 7.90 dB and 5.56 dB on *vocal* and *other* sources, respectively. The CDE-HTCN achieves the best SDR metric exceeding G-MSS 4%, but has the amount of parameters times 7.4 of that of G-MSS.

According to their reports, we also conducted comparisons with the other five methods on the SAR and SIR metrics, as shown in Table 3. Remarkably, our proposed G-MSS achieves the highest mean SIR score among the compared methods. In general, we place a greater emphasis on the SDR evaluation metric.

6 Conclusion

In this paper, we propose a lightweight Music Source Separation network combined with a Graph convolutional network (G-MSS), which includes an *Encoder* and four *Decoder*s, each *Decoder* outputs a target music source. The experimental results show that the GCN_A designed in the G-MSS can effectively improve the separation performance by paying more attention to frequency information. In addition, the simultaneous output of four target music sources by multiple *Decoder*s can reduce the interference between target music sources and thus improve the separation performance. Meanwhile, we verify that using T-F domain loss has better separation performance than using time domain loss. Comparing with the other 13 methods on the MUSDB18 dataset, our proposed G-MSS achieves comparable separation performance while maintaining the lower amount of parameters.

Acknowledgements. This work is supported by the Multi-lingual Information Technology Research Center of Xinjiang (ZDI145-21).

References

1. Mesaros, A., Virtanen, T.: Automatic recognition of lyrics in singing. EURASIP J. Audio Speech Music Process. **1–11**, 2010 (2010)
2. Rosner, A., Kostek, B., Schuller, B.: Classification of music genres based on music separation into harmonic and drum components. In: Archives of Acoustics, pp. 629–638 (2014)
3. Dittmar, C., Cano, E., Abeßer, J., Grollmisch, S.: Music information retrieval meets music education. In: Dagstuhl Follow-Ups, vol. 3. Schloss Dagstuhl-Leibniz-Zentrum fuer Informatik (2012)
4. Reimer, B.: A Philosophy of Music Education: Advancing the Vision. State University of New York Press, New York (2022)
5. Luo, Y., Mesgarani, N.: Conv-tasnet: surpassing ideal time-frequency magnitude masking for speech separation. IEEE/ACM Trans. Audio Speech Lang. Process. **27**(8), 1256–1266 (2019)
6. Li, K., Yang, R., Hu, X.: An efficient encoder-decoder architecture with top-down attention for speech separation. arXiv preprint arXiv:2209.15200 (2022)
7. Yip, J.Q., et al.: Aca-net: towards lightweight speaker verification using asymmetric cross attention. arXiv preprint arXiv:2305.12121 (2023)
8. Macartney, C., Weyde, T.: Improved speech enhancement with the wave-u-net. arXiv preprint arXiv:1811.11307 (2018)
9. Defossez, A., Synnaeve, G., Adi, Y.: Real time speech enhancement in the waveform domain. arXiv preprint arXiv:2006.12847 (2020)

10. Wang, L., Wei, W., Chen, Y., Hu, Y.: D2 net: a denoising and dereverberation network based on two-branch encoder and dual-path transformer. In: 2022 Asia-Pacific Signal and Information Processing Association Annual Summit and Conference (APSIPA ASC), pp. 1649–1654. IEEE (2022)
11. Stoller, D., Ewert, S., Dixon, S.: Wave-u-net: a multi-scale neural network for end-to-end audio source separation. arXiv preprint arXiv:1806.03185 (2018)
12. Défossez, A., Usunier, N., Bottou, L., Bach, F.: Music source separation in the waveform domain. arXiv preprint arXiv:1911.13254 (2019)
13. Kim, J., Kang, H.-G.: Contrastive learning based deep latent masking for music source separation. In: Proceedings of INTERSPEECH, vol. 2023, pp. 3709–3713 (2023)
14. Koyama, Y., Murata, N., Uhlich, S., Fabbro, G., Takahashi, S., Mitsufuji, Y.: Music source separation with deep equilibrium models. In: International Conference on Acoustics, Speech and Signal Processing (ICASSP), pp. 296–300. IEEE (2022)
15. Takahashi, N., Mitsufuji, Y.: D3net: densely connected multidilated densenet for music source separation. arXiv preprint arXiv:2010.01733 (2020)
16. Choi, W., Kim, M., Chung, J., Jung, S.: Lasaft: latent source attentive frequency transformation for conditioned source separation. In: International Conference on Acoustics, Speech and Signal Processing (ICASSP), pp. 171–175. IEEE (2021)
17. Li, T., Chen, J., Hou, H., Li, M.: Sams-net: a sliced attention-based neural network for music source separation. In: 2021 12th International Symposium on Chinese Spoken Language Processing (ISCSLP), pp. 1–5. IEEE (2021)
18. Stöter, F.-R., Uhlich, S., Liutkus, A., Mitsufuji, Y.: Open-unmix-a reference implementation for music source separation. J. Open Source Softw. $\mathbf{4}$(41), 1667 (2019)
19. Sawata, R., Uhlich, S., Takahashi, S., Mitsufuji, Y.: All for one and one for all: Improving music separation by bridging networks. In: International Conference on Acoustics, Speech and Signal Processing (ICASSP), pp. 51–55. IEEE (2021)
20. Huang, G., Liu, Z., Van Der Maaten, L., Weinberger, K.Q.: Densely connected convolutional networks. In: Proceedings of the IEEE Conference on Computer Vision and Pattern Recognition, pp. 4700–4708 (2017)
21. Ying, H., Chen, Y., Yang, W., He, L., Huang, H.: Hierarchic temporal convolutional network with cross-domain encoder for music source separation. IEEE Signal Process. Lett. $\mathbf{29}$, 1517–1521 (2022)
22. Luo, Y., Yu, J.: Music source separation with band-split rnn. IEEE/ACM Trans. Audio Speech Lang. Process. (2023)
23. Vaswani, A., et al.: Attention is all you need. Adv. Neural Inf. Process. Syst. $\mathbf{30}$ (2017)
24. Devlin, J., Chang, M.-W., Lee, K., Toutanova, K.: Bert: bidirectional encoder representations from transformers (2016)
25. Lewis, M., et al.: Bart: denoising sequence-to-sequence pre-training for natural language generation, translation, and comprehension. arXiv preprint arXiv:1910.13461 (2019)
26. Dosovitskiy, A., et al.: An image is worth 16×16 words: transformers for image recognition at scale (2020). arXiv preprint arXiv:2010.11929
27. Liu, Z., et al.: Swin transformer: hierarchical vision transformer using shifted windows. In: Proceedings of the IEEE/CVF International Conference on Computer Vision, pp. 10012–10022 (2021)
28. Subakan, C., Ravanelli, M., Cornell, S., Bronzi, M., Zhong, J.: Attention is all you need in speech separation. In: International Conference on Acoustics, Speech and Signal Processing (ICASSP), pp. 21–25 (2021)

29. Rouard, S., Massa, F., Défossez, A.: Hybrid transformers for music source separation. In: International Conference on Acoustics, Speech and Signal Processing (ICASSP), pp. 1–5. IEEE (2023)

30. Luo, Y., Chen, Z., Yoshioka, T.: Dual-path rnn: efficient long sequence modeling for time-domain single-channel speech separation. In: International Conference on Acoustics, Speech and Signal Processing (ICASSP), pp. 46–50. IEEE (2020)

31. Chen, J., Mao, Q., Liu, D.: Dual-path transformer network: direct context-aware modeling for end-to-end monaural speech separation. arXiv preprint arXiv:2007.13975 (2020)

32. Kipf, T.N., Welling, M.: Semi-supervised classification with graph convolutional networks. arXiv preprint arXiv:1609.02907 (2016)

33. Wang, T., Pan, Z., Ge, M., Yang, Z., Li, H.: Time-domain speech separation networks with graph encoding auxiliary. IEEE Signal Process. Lett. **30**, 110–114 (2023)

34. Tzirakis, P., Kumar, A., Donley, J.: Multi-channel speech enhancement using graph neural networks. In: International Conference on Acoustics, Speech and Signal Processing (ICASSP), pp. 3415–3419. IEEE (2021)

35. Hu, Y., Tang, Y., Huang, H., He, L.: A graph isomorphism network with weighted multiple aggregators for speech emotion recognition. arXiv preprint arXiv:2207.00940 (2022)

36. Shirian, A., Guha, T.: Compact graph architecture for speech emotion recognition. In: International Conference on Acoustics, Speech and Signal Processing (ICASSP), pp. 6284–6288. IEEE (2021)

37. Hu, Y., Zhu, X., Li, Y., Huang, H., He, L.: A multi-grained based attention network for semi-supervised sound event detection. arXiv preprint arXiv:2206.10175 (2022)

38. Rafii, Z., Liutkus, A., Stöter, F.R., Mimilakis, S.I., Bittner, R.: Musdb18-a corpus for music separation (2017)

39. Vincent, E., Gribonval, R., Févotte, C.: Performance measurement in blind audio source separation. IEEE Trans. Audio Speech Lang. Process. **14**(4), 1462–1469 (2006)

40. Stöter, F.-R., Liutkus, A., Ito, N.: The 2018 signal separation evaluation campaign. In: Deville, Y., Gannot, S., Mason, R., Plumbley, M.D., Ward, D. (eds.) LVA/ICA 2018. LNCS, vol. 10891, pp. 293–305. Springer, Cham (2018). https://doi.org/10.1007/978-3-319-93764-9_28

41. Samuel, D., Ganeshan, A., Naradowsky, J.: Meta-learning extractors for music source separation. In: International Conference on Acoustics, Speech and Signal Processing (ICASSP), pp. 816–820. IEEE (2020)

Joint Time-Domain and Frequency-Domain Progressive Learning for Single-Channel Speech Enhancement and Recognition

Gongzhen Zou[1], Jun Du[1(✉)], Shutong Niu[1], Hang Chen[1], Yuling Ren[2], Qinglong Li[2], Ruibo Liu[2], and Chin-Hui Lee[3]

[1] University of Science and Technology of China, Hefei, China
jundu@ustc.edu.cn
[2] China Mobile Online Services Company Limited, Zhengzhou, China
[3] Georgia Institute of Technology, Atlanta, USA

Abstract. Single-channel speech enhancement for automatic speech recognition (ASR) has been extensively researched. Traditional methods usually directly learn clean target, which may introduce speech distortions and limit ASR performance. Meanwhile, these methods usually focus on either the time or frequency domain, ignoring their potential connections. To tackle these problems, we propose a joint time and frequency domain progressive learning (TFDPL) method for speech enhancement and recognition. TFDPL leverages information from both domains to estimate frequency masks and waveforms, and further combines the information from both domains through a fusion loss, gradually predicting less-noisy and cleaner targets. Experimental results show that TFDPL outperforms traditional methods in ASR and perceptual metrics. TFDPL achieves relative reductions of 43.83% and 36.03% in word error rate for its intermediate outputs on the CHiME-4 real test set using two different acoustic models and certain improvements in PESQ and STOI metrics for clean output on the simulated test set.

Keywords: automatic speech recognition · speech enhancement · joint time and frequency domain · progressive learning

1 Introduction

With the advancement of deep learning, automatic speech recognition (ASR) [1] has made significant progress and has been widely applied in our daily lives [2]. However, in complex acoustic environments, speech may be interfered with by various sources of noise, leading to degradation in ASR performance.

Speech enhancement (SE) is a critical technology in speech processing that aims to improve the quality and intelligibility of corrupted speech [3]. Moreover, it can be utilized as a front-end system to enhance the robustness of ASR systems [4]. In recent years, supervised speech enhancement techniques based on deep neural networks have been widely studied and established as the mainstream

ⓒ The Author(s), under exclusive license to Springer Nature Singapore Pte Ltd. 2024
J. Jia et al. (Eds.): NCMMSC 2023, CCIS 2006, pp. 37–52, 2024.
https://doi.org/10.1007/978-981-97-0601-3_4

approach [5]. These methods can be categorized into two classes based on the domain: frequency-domain methods and time-domain methods.

Frequency-domain methods often utilize the Short-Time Fourier Transform (STFT) to convert the original waveform into a time-frequency spectrogram, which serves as the input to the neural network. These methods aim to predict frequency-domain masks or features, such as Ideal Ratio Mask (IRM) [6] or log-power spectra (LPS) [7]. However, prediction of these methods often discards the clean phase information, which can be detrimental for speech recognition [8]. Additionally, certain prediction targets like IRM are limited by their own assumptions and cannot perfectly reconstruct the clean speech, thus limiting the performance of frequency-domain methods. In recent years, there has been increasing interest in time-domain-based speech enhancement methods [9–11]. These methods directly process the raw waveform to overcome challenges associated with phase estimation and have a higher theoretical performance ceiling.

Most speech enhancement methods aim to improve the quality and intelligibility of corrupted speech. During training, these methods often utilize targets such as IRM, clean LPS, or clean waveform. However, using these targets can sometimes result in excessive suppression and distortion, which may have a negative impact on ASR performance [12]. To address this issue, researchers have proposed methods to mitigate over-suppression and improve ASR performance. [12] introduced an asymmetric loss function to improve speech preservation, while [13] introduced a progressive learning-based speech enhancement network that gradually improves the Signal-to-Noise Ratio (SNR) until learning clean spectral features, where the intermediate target can effectively preserve speech information. [14] and [15] proposed progressive learning (PL) methods in the frequency and time domains, respectively, and demonstrated the effectiveness of intermediate targets in improving ASR performance. In [16], the authors demonstrated that all audio-only object-oriented progressive learning models outperform their audio-only counterparts in speech enhancement. These findings highlight the advantages of progressive learning methods in ASR back-end and speech perceptual quality. On the other hand, previous research in speech enhancement has primarily focused on separate modeling of either time-domain or frequency-domain information. However, due to the complementary nature of the latent information in these two domains, integrating them can enhance the performance of the models [17,18].

In this paper, we propose a novel joint time-domain and frequency-domain progressive learning approach (TFDPL) for single-channel speech enhancement and recognition. TFDPL progressively predicts less-noisy and clearer speech, simultaneously estimates time-frequency masks and waveform using information from both the time and frequency domains, and further combines these two prediction targets through a fusion loss. Experimental results demonstrate that TFDPL outperforms traditional methods in ASR and perceptual metrics. On the CHiME-4 real test set, TFDPL's intermediate output achieves relative word error rate (WER) reductions of 43.83% and 36.03% compared to the untreated noisy speech, respectively, using two different acoustic models without retraining. The final results also demonstrate the best PESQ and STOI scores on the simulated test set.

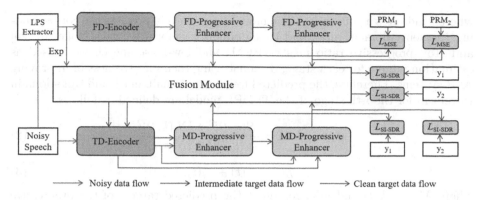

Fig. 1. The overview of proposed joint time-domain and frequency-domain progressive learning network.

2 Progressive Learning with Joint Time Domain and Frequency Domain

In this section, we will provide a detailed description of the TFDPL model. We divide the prediction of clean speech into two stages, aiming to progressively predict intermediate target speech with a 10 dB improvement in SNR relative to the noisy speech and the final clean speech. The overview of the TFDPL model is shown in Fig. 1.

TFDPL model consists of three modules: the progressive frequency-domain masking module, the progressive mix-domain module, and the fusion module. The TFDPL model takes noisy time-domain signals as input. First, the LPS features of the signal are extracted and normalized [19]. Then, the normalized features are fed into the progressive frequency-domain masking module to estimate progressive masks. The estimated masks are multiplied with the original spectrogram and reconstructed back to waveform signals using inverse STFT (ISTFT). The reconstructed signals, along with the original noisy speech, are then fed into the progressive mix-domain module. In addition, we propose a novel fusion strategy, where the fusion module extracts the corresponding LPS features from the estimated targets of the two modules at the same stage of progressive learning, and weights and reconstructs them into fused speech to better utilize frequency and time domain information.

2.1 Problem Formulation

For single-channel speech enhancement, we have a noisy speech signal denoted as y, which consists of a combination of the clean target speech signal s and background noise signal n.

$$y(t) = s(t) + n(t) \tag{1}$$

where t indicates a time index. The model progressively predicts the corresponding frequency-domain mask and time-domain waveform for each stage. In [20] and [15], progressive ratio masks (PRM) and low-noise speech were used as estimation targets for each stage, demonstrating their effectiveness in improving ASR systems. Therefore, the predicted frequency-domain mask and time-domain waveform for the i-th stage of the TFDPL model are defined as follows:

$$M_{\mathrm{PRM}_i}(k, l) = \frac{|S(k, l)|^2 + |N_i(k, l)|^2}{|S(k, l)|^2 + |N(k, l)|^2} \tag{2}$$

$$y_i(t) = s(t) + n_i(t) \tag{3}$$

where $M_{\mathrm{PRM}_i}(k, l)$ and $y_i(t)$ represent the predicted targets of the progressive frequency-domain masking module and the progressive mix-domain module in the i-th stage, respectively. $n_i(t)$ represents the residual noise in the i-th stage, while $S(k, l)$, $N_i(k, l)$ and $N(k, l)$ represent the STFT of the clean speech, residual noise in the i-th stage, and input noise, respectively. k and l are indices representing frames and frequency bins, respectively.

Indeed, it can be observed that when $N_i(k, l)$ and $n_i(t)$ are both equal to 0, M_{PRM_i} and y_i correspond to the traditional IRM and the clean speech s, respectively. They serve as the clean targets for the progressive frequency-domain masking module and the progressive mix-domain module.

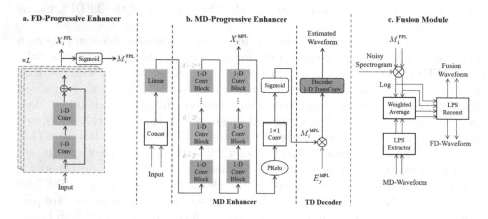

Fig. 2. The detailed design of the components in the TFDPL model. (a) Frequency-domain progressive enhancer. (b) Mix-domain progressive enhancer. (c) Fusion module.

2.2   Joint Time and Frequency Domain Progressive Learning

Progressive Frequency-Domain Masking Module. The progressive frequency-domain masking module consists of two components: the frequency-domain (FD) encoder and the frequency-domain (FD) progressive enhancer (shown in Fig. 2(a)). The data flow and parameters of this module are denoted as FPL. In [16], each stage of progressive learning is composed of L Blocks, each consisting of a 1D convolutional layer with residual connection, a ReLU activation, and a batch normalization. We employ 5 Blocks and 2 conformer layers as the FD encoder, denoted as $\mathcal{F}^{\mathrm{FPL}}_{\mathrm{encoder}}(\cdot)$.

First, the LPS features of the input signal y are extracted and then fed into the FD encoder:

$$E^{\mathrm{FPL}} = \mathcal{F}^{\mathrm{FPL}}_{\mathrm{encoder}}(\mathcal{F}^{\mathrm{LPS}}(y), \Lambda^{\mathrm{FPL}}_{\mathrm{encoder}}) \tag{4}$$

where $\mathcal{F}^{\mathrm{LPS}}(\cdot)$ represents LPS extractor. $\Lambda^{\mathrm{FPL}}_{\mathrm{encoder}}$ and E^{FPL} represent the parameter sets of the FD encoder and the encoded features.

$\mathcal{F}^{\mathrm{FPL}}_{\mathrm{enhancer}_i}(\cdot)$ denotes subsequent FD progressive enhancer composed of 5 Blocks, which is used to predict the intermediate representation of that stage:

$$X^{\mathrm{FPL}}_i = \mathcal{F}^{\mathrm{FPL}}_{\mathrm{enhancer}_i}(X^{\mathrm{FPL}}_{i-1}, \Lambda^{\mathrm{FPL}}_{\mathrm{enhancer}_i}) \tag{5}$$

where $\Lambda^{\mathrm{FPL}}_{\mathrm{enhancer}_i}$ and X^{FPL}_i represent the parameter set and the intermediate representation in the i-th stage, respectively. In the case of the first stage, X^{FPL}_0 is equivalent to E^{FPL}.

Then, the mask M^{FPL}_i for the i-th stage is obtained by applying a sigmoid activation.

$$M^{\mathrm{FPL}}_i = \sigma(X^{\mathrm{FPL}}_i) \tag{6}$$

where M^{FPL}_i is a mask with values ranging from 0 to 1, we expect to use the mask to obtain enhanced speech in the i-th stage with a specified relative increase in SNR compared to the noisy input:

$$\hat{y}^{\mathrm{FPL}}_i = \mathcal{F}^{\mathrm{Reconst}}_{\mathrm{PRM}}(y, M^{\mathrm{FPL}}_i, W_{\mathrm{istft}}) \tag{7}$$

$\mathcal{F}^{\mathrm{Reconst}}_{\mathrm{PRM}}(\cdot)$ represents the waveform reconstruction module, which involves multiplying the extracted spectrum of the input noisy speech y with the predicted mask, and then using the y phase information to reconstruct the waveform using ISTFT. $\hat{y}^{\mathrm{FPL}}_i$ represents the waveform reconstructed by the progressive frequency-domain masking module in the i-th stage.

Progressive Mix-Domain Module. The progressive mix-domain module consists of two components: a time-domain (TD) encoder, and a mix-domain (MD) progressive enhancer (shown in Fig. 2(b)). The data flow and parameters of this module are denoted as MPL. The MD progressive enhancer consists of three sub-parts: a linear layer, a mix-domain (MD) enhancer, and a time-domain (TD) decoder. We introduced a linear layer to fuse the encoded features of the time-domain waveform and the waveform reconstructed based on the progressive frequency-domain masking module. Each MD enhancer consists of two stacks of n 1D-ConvBlocks, with increasing dilation factors. The structure of the 1D-ConvBlocks is the same as in Conv-TasNet [21], but without skip connections.

The trainable TD encoder takes noisy speech y and progressive frequency-domain masking module estimated speech $\hat{y}_i^{\text{FPL}}$ as inputs, where i takes a value of 1 or 2, resulting in high-dimensional encoded features.

$$(E_y^{\text{MPL}}, E_{\hat{y}_1^{\text{FPL}}}^{\text{MPL}}, E_{\hat{y}_2^{\text{FPL}}}^{\text{MPL}}) = \mathcal{F}_{\text{encoder}}^{\text{MPL}}((y, \hat{y}_1^{\text{FPL}}, \hat{y}_2^{\text{FPL}}), \Lambda_{\text{encoder}}^{\text{MPL}}) \tag{8}$$

where $\mathcal{F}_{\text{encoder}}^{\text{MPL}}(\cdot)$ represents TD encoder, which consists of a 1D convolutional layer. E_y^{MPL}, $E_{\hat{y}_1^{\text{FPL}}}^{\text{MPL}}$, and $E_{\hat{y}_2^{\text{FPL}}}^{\text{MPL}}$ denote the encoded high-level features of the input noisy speech y, $\hat{y}_1^{\text{FPL}}$, and $\hat{y}_2^{\text{FPL}}$, respectively. $\Lambda_{\text{encoder}}^{\text{MPL}}$ represents the parameter set of the TD encoder. Next, we concatenate the features E_y^{MPL} and $E_{\hat{y}_1^{\text{FPL}}}^{\text{MPL}}$, and pass the concatenated feature through a linear layer for information fusion and dimensionality reduction. This allows us to obtain a feature that contains both time-domain and frequency-domain information.

$$E_1^{\text{MPL}} = \mathcal{F}_{\text{linear}_1}(Concat(E_y^{\text{MPL}}, E_{\hat{y}_1^{\text{FPL}}}^{\text{MPL}}), \Lambda_{\text{linear}_1}) \tag{9}$$

where $Concat$ represents the concatenation of two vectors along the feature dimension. E_1^{MPL} represents the fused features in the first stage, which is then fed into the first MD enhancer.

$$= \mathcal{F}_{\text{enhancer}_1}^{\text{MPL}}(E_1^{\text{MPL}}, \Lambda_{\text{enhancer}_1}^{\text{MPL}}) \tag{10}$$

where $\mathcal{F}_{\text{enhancer}_1}^{\text{MPL}}(\cdot)$ and $\Lambda_{\text{enhancer}_1}^{\text{MPL}}$ represent the first MD enhancer and the corresponding parameter set. X_1^{MPL} and M_1^{MPL} denote the mix-domain module intermediate representation and masks in the first stage. We concatenate X_1^{MPL} with $X_{\hat{y}_2^{\text{FPL}}}^{\text{MPL}}$ and feed it into the second linear layer to extract fusion information for the second stage, which is then passed to the second MD progressive enhancer.

$$E_2^{\text{MPL}} = \mathcal{F}_{\text{linear}_2}(Concat(X_1^{\text{MPL}}, E_{\hat{y}_2^{\text{FPL}}}^{\text{MPL}}), \Lambda_{\text{linear}_2}) \tag{11}$$

$$= \mathcal{F}_{\text{enhancer}_2}^{\text{MPL}}(E_2^{\text{MPL}}, \Lambda_{\text{enhancer}_2}^{\text{MPL}}) \tag{12}$$

where E_2^{MPL} and M_2^{MPL} represent the fusion features and masks for the second stage. Finally, we perform element-wise multiplication between the obtained masks (M_1^{MPL} and M_2^{MPL}) and the E_y^{MPL} separately, and pass each result through their respective TD decoder to obtain the estimated intermediate targets and clean targets from the mix-domain module.

$$\hat{y}_i^{\text{MPL}} = \mathcal{F}_{\text{decoder}_i}^{\text{MPL}}(E_y^{\text{MPL}}, M_i^{\text{MPL}}, \Lambda_{\text{decoder}_i}^{\text{MPL}}) \tag{13}$$

where $\mathcal{F}_{\text{decoder}_i}^{\text{MPL}}(\cdot)$ and $\Lambda_{\text{decoder}_i}^{\text{MPL}}$ represent the i-th TD decoder and its corresponding parameter set, and $\hat{y}_i^{\text{MPL}}$ represents the waveform estimated by the progressive mix-domain module in the i-th stage.

Fusion Module. This module connects the progressive frequency-domain masking module and progressive mix-domain module, as shown in Fig. 2(c) for more details. First, the fusion module receives the masking M_i^{FPL} estimated by the progressive frequency-domain masking module and the spectrum and phase of the noisy speech y to obtain the reconstructed speech $\hat{y}_i^{\text{FPL}}$ through ISTFT. Afterward, we can input $\hat{y}_i^{\text{FPL}}$ into the progressive mix-domain module to fuse the information from both the time-domain and frequency-domain, and obtain its predicted waveform $\hat{y}_i^{\text{MPL}}$.

Finally, we propose a novel fusion strategy that combines the outputs of the progressive frequency-domain masking module and progressive mix-domain module, aiming to further exploit the complementarity of different domain information:

$$\text{LPS}_i^{\text{fusion}} = \lambda * \mathcal{F}^{\text{LPS}}(\hat{y}_i^{\text{MPL}}) + (1 - \lambda) * \mathcal{F}^{\text{LPS}}(\hat{y}_i^{\text{FPL}}) \tag{14}$$

$$\hat{y}_i^{\text{fusion}} = \mathcal{F}_{\text{LPS}}^{\text{Reconst}}(\hat{y}_i^{\text{MPL}}, \text{LPS}_i^{\text{fusion}}, W_{\text{istft}}) \tag{15}$$

By employing a weighted fusion approach, we acquire the fused LPS feature $\text{LPS}_i^{\text{fusion}}$, with a weight parameter λ ranging from 0 to 1. The function $\mathcal{F}_{\text{LPS}}^{\text{Reconst}}(\cdot)$ denotes the waveform reconstruction based on the LPS features and the phase of $\hat{y}_i^{\text{MPL}}$. $\hat{y}_i^{\text{fusion}}$ represents the waveform obtained through the fusion strategy.

2.3   Multi-target Loss

We propose a multi-task learning approach to train TFDPL. In this approach, we utilize a multi-scale scale-invariant signal-to-distortion ratio (SI-SDR) loss [22] for the progressive mix-domain module and fusion module. Additionally, we use a minimum mean squared error (MSE) loss for the progressive frequency-domain masking module.

$$\mathcal{L}_i^{\text{FPL}} = \mathcal{L}_{\text{MSE}}(M_i^{\text{FPL}}, M_{\text{PRM}_i}) \tag{16}$$

$$\mathcal{L}_i^{\text{MPL}} = \mathcal{L}_{\text{SI-SDR}}(\hat{y}_i^{\text{MPL}}, y_i) \tag{17}$$

$$\mathcal{L}_i^{\text{fusion}} = \mathcal{L}_{\text{SI-SDR}}(\hat{y}_i^{\text{fusion}}, y_i) \tag{18}$$

The final optimization objective of TFDPL is a linear combination of the three mentioned losses.

$$\mathcal{L}^{\text{TFDPL}} = \sum_i \alpha_i * \mathcal{L}_i^{\text{FPL}} + \beta_i * \mathcal{L}_i^{\text{MPL}} + \gamma_i * \mathcal{L}_i^{\text{fusion}} \tag{19}$$

3   Experiments and Analysis

3.1   Data Corpus

Clean speech is obtained from the WSJ0 SI-84 dataset [23], which consists of 7,138 utterances from different speakers. We randomly selected 7,000 utterances for training, 65 for validation, and 73 for testing. The noise data used to generate noisy-clean pairs is sourced from the CHIME-4 noise dataset [24]. During the training process, we employed an online augmentation strategy where speech and noise pairs were randomly selected and randomly segmented into durations ranging from 4 to 6 s. The SNR was varied between –5 dB and 5 dB to generate noisy speech. We simulated a test set to evaluate the perceived quality of the enhanced speech at five different SNR levels (–10 dB, –5 dB, 0 dB, 5 dB, 10 dB) using 73 clean speech utterances from the test set and three training-time unseen noises from the NOISEX-92 corpus [25]: Destroyer Engine, Factory1, and Speech Babble. Additionally, we conducted an ASR performance evaluation of our framework on the CHiME-4 real test set, which includes 1,320 real recordings in four different conditions: bus (BUS), cafe (CAF), pedestrian area (PED), and street (STR).

3.2 Implementation Details

The speech waveform was sampled at a frequency of 16 kHz. We applied a 32-ms Hanning window with 16-ms overlap to extract audio frames. Then, a 512-point STFT was used to compute the spectrum of each frame, resulting in 257-dimensional LPS features. Before feeding the features into the neural network, they were normalized using global mean and variance [19].

For the TFDPL model, each MD enhancer is composed of two stacks, each containing 8 1D-ConvBlocks. The remaining hyper-parameters setting is similar to the original Conv-TasNet [21]. Besides, we do not use skip connection in MD enhancer. We used PyTorch to train the model, with an initial learning rate set to 5e–4. The batch size was 12. If the loss on the validation set did not decrease after an epoch, the learning rate was halved. Adam [26] is used as the optimizer. For the loss configuration, we use $\alpha_1 = 2.3$, $\alpha_2 = 1.5$, and β_i and γ_i are set to 1 to balance the training loss. For the fusion configuration, we set $\lambda = 0.5$.

We trained four baseline models for comparison with our proposed method. The first model has the same network structure as the progressive frequency-domain masking module of TFDPL, denoted as FDPL. The second model is denoted as TDPL, we set the stack of the progressive enhancer in TDPL to contain 8 1D-ConvBlocks, because we found that this has better performance, and the rest of the hyperparameters are set similarly to [15]. The third model, denoted PL-ANSE [14], combines progressive learning with the traditional IMCRA [27] algorithm. The fourth is the traditional speech enhancement Conv-TasNet, denoted as TDSE.

We evaluated our method on two different ASR systems. The first one is an official ASR system [24], referred to as $ASR(1)$, where the acoustic model is trained using the DNN-HMM architecture with sMBR criteria [28]. The second system is trained with LF-MMI using TDNN and is referred to as $ASR(2)$. Both ASR systems used a 5-gram Kneser-Ney (KN) smoothed language model for the first-pass decoding [29], and scoring was performed using an RNN-based language model.

3.3 Evaluation Metrics

To evaluate the perceived quality of the enhanced speech, we employed the perceptual evaluation of speech quality (PESQ) [30] and short-time objective intelligibility (STOI) [31] metrics. Higher values in both metrics indicate better performance. Additionally, we employed word error rate (WER) to assess the model's improvement on the ASR system, where lower values are better. The intermediate and clean outputs of the model render great service to ASR back-end and human listener, respectively.

3.4 Results and Analysis

In this section, we will compare the performance of the proposed model with other models in terms of ASR back-end and perceived quality of the enhanced speech. Explanation of terms in Tables 1, 2, 3, 4 and 5: Noisy represents the noisy speech without any enhancement. D-Fusion represents divide fusion, which is the fusion output obtained by applying the fusion strategy proposed in Sect. 2.2 to separately trained FDPL and TDPL models. J-Fusion represents joint fusion, which indicates the output of the fusion module in TFDPL.

Table 1. WER(%) comparison of different targets for different methods on CHiME-4 real test set with $ASR(1)$ and $ASR(2)$.

Target	Model	$ASR(1)$	$ASR(2)$
Noisy	–	23.84	13.46
–	PL-ANSE	18.57	12.48
Clean	TDSE	24.55	21.69
+10 dB	FDPL	16.26	10.38
	TDPL	15.18	9.74
	D-Fusion	14.41	9.12
Clean	FDPL	17.76	11.54
	TDPL	26.43	25.16
	D-Fusion	14.83	10.64

Analysis on Recognition Performance. Table 1 presents the results of different speech enhancement methods in ASR systems. PL-ANSE [14] combines multiple targets of progressive learning with IMCRA [27] algorithm to estimate speech. We can observe that the intermediate target (+10 dB) of progressive learning can always improve the performance of ASR, while its clean target or TDSE directly estimating the clean target may degrade the performance of ASR. Studies [20] and [15] have demonstrated separately that the intermediate targets in progressive frequency-domain and time-domain models effectively enhance the performance of ASR systems. Our experiments validate this finding. Additionally, the ASR performance of the intermediate targets of FDPL and TDPL is better than that of PL-ANSE and TDSE, so we choose FDPL and TDPL for subsequent fusion experiments. When FDPL and TDPL results are fused, the performance of intermediate target fusion still exceeds the performance of clean target fusion on two different ASR backends and improves ASR performance. Due to the superior performance of the intermediate targets of the progressive learning method in ASR, subsequent ASR experiments are based on the intermediate targets. Furthermore, the fused results show improvements compared to the best results of the individual models before fusion, providing strong evidence for the complementary nature of the frequency and time domains and indicating the effectiveness of our fusion strategy. This is also the motivation behind our joint training.

Table 2. WER(%) comparison of different speech enhancement methods in various environments on CHiME-4 real test set with $ASR(1)$.

Model	Domain	BUS	CAF	PED	STR	AVG
Noisy	–	36.55	24.73	19.92	14.16	23.84
FDPL	*Freq.*	24.36	17.69	13.30	9.69	16.26
TDPL	*Time*	21.18	16.16	13.32	10.05	15.18
D-Fusion	*Time & Freq.*	21.05	15.39	12.01	9.19	14.41
TFDPL	*Freq.*	21.63	14.94	11.79	9.28	14.41
	Time & Freq.	20.21	14.59	12.05	9.19	14.01
J-Fusion	*Time & Freq.*	**19.42**	**14.40**	**11.04**	**8.70**	**13.39**

Table 3. WER(%) comparison of different speech enhancement methods in various environments on CHiME-4 real test set with $ASR(2)$.

Model	Domain	BUS	CAF	PED	STR	AVG
Noisy	–	21.16	13.39	10.41	8.87	13.46
FDPL	*Freq.*	16.41	10.44	8.48	6.20	10.38
TDPL	*Time*	14.41	9.60	8.13	6.84	9.74
D-Fusion	*Time & Freq.*	13.74	9.10	7.60	6.05	9.12
TFDPL	*Freq.*	13.98	9.66	7.94	6.05	9.41
	Time & Freq.	12.99	9.54	7.75	6.54	9.21
J-Fusion	*Time & Freq.*	**12.75**	**8.52**	**7.21**	**5.98**	**8.61**

Tables 2 and 3 display the WER for different speech enhancement methods under various environmental conditions on the CHiME-4 real test set, using $ASR(1)$ and $ASR(2)$. The observation reveals that the intermediate targets of progressive learning have improved ASR performance in various environmental conditions. By comparing the WER in various environments, FDPL and TDPL have demonstrated their advantages in low-noise and challenging conditions, respectively, highlighting the complementarity of frequency-domain and time-domain information. We further fused the prediction results of TDPL and FDPL, and it led to improved ASR performance in all environments, highlighting the robustness of time and frequency domain information fusion.

Regarding the TFDPL model, its progressive frequency-domain masking module shows significant improvements compared to FDPL, achieving relative improvements of 11.38% and 9.34% on the two acoustic models, respectively, with no additional parameters compared to FDPL. As for the progressive mix-domain module, it incorporates information from the frequency-domain leading to performance improvements relative to TDPL in all environments for two acoustic models, with average relative improvements of 7.71% and 5.44%, respectively. The performance of both the progressive frequency-domain masking module and progressive mix-domain module in TFDPL has improved, indicating that there is a mutually reinforcing effect between the time-domain and frequency-domain in TFDPL, thus confirming the effectiveness of TFDPL.

Finally, the fusion of TFDPL model results achieved relative improvements of 43.83% and 36.03% compared to noisy speech in two acoustic models, reaching the best performance. Compared to the fusion results of separately trained FDPL and TDPL, TFDPL achieved relative improvements of 7.08% and 5.59% in two acoustic models, and relative improvements of 11.79% and 11.60% compared to TDPL. In Fig. 3, we selected a representative utterance from a real BUS environment to visually compare the performance of FDPL, TDPL, and TFDPL. In the blue boxed regions in Fig. 3(b) and Fig. 3(c), the target speech of FDPL and TDPL respectively experienced excessive suppression and distortion, resulting in substitution errors in the corresponding ASR results. TFDPL effectively combines information from both the time and frequency domain, thereby avoiding such errors. The experimental results strongly demonstrate the effectiveness of TFDPL's intermediate targets in ASR.

(a) Noisy

TODAY NINETY PERCENT OF THE FOUR BILLION MILLION DOLLARS OF B. A. S.
F. SALES IN AND THE U. S. IS HAS PRODUCED THERE

(b) FDPL

TODAY NINETY PERCENT OF THE FOUR BILLION MILLION DOLLARS OF B. A. S.
F. SALES IN AND THE U. S. IS PRODUCED THERE AND

(c) TDPL

TODAY NINETY PERCENT OF THE FOUR BILLION DOLLARS OF B. A. S. F. SALES
IN AND THE U. S. IS PRODUCED THERE

(d) TFDPL

Fig. 3. Spectrograms and ASR Results of FDPL, TDPL, and TFDPL. (a) Noisy speech,
(b) Intermediate output of the FDPL model, (c) Intermediate output of the TDPL
model, (d) Intermediate output of the TFDPL model fusion module.

Table 4. PESQ comparison on different speech enhancement methods at several SNRs.

Metrics		PESQ					
Model	Domain	−10	−5	0	5	10	*avg.*
Noisy	−	1.38	1.52	1.79	2.11	2.44	1.85
TDSE	*Time*	1.57	2.27	2.76	3.11	3.39	2.62
FDPL	*Freq.*	1.48	1.99	2.42	2.75	3.02	2.33
TDPL	*Time*	1.59	2.28	2.78	3.14	3.41	2.64
D-Fusion	*Time & Freq.*	1.66	2.33	2.84	3.20	3.45	2.70
TFDPL	*Freq.*	1.54	2.06	2.47	2.79	3.05	2.38
	Time & Freq.	1.41	2.11	2.67	3.04	3.36	2.52
J-Fusion	*Time & Freq.*	**1.75**	**2.42**	**2.94**	**3.27**	**3.50**	**2.78**

Table 5. STOI(%) comparison on different speech enhancement methods at several SNRs.

Metrics		STOI(%)					
Model	Domain	−10	−5	0	5	10	*avg.*
Noisy	−	49.53	60.20	72.38	83.10	90.46	71.13
TDSE	*Time*	62.40	81.77	90.19	93.99	95.99	84.87
FDPL	*Freq.*	58.35	74.10	84.76	90.99	94.53	80.55
TDPL	*Time*	62.66	82.21	90.57	94.25	96.17	85.17
D-Fusion	*Time & Freq.*	64.77	82.26	90.47	94.30	96.29	85.62
TFDPL	*Freq.*	59.63	75.22	85.65	91.48	94.75	81.35
	Time & Freq.	62.84	82.46	90.99	94.62	96.43	85.47
J-Fusion	*Time & Freq.*	**65.53**	**82.99**	**91.12**	**94.69**	**96.46**	**86.16**

Analysis on Perceptual Quality Metrics. Table 4 and 5 presents the average PESQ and STOI comparisons of different models' clean targets across five SNR levels and three types of unknown noise. The PESQ and STOI of TDPL's clean target are superior to TDSE's clean target, demonstrating the effectiveness of progressive learning methods in perceptual quality. The fusion results of TFDPL achieved the best PESQ and STOI scores across all SNRs. The PESQ and STOI of TDPL's clean target are superior to TDSE's clean target, demonstrating the effectiveness of progressive learning methods in perception. The fusion results of FDPL and TDPL outperformed their individual models, indicating the complementary nature of time and frequency domain information in improving perceptual quality. Particularly, at SNR of −10 dB, the STOI improvement relative to TDPL was 2.11, demonstrating the robustness of the time and frequency domain fusion in challenging environments.

For the progressive frequency-domain masking module of TFDPL, it still maintains improvement over FDPL. In the progressive mix-domain module, it shows an improvement in STOI compared to TDPL but a decrease in PESQ. However, the fusion of the two yields the best performance, with an improvement of 0.08 in PESQ and 0.76 in STOI compared to the fusion results of separately trained models, demonstrating its effectiveness in perceptual quality.

4   Conclusion

In this paper, we propose a TFDPL method for speech enhancement and recognition. TFDPL progressively predicts less-noisy and clearer speech, while estimating time-frequency masks and waveforms using information in the time and frequency domains, and further combines these two prediction targets through a fusion loss. Finally, the mutually beneficial effect of time-domain and frequency-domain is achieved. The experimental results demonstrate that TFDPL outperforms both time-domain and frequency-domain progressive learning methods, as well as their fusion results, in both ASR and human listener tasks. On the CHiME-4 real test set, TFDPL's intermediate output achieves relative WER reductions of 43.83% and 36.03% compared to the untreated noisy speech, under two different acoustic models. The clean output also demonstrates the best PESQ and STOI scores on the simulated test set. The positive experimental results demonstrate the effectiveness of the TFDPL approach.

References

1. Levinson, S.E., Rabiner, L.R., Sondhi, M.M.: An introduction to the application of the theory of probabilistic functions of a markov process to automatic speech recognition. Bell Syst. Tech. J. **62**(4), 1035–1074 (1983)
2. Zhang, Y., Chan, W., Jaitly, N.: Very deep convolutional networks for end-to-end speech recognition. In: IEEE International Conference on Acoustics, Speech and Signal Processing (ICASSP), pp. 4845–4849. IEEE (2017)
3. Loizou, P.C.: Speech Enhancement: Theory and Practice. CRC Press, Boca Raton (2013)
4. Tu, Y., Du, J., Sun, L., Ma, F., Lee, C.H.: On design of robust deep models for chime-4 multi-channel speech recognition with multiple configurations of array microphones. In: INTERSPEECH, pp. 394–398 (2017)
5. Wang, D.L., Chen, J.: Supervised speech separation based on deep learning: an overview. IEEE/ACM Trans. Audio Speech Lang. Process. **26**(10), 1702–1726 (2018)
6. Wang, D.L.: On ideal binary mask as the computational goal of auditory scene analysis. In: Speech Separation by Humans and Machines, pp. 181–197. Springer, Heidelberg (2005). https://doi.org/10.1007/0-387-22794-6_12
7. Yong, X., Jun, D., Dai, L.-R., Lee, C.-H.: An experimental study on speech enhancement based on deep neural networks. IEEE Signal Process. Lett. **21**(1), 65–68 (2013)
8. Paliwal, K., Wójcicki, K., Shannon, B.: The importance of phase in speech enhancement. Speech Commun. **53**(4), 465–494 (2011)

9. Kinoshita, K., Ochiai, T., Delcroix, M., Nakatani, T.: Improving noise robust automatic speech recognition with single-channel time-domain enhancement network. In: ICASSP 2020–2020 IEEE International Conference on Acoustics, Speech and Signal Processing (ICASSP), pp. 7009–7013. IEEE (2020)
10. Szu-Wei, F., Tao-Wei Wang, Yu., Tsao, X.L., Kawai, H.: End-to-end waveform utterance enhancement for direct evaluation metrics optimization by fully convolutional neural networks. IEEE/ACM Trans. Audio Speech Lang. Process. **26**(9), 1570–1584 (2018)
11. Pandey, A., Wang, D.L.: Dense CNN with self-attention for time-domain speech enhancement. IEEE/ACM Trans. Audio Speech Lang. Process. **29**, 1270–1279 (2021)
12. Wang, Q., et al.: Voicefilter-lite: streaming targeted voice separation for on-device speech recognition. arXiv preprint arXiv:2009.04323 (2020)
13. Yan-Hui, T., Jun, D., Gao, T., Lee, C.-H.: A multi-target snr-progressive learning approach to regression based speech enhancement. IEEE/ACM Trans. Audio Speech Lang. Process. **28**, 1608–1619 (2020)
14. Nian, Z., Tu, Y.H., Du, J., Lee, C.H.: A progressive learning approach to adaptive noise and speech estimation for speech enhancement and noisy speech recognition. In: ICASSP 2021–2021 IEEE International Conference on Acoustics, Speech and Signal Processing (ICASSP), pp. 6913–6917. IEEE (2021)
15. Nian, Z., Du, J., Yeung, Y.T., Wang, R.: A time domain progressive learning approach with snr constriction for single-channel speech enhancement and recognition. In: ICASSP 2022–2022 IEEE International Conference on Acoustics, Speech and Signal Processing (ICASSP), pp. 6277–6281. IEEE (2022)
16. Zhang, C.Y., Chen, H., Du, J., Yin, B.C., Pan, J., Lee, C.H: Incorporating visual information reconstruction into progressive learning for optimizing audio-visual speech enhancement. In: ICASSP 2023–2023 IEEE International Conference on Acoustics, Speech and Signal Processing (ICASSP), pp. 1–5. IEEE (2023)
17. Dang, F., Hu, Q., Zhang, P., Yan, Y.: Forknet: simultaneous time and time-frequency domain modeling for speech enhancement. arXiv preprint arXiv:2305.08292 (2023)
18. Tang, C., Luo, C., Zhao, Z., Xie, W., Zeng, W.: Joint time-frequency and time domain learning for speech enhancement. In: Proceedings of the Twenty-Ninth International Conference on International Joint Conferences on Artificial Intelligence, pp. 3816–3822 (2021)
19. Yong, X., Jun, D., Dai, L.-R., Lee, C.-H.: A regression approach to speech enhancement based on deep neural networks. IEEE/ACM Trans. Audio Speech Lang. Process. **23**(1), 7–19 (2014)
20. Zhou, N., Jun, D., Yan-Hui, T., Gao, T., Lee, C.-H.: A speech enhancement neural network architecture with snr-progressive multi-target learning for robust speech recognition. In: Asia-Pacific Signal and Information Processing Association Annual Summit and Conference (APSIPA ASC), pp. 873–877. IEEE (2019)
21. Luo, Y., Mesgarani, N.: Conv-tasnet: surpassing ideal time-frequency magnitude masking for speech separation. IEEE/ACM Trans. Audio Speech Lang. Process. **27**(8), 1256–1266 (2019)
22. Roux, J.L., Wisdom, S., Erdogan, H., Hershey, J.R.: SDR-half-baked or well done?. In: ICASSP 2019–2019 IEEE International Conference on Acoustics, Speech and Signal Processing (ICASSP), pp. 626–630. IEEE (2019)
23. Paul, D.B., Baker, J.: The design for the wall street journal-based CSR corpus. In: Speech and Natural Language: Proceedings of a Workshop Held at Harriman, New York, 23–26 February 1992 (1992)

24. Vincent, E., Watanabe, S., Nugraha, A.A., Barker, J., Marxer, R.: An analysis of environment, microphone and data simulation mismatches in robust speech recognition. Comput. Speech Lang. **46**, 535–557 (2017)
25. Varga, A., Steeneken, H.J.M.: Assessment for automatic speech recognition: Ii. noisex-92: a database and an experiment to study the effect of additive noise on speech recognition systems. Speech Commun. **12**(3), 247–251 (1993)
26. Kingma, D.P., Ba, J.: Adam: a method for stochastic optimization. arXiv preprint arXiv:1412.6980 (2014)
27. Cohen, I.: Noise spectrum estimation in adverse environments: improved minima controlled recursive averaging. IEEE Trans. Speech Audio Process. **11**(5), 466–475 (2003)
28. Saon, G., Soltau, H.: A comparison of two optimization techniques for sequence discriminative training of deep neural networks. In: 2014 IEEE International Conference on Acoustics, Speech and Signal Processing (ICASSP), pp. 5567–5571. IEEE (2014)
29. Kneser, R., Ney, H.: Improved backing-off for m-gram language modeling. In: 1995 International Conference on Acoustics, Speech, and Signal Processing, vol. 1, pp. 181–184. IEEE (1995)
30. Rix, A.W., Beerends, J.G., Hollier, M.P., Hekstra, A.P.: Perceptual evaluation of speech quality (pesq)-a new method for speech quality assessment of telephone networks and codecs. In: 2001 IEEE International Conference on Acoustics, Speech, and Signal Processing. Proceedings (Cat. No. 01CH37221), vol. 2, pp. 749–752. IEEE (2001)
31. Taal, C.H., Hendriks, R.C., Heusdens, R., Jensen, J.: An algorithm for intelligibility prediction of time-frequency weighted noisy speech. IEEE Trans. Audio Speech Lang. Process. **19**(7), 2125–2136 (2011)

A Study on Domain Adaptation for Audio-Visual Speech Enhancement

Chenxi Wang[1], Hang Chen[1], Jun Du[1(✉)], Chenyue Zhang[1], Yuling Ren[2], Qinglong Li[2], Ruibo Liu[2], and Chin-Hui Lee[3]

[1] University of Science and Technology of China, Hefei, China
jundu@ustc.edu.cn
[2] China Mobile Online Services Company Limited, Zhengzhou, China
[3] Georgia Institute of Technology, Atlanta, USA

Abstract. This paper presents the DA-AVSE system developed for the ASRU 2023 Audio-Visual Speech Enhancement (AVSE) Challenge. We initially employed three well-established AVSE models: MEASE, MTMEASE, and PLMEASE. These models demonstrated effectiveness even without utilizing matched data for training. To further enhance the performance, we introduced a domain adaptation method. More specifically, we utilized pseudo-labels generated by the models above in conjunction with the official baseline to fine-tune each model. Through extensive experiments, we observed that our method significantly improved the models' generalization to the target test set, regardless of whether the training and testing conditions matched. Additionally, we implemented a multi-model fusion strategy to enhance the overall model performance further. Our system exhibited significant improvements in all objective metrics, including PESQ, STOI, and SiSDR, compared to almost all competing teams. As a result, our system ranked the 2nd place in the objective metrics comparison for track 1.

Keywords: Speech enhancement · domain adaptation · audio-visual

1 Introduction

There are many speech-related applications, such as automatic speech recognition [1], hearing aids [2], video conferencing [3] in reality. But speech is often interfered with by noise in the daily acoustic environment, which has a bad impact on the performance of the applications. Therefore, speech enhancement to reduce noise while maintaining speech quality and intelligibility is important research with great practical application value.

The research on speech enhancement has been developed for decades. Traditional speech enhancement methods are based on statistical signal processing, including spectral subtraction [4], Wiener filtering [5] and minimum mean squared error (MMSE) estimation [6]. In recent years, deep learning has boomed and achieved promising results in speech enhancement [7,8]. Deep learning-based

© The Author(s), under exclusive license to Springer Nature Singapore Pte Ltd. 2024
J. Jia et al. (Eds.): NCMMSC 2023, CCIS 2006, pp. 53–65, 2024.
https://doi.org/10.1007/978-981-97-0601-3_5

speech enhancement methods recover clean speech from noisy speech via a deep neural network. They can better deal with non-stationary noise in real acoustic scenes than traditional methods. The deep learning-based methods can be divided into the audio-only speech enhancement (AOSE) method and the audio-visual speech enhancement (AVSE) method. It's difficult for the AOSE method to handle scenarios with multiple speakers, but the AVSE method can do better in this case because visual information can assist speech enhancement [9].

Many previous papers [10,11] have pointed out that the AVSE models bring significant improvements compared to audio-only algorithms, especially in low SNRs. Since the visual cues, such as facial and lip movements, are immune to acoustic conditions. Ephrat et al. [12] proposed an AVSE model for complex ratio mask estimation to separate speech from overlapping speech and background noises. Hou et al. [13] used a multimodal deep convolutional neural network that received both noisy audio and lip images as input and generated enhanced audio and lip images as output, fully leveraging visual information. Chen et al. [14] adopted a multi-modal embedding extracted from a pre-trained articulation place classifier to avoid performance distortion at high SNRs, and realized a significant improvement. [15] designed a lighter model structure by removing the visual feature extraction network with visual data compression, and experimental results confirmed that it provides better performance than the audio-only and AVSE systems without visual data compression. [16] fused audio-visual features layer by layer and introduced the channel and spectral attention mechanisms to pay more attention to informative regions of the fused AV feature maps.

Data-driven speech enhancement methods typically use many noisy and clean speech pairs for training. In the acoustic environment of practical application scenarios, noise types are always not included in the training, which degrades speech enhancement performance. To improve the generalization ability of the speech enhancement models for unseen acoustic environments, the usual practice is to increase the types of acoustic conditions during training as much as possible. However,it is not feasible to include all possible conditions during training, so the mismatch between training and testing of data-driven models has always been existing.

Domain adaptation can be used as a solution to the above problem. It has been extensively studied in computer vision in recent years [17,18], but it is uncommon in speech signal processing. Domain adaptation aims to transfer the knowledge learned from the source domain to the target domain. Domain adaptation can be used to address the degradation of performance when migrating speech enhancement models to unseen acoustic environments. [19] proposed a domain adversarial training method by utilizing unlabelled target domain noisy speech to extract noise-invariant features, which improved the speech enhancement performance in the unseen target domain. [20] designed a cross-task transfer learning method using paired senone classifiers to align speech signals in the source and target domains and transfer knowledge from the source domain to the target domain through multi-task learning. [21] used a Relativistic Discriminator and Multi-Kernel Maximum Mean Discrepancy (MK-MMD) to align the speech distributions between the source and target domains, thereby improving the speech enhancement model's performance on the target domain.

The Audio-Visual Speech Enhancement (AVSE) Challenge 2023 aims to explore novel approaches to audio-visual speech enhancement. The objective evaluation indicators of the challenge include PESQ, STOI and SISDR. The challenge is divided into two tracks: track 1, which prohibits the use of additional data, and track 2, which allows it. We participated in track 1. In our work, we first use the official training set released by the challenge to train three models, namely MEASE [14], MTMEASE [22] and PLMEASE, which is a model that applies the audio-visual progressive learning framework proposed in [23] to MEASE. These three models achieve competitive results on the official evaluation set. To improve the performance of our system on the evaluation set, we propose a simple and efficient domain adaptation method called Multi-Model Mixture Pseudo-Label Domain Adaptation (MMMP-DA). The method jointly uses pseudo-labels generated by multiple models to fine-tune the models. Specifically, we generate pseudo-labels for the unlabeled evaluation set using three trained models and the officially trained baseline [24]. Then a small-scale dataset is simulated with pseudo-labels to fine-tune the four models. The experimental results demonstrate that our MMMP-DA method can effectively leverage the complementarity among multiple models and enhance their adaptation ability to the evaluation set, regardless of whether the training and evaluation environments match.

Furthermore, we use a multi-model fusion strategy to fuse the enhanced speech predicted by all models, reducing the prediction bias of individual models and resulting in more robust and reliable predictions. After applying our MMMP-DA method and the multi-model fusion strategy, our DA-AVSE submission system achieves 1.77 PESQ, 71.23% STOI and 7.68 SISDR on the evaluation set of the AVSE Challenge 2023, which ranked 2nd in track 1.

2 Proposed System

2.1 Employed Model

We use four models in the challenge: the official baseline model, MEASE, MTMEASE and PLMEASE. The input of the official baseline model includes the magnitude spectrum and the video frames, and the output of the model is the mask of the magnitude spectrum. The MEASE model was proposed in [14], whose full name is multimodal embedding aware speech enhancement model. Its structure diagram is shown in Fig. 1. The MEASE includes a multimodal embedding extractor and an embedding-aware speech enhancement network. There are two branches inside the multimodal embedding extractor, which extract audio embedding and visual embedding respectively. The input of the audio branch is the Mel Filter Bank (FBANK) feature of noisy speech, and the audio embedding is obtained through a 1D convolutional layer, a batch normalization, a ReLU activation and an 18-layer variant of ResNet [25] in which all 2D convolutions are replaced with 1D convolutions. The input of the visual branch is the lip frames cropped from the video, and the visual embedding is obtained through a 3D convolutional layer, a batch normalization, a ReLU activation, a 3D max-pooling layer and an 18-layer ResNet. Then the audio and visual embeddings are concatenated over channel dimension and fused through the fusion module containing a 2-layer Bidirectional Gated Recurrent Unit (BiGRU) to obtain the multimodal embedding.

The 1D ConvBlock in the embedding-aware speech enhancement network includes a 1D convolution layer with a residual connection, a ReLU activation, and a batch normalization. The embedding-aware speech enhancement network takes a concatenation of multimodal embedding and noisy log power spectrum (LPS) [26] as input and predicts the ideal ratio mask (IRM) [27] of the noisy speech.

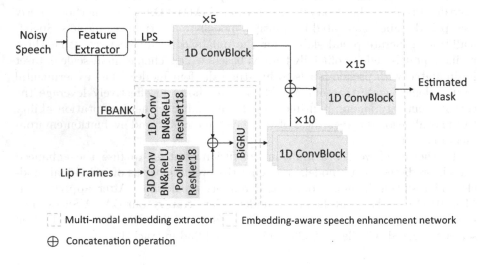

Fig. 1. Schematic diagram of MEASE.

The multimodal embedding extractor is pretrained with the articulation place classification task [14] and kept frozen during the embedding-aware speech enhancement network training. The forward propagation process of the enhancement network for a given sample (A, V), where A represents the noisy speech and V represents the corresponding video, is represented by Eq. (1), where W refers to the model parameters. The enhancement network is trained by minimizing the mean square error (MSE) loss between the predicted ideal ratio mask $\dot{M}$ and the target ideal ratio mask M, as shown in Eq. (2). During training, the network's parameters are updated by backpropagating the loss, which is represented by Eq. (3) and involves the learning rate lr.

$$\dot{M} = \mathcal{F}(A, V, W) \tag{1}$$

$$L_{\text{training}} = \sum \left\| \dot{M} - M \right\|_2^2 \tag{2}$$

$$W \leftarrow W - lr \frac{\partial \mathcal{L}_{\text{training}}}{\partial W} \tag{3}$$

Building on the work of MEASE, [22] considered the differences between audio and visual modalities. Specifically, audio contains more and finer acoustic information compared to video, which is less and rougher. Therefore, to prevent a loss of acoustic details, a finer classification task was used to extract audio embeddings. The phone label was found to be more suitable than the articulation place label due to its finer granularity. Expanding on this idea, [22] proposed a multi-task pre-training method for the multimodal embedding extractor, using both the phone and articulation place labels as training targets to extract more effective multimodal embeddings. The MEASE model that applies this multi-task pre-training method is called multi-task multimodal embedding aware speech enhancement (MTMEASE). The forward propagation and loss backpropagation methods of the enhancement network in the MTMEASE are the same as those in the enhancement network of the MEASE.

In [23], a mask-based audio-visual progressive learning speech enhancement (AVPL) model was proposed to address the problem of a large signal-to-noise ratio (SNR) gap between the learning target and input noisy speech. The model accomplishes this by dividing the mapping between noisy and clean speech into multiple stages, gradually narrowing the SNR gap at each stage. The first stage of AVPL takes a concatenation of the pre-trained visual embedding and the noisy LPS feature as input. In each subsequent stage, the visual embedding and the representation outputted by the previous stage are used as input. The final stage of AVPL predicts the IRM of the noisy speech. The AVPL is able to suppress more noise while preserving more spectral information. We replaced the input visual embeddings of AVPL with multimodal embeddings, and named the resulting model the progressive learning multimodal embedding aware speech enhancement (PLMEASE) model. The motivation for this change is to take advantage of the complementarity of audio and video modalities in multimodal embedding to further improve the performance of speech enhancement. The forward and backward propagation processes of PLMEASE are the same as those of AVPL described in [23].

2.2   Proposed Multi-Model Mixture Pseudo-Label Domain Adaptation (MMMP-DA) Method

Our method is model-agnostic and can be applied to multiple models. We regard the labeled training set as the source domain and the unlabeled evaluation set as the target domain. The models are trained on the source domain before applying our MMMP-DA method. The structure of the MMMP-DA method is shown in Fig. 2. Assuming a total of K AVSE models, the evaluation set has Q unlabeled noisy speech and the corresponding videos. We use $A = \{A_i, 0 \leq i < Q\}$ to represent the noisy speech feature set of the evaluation set, and $V = \{V_i, 0 \leq i < Q\}$ to represent the visual feature set. The steps of our method are summarized as follows:

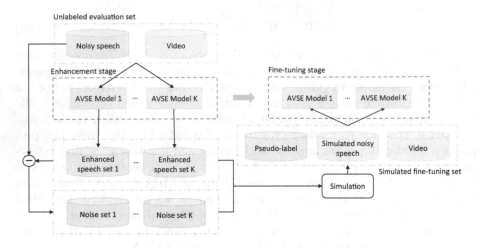

Fig. 2. The proposed MMMP-DA method.

Step 1: The first step is the enhancement stage. The K trained source AVSE models predict pseudo-labels on the target unlabeled evaluation set, respectively. For the i-th sample (A_i, V_i), the predicted mask of the j-th model $\dot{M}_i^j$ can be obtained using Eq. (4), where W_j represents the parameters of the j-th model.

$$\dot{M}_i^j = \mathcal{F}^j(A_i, V_i, W_j) \tag{4}$$

The enhanced speech feature and noise feature of the i-th sample generated by the j-th model can be obtained by Eq. (5) and Eq. (6), respectively.

$$\hat{E}_i^j = A_i \odot \dot{M}_i^j \tag{5}$$

$$\hat{N}_i^j = A_i \odot (1 - \dot{M}_i^j) \tag{6}$$

Each model can get an enhanced speech feature set and a noise feature set of evaluation set. We collect all the enhanced speech feature sets from the K models into a set represented in Eq. (7), and we collect the corresponding noise feature sets into another set represented as shown in Eq. (8).

$$\hat{E} = \left\{ \hat{E}_0^0, \cdots, \hat{E}_{Q-1}^0, \cdots, \hat{E}_0^{K-1}, \cdots, \hat{E}_{Q-1}^{K-1} \right\} \tag{7}$$

$$\hat{N} = \left\{ \hat{N}_0^0, \cdots, \hat{N}_{Q-1}^0, \cdots, \hat{N}_0^{K-1}, \cdots, \hat{N}_{Q-1}^{K-1} \right\} \tag{8}$$

Step 2: The second step is random simulation, where we create a dataset with a total duration of 20 h. For each sample, we randomly select an enhanced speech feature $\hat{E}_i^j$ from the set $\hat{E}$ and a noise feature $\hat{N}_u^v$ from the set $\hat{N}$. We then randomly select a value from the set signal-to-noise ratio (SNR) range as the SNR and use it to calculate the noise adjustment factor α during the simulation. We denote the simulated sample as $\tilde{A}_i^u$, and the simulation formula is shown in

Eq. (9). The corresponding visual feature is V_i. The label of this sample can be calculated with Eq. (10).

$$\tilde{A}_i^u = \hat{E}_i^j + \alpha \hat{N}_u^v \qquad (9)$$

$$\tilde{M}_i^u = \frac{\left\| \hat{E}_i^j \right\|_2^2}{\left\| \hat{E}_i^j \right\|_2^2 + \left\| \alpha \hat{N}_u^v \right\|_2^2} \qquad (10)$$

Step 3: The third step involves fine-tuning models using the simulated dataset separately. During the fine-tuning stage, we employ the k-fold cross-validation method [28] which divides the simulated dataset into k subsets for a total of k iterations. We leave one fold for testing each iteration and use the remaining k−1 folds to train the model. This method allowed us to fully use the data, avoid overfitting issues caused by insufficient data, and improve the model's generalization ability. When the input is the simulated sample $(\tilde{A}_i^u, V_i)$, the output mask $\hat{M}_i^j$ of the j-th model is calculated using Eq. (11).

$$\hat{M}_{i,u}^j = \mathcal{F}^j(\tilde{A}_i^u, V_i, W_j) \qquad (11)$$

The loss function we use during fine-tuning is the MSE loss between the predicted mask and the target mask, as shown in Eq. (12). The update formula of the model parameters is shown in Eq. (13), where lr is the learning rate.

$$L_{\text{finetuning}} = \sum \left\| \hat{M}_{i,u}^j - \tilde{M}_i^u \right\|_2^2 \qquad (12)$$

$$W \leftarrow W - lr \frac{\partial \mathcal{L}_{\text{finetuning}}}{\partial W} \qquad (13)$$

3 Experiments

3.1 Datasets

The AVSE Challenge 2023 provides 113 h of training set and 8.5 h of development set. Audio tracks of interferers are composed of a single competing speaker or a noise source in the following ranges: −15 dB to 5 dB (competing speaker) and −10 dB to 10 dB (noise) [24]. The videos of the target speakers and the competing speakers in the training set are selected from the LRS3 dataset [29]. Noise data mainly comes from Clarity Challenge (First edition) [30], Freesound [31], and DNS Challenge (Second edition) [32]. All audio files are monaural speech with a 16 kHz sampling frequency and 16 bits of bit depth. The target speakers, competing speakers and noise files of the training and development set are all disjoint but share the same noise categories.

The official evaluation set has 1,389 extracted sentences from 30 speakers. Approximately half of the mixed speech in the evaluation set has a competing speaker scenario while the other half has noise. There are six competing speakers.

The noise types used in the evaluation set are a subset of the noise types used in the training and development sets.

To evaluate the generalization ability of our MMMP-DA method, we create an acoustic environment that does not match the evaluation set. Specifically, we build an audio-visual dataset based on the benchmark released by the Multimodal Information-based Speech Processing (MISP) 2021 Challenge [33]. The MISP2021 Challenge Audio-Visual Speech Recognition (AVSR) dataset contains 122.53 h of audio-visual data recorded in a real-home TV room, with the language being Chinese. We utilize near-field audio and corresponding mid-field video from this dataset and mix it with MISP home-scene noise at 6 levels of SNRs (−15 dB, −10 dB, −5 dB, 0 dB, 5 dB, and 10 dB) to create a 132-h training set and a 10-h development set. We named this dataset as the MISP home-noise dataset. The acoustic environment of this dataset is significantly mismatched from that of the AVSE Challenge 2023 evaluation set, as the noise and language types are completely different.

3.2   Experimental Settings

For the MEASE, MTMEASE, and PLMEASE models, we utilize a 25-ms Hanning window and an overlap of 10 ms to extract audio frames during the audio preprocessing phase. The spectra of each frame is computed by a 400-point short-time Fourier transform. The videos are resampled to 25 fps and the lip area with a size of 98×98 pixels is cropped from each video frame. We compute 201-dimensional LPS features as the input of the enhancement network, 40-dimensional FBANK features and lip frames as input to the multimodal embedding extractor.

The objective metrics used on the challenge leaderboard include perceptual evaluation of speech quality (PESQ) [34], short-time objective intelligibility (STOI) [35] and Scale-Invariant Signal-to-Distortion Ratio (SISDR) [36]. Higher is better on three metrics.

The training methods of the models MEASE, MTMEASE and PLMEASE are similar. Firstly, we train the multimodal embedding extractor for 100 epochs with the Adam optimizer. The initial learning rate is 3e−4, decreasing on log scale after 30 epochs. After training the multimodal embedding extractor, we freeze it and train the speech enhancement network for 100 epochs using the Adam optimizer. The initial learning rate is set to 1e−3 and halved if the validation loss does not decrease for three epochs.

During random simulation, we generate a dataset of 20 h of noisy-clean pairs with SNRs ranging from −10 dB to 0 dB. In the fine-tuning stage, the number of cross-validation folds is set to 10. We use the Adam optimizer with an initial learning rate of 3e−4, which is halved if the validation loss fails to improve after three consecutive epochs.

3.3   Results of the MEASE, MTMEASE and PLMEASE in Matched Scenario

In this section, we present the results of the MEASE, MTMEASE and PLMEASE models on the evaluation set after being trained on the official training set, which represents a matched condition between training and test data. As evidenced quantitatively in Table 1, the three aforementioned models demonstrated competitive performance across all evaluation metrics.

Table 1. Performance of MEASE, MTMEASE and PLMEASE on the evaluation set in the matched scenario.

Model	PESQ	STOI (%)	SISDR
Noisy	1.14	44.10	−5.07
Baseline	1.41	55.63	3.67
MEASE	1.60	67.66	5.34
MTMEASE	1.61	67.86	5.46
PLMEASE	1.56	66.28	4.97

3.4   Performance Analysis of the Proposed MMMP-DA Method

To demonstrate the effectiveness of our proposed method, we compare the performance of our MMMP-DA method with a common single-model pseudo-label domain adaptation method, denoted as SMP-DA. The SMP-DA method only uses pseudo-labels and noise generated by a single model to simulate the fine-tuned dataset, which means that domain adaptation is only performed on a single model. We apply the SMP-DA and proposed MMMP-DA methods on four models: the official baseline, MEASE, MTMEASE and PLMEASE. The results are shown in Table 2.

Table 2. Performance comparison of SMP-DA and MMMP-DA methods on the baseline, MEASE, MTMEASE and PLMEASE models on the evaluation set in matched scenario.

Model	SMP-DA			MMMP-DA		
	PESQ	STOI(%)	SISDR	PESQ	STOI(%)	SISDR
Baseline	1.60	64.26	8.01	1.63	65.51	8.43
MEASE	1.64	69.09	5.79	1.68	70.30	6.26
MTMEASE	1.65	69.25	5.85	1.68	70.30	6.24
PLMEASE	1.61	68.54	5.67	1.64	69.25	5.93

By comparing the results presented in Tables 1 and 2, we can see that our MMMP-DA method consistently improves the performance of all models, indicating that our method is robust and effective. For example, the SISDR of the baseline model increased from 3.67 to 8.43, and the STOI of the PLMEASE model increased from 66.28% to 69.25%.

Comparing the results of the SMP-DA and MMMP-DA methods in Table 2, it is clear that the MMMP-DA method consistently outperforms the SMP-DA method. These findings suggest that leveraging prior knowledge from multiple models through the MMMP-DA method is more effective in facilitating domain adaptation than using a single model, as in the SMP-DA method. Specifically, the MMMP-DA method can leverage the complementarity among multiple models and capture a wider range of variations in the target evaluation set, enabling effective domain adaptation across multiple models. In contrast, using a single model for domain adaptation may lead to limited effectiveness in improving its performance due to the model's inability to capture all the variations in the target evaluation set.

To demonstrate the validity of our MMMP-DA method in mismatched scenarios, we evaluate the performance of the MEASE model with and without the MMMP-DA method applied in a mismatched scenario using the MISP home-noise dataset as the training set. We also set up a fine-tuning scenario training with the MISP home-noise dataset and fine-tuning on the official development set to compare with the MMMP-DA method. The number of cross-validation folds when fine-tuning with the development set and the learning rate setting are the same as that of our MMMP-DA method. We present the performance comparison on the evaluation set in Fig. 3.

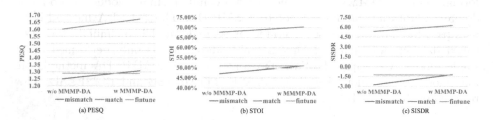

Fig. 3. The performance comparison of MEASE on the evaluation set. "w/o MMMP-DA" stands for MEASE without applying our MMMP-DA method. "w MMMP-DA" stands for MEASE applying our MMMP-DA method. "mismatch" represents that MEASE is trained in the mismatched scenario, that is, the MISP home-noise dataset. "match" represents that MEASE is trained in the matched scenario which is the official training set. "finetune" means that the MEASE is fine-tuned with a small amount of matching data after training in mismatched scenario.

Figure 3 shows that the MMMP-DA method steadily improves the performance of the MEASE model in all three evaluation metrics, regardless of whether the environments are matched or mismatched. In the case of mismatched environments, our method enables the model to achieve comparable or even better performance than the model fine-tuned with the matched development set. This suggests that the MMMP-DA method can effectively improve the model's generalization and make it more suitable for the evaluation set, even when labeled real matching data is not available during the fine-tuning stage.

3.5 Objective Results on AVSE Challenge 2023

In this section, we introduce the method we use in the challenge. We first use the official training set to train the MEASE, MTMEASE and PLMEASE models, and then we jointly apply our MMMP-DA method on these three models and the official trained baseline provided by the challenge. Finally, we fuse the prediction results of the evaluation set from the four models, that is, average the predicted time domain waveforms to obtain the final predicted waveform. This operation can effectively reduce the bias and variance of a single model, thereby improving the stability and accuracy of predictions.

Table 3. Performance comparison of our system and some other competing systems on the evaluation set.

System	PESQ	STOI (%)	SISDR
Baseline	1.41	55.63	3.67
BioASP_CITI	1.41	53.78	3.61
AVSE02	1.61	68.21	8.81
Ict_avsu	1.66	67.67	6.73
Merl	2.71	83.76	14.43
DA-AVSE	1.77	71.23	7.68

Table 3 presents a comparison of the performance of our DA-AVSE system with some other participating systems in the challenge track 1. The table shows that our DA-AVSE system outperforms almost all other competing systems regarding all three evaluation metrics. Specifically, our system achieves a PESQ of 1.77, an STOI of 71.23%, a SISDR of 7.68 dB. These results demonstrate the effectiveness of the MMMP-DA method and the model fusion strategy.

4   Conclusions

This paper presents our submission to the AVSE Challenge 2023 track 1, where we introduced three AVSE models: MEASE, MTMEASE, and PLMEASE, yielding competitive results. Additionally, we propose a domain adaptation method called MMMP-DA to enhance the model's generalization capability to the target test set. By applying the MMMP-DA method to the official baseline, as well as the MEASE, MTMEASE, and PLMEASE models, and employing a multi-model fusion strategy, we further improve the overall model performance. Our final results on the evaluation set demonstrate notable achievements, ranked 2nd on track 1 in terms of objective metrics.

References

1. Li, C.-Y., Vu, N.T.: Improving speech recognition on noisy speech via speech enhancement with multi-discriminators CycleGAN. In: 2021 IEEE Automatic Speech Recognition and Understanding Workshop (ASRU), pp. 830–836 (2021)
2. Panahi, I., Kehtarnavaz, N., Thibodeau, L.: Smartphone-based noise adaptive speech enhancement for hearing aid applications. In: 2016 38th Annual International Conference of the IEEE Engineering in Medicine and Biology Society (EMBC), pp. 85–88 (2016)
3. Cutler, R., et al.: Multimodal active speaker detection and virtual cinematography for video conferencing. In: ICASSP 2020, pp. 4527–4531 (2020)
4. Boll, S.: Suppression of acoustic noise in speech using spectral subtraction. IEEE Trans. Acoust. Speech Signal Process. **27**(2), 113–120 (1979)
5. Lim, J., Oppenheim, A.: All-pole modeling of degraded speech. IEEE Trans. Acoust. Speech Signal Process. **26**(3), 197–210 (1978)
6. Ephraim, Y., Malah, D.: Speech enhancement using a minimum mean-square error log-spectral amplitude estimator. IEEE Trans. Acoust. Speech Signal Process. **33**(2), 443–445 (1985)
7. Wang, Y., Narayanan, A., Wang, D.L.: On training targets for supervised speech separation. IEEE/ACM Trans. Audio Speech Lang. Process. **22**(12), 1849–1858 (2014)
8. Yong, X., Jun, D., Dai, L.-R., Lee, C.-H.: A regression approach to speech enhancement based on deep neural networks. IEEE/ACM Trans. Audio Speech Lang. Process. **23**(1), 7–19 (2015)
9. Shetu, S.S., Chakrabarty, S., Habets, E.A.P.: An empirical study of visual features for DNN based audio-visual speech enhancement in multi-talker environments. In: ICASSP 2021, pp. 8418–8422 (2021)
10. Sumby, W.H., Pollack, I.: Visual contribution to speech intelligibility in noise. J. Acoust. Soc. Am. **26**(2), 212–215 (1954)
11. McGurk, H., MacDonald, J.: Hearing lips and seeing voices. Nature **264**(5588), 746–748 (1976)
12. Ephrat, A., et al.: Looking to listen at the cocktail party: a speaker-independent audio-visual model for speech separation. arXiv preprint arXiv:1804.03619 (2018)
13. Hou, J.-C., Wang, S.-S., Lai, Y.-H., Tsao, Y., Chang, H.-W., Wang, H.-M.: Audio-visual speech enhancement using multimodal deep convolutional neural networks. IEEE Trans. Emerg. Top. Comput. Intell. **2**(2), 117–128 (2018)
14. Chen, H., Du, J., Hu, Y., Dai, L.-R., Yin, B.-C., Lee, C.-H.: Correlating subword articulation with lip shapes for embedding aware audio-visual speech enhancement. Neural Netw. **143**, 171–182 (2021)
15. Chuang, S.-Y., Tsao, Y., Lo, C.-C., Wang, H.-M.: Lite audio-visual speech enhancement. arXiv preprint arXiv:2005.11769 (2020)
16. Xu, X., et al.: MFFCN: multi-layer feature fusion convolution network for audio-visual speech enhancement. arXiv preprint (2021)
17. Mansour, Y., Mohri, M., Rostamizadeh, A.: Domain adaptation: learning bounds and algorithms. CoRR, abs/0902.3430 (2009)
18. Tzeng, E., Hoffman, J., Saenko, K., Darrell, T.: Adversarial discriminative domain adaptation. In: 2017 IEEE Conference on Computer Vision and Pattern Recognition (CVPR), pp. 2962–2971 (2017)
19. Liao, C.-F., Tsao, Y., Lee, H.-Y., Wang, H.-M.: Noise adaptive speech enhancement using domain adversarial training. In: Proceedings of the Interspeech 2019, pp. 3148–3152 (2019)

20. Wang, S., Li, W., Siniscalchi, S.M., Lee, C.-H.: A cross-task transfer learning approach to adapting deep speech enhancement models to unseen background noise using paired senone classifiers. In: ICASSP 2020, pp. 6219–6223 (2020)
21. Cheng, J., Liang, R., Liang, Z., Zhao, L., Huang, C., Schuller, B.: A deep adaptation network for speech enhancement: combining a relativistic discriminator with multi-kernel maximum mean discrepancy. IEEE/ACM Trans. Audio Speech Lang. Process. **29**, 41–53 (2021)
22. Wang, C., Chen, H., Du, J., B., Pan, J.: Multi-task joint learning for embedding aware audio-visual speech enhancement. In: 2022 13th International Symposium on Chinese Spoken Language Processing (ISCSLP), pp. 255–259 (2022)
23. Zhang, C.-Y., Chen, H., Du, J., Yin, B.-C., Pan, J., Lee, C.-H.: Incorporating visual information reconstruction into progressive learning for optimizing audio-visual speech enhancement. In: ICASSP 2023, pp. 1–5 (2023)
24. Blanco, A.L.A., et al.: AVSE challenge: audio-visual speech enhancement challenge. In: 2022 IEEE Spoken Language Technology Workshop (SLT), pp. 465–471 (2023)
25. He, K., Zhang, X., Ren, S., Sun, J.: Deep residual learning for image recognition. In: Proceedings of the IEEE Conference on Computer Vision and Pattern Recognition, pp. 770–778 (2016)
26. Du, J., Huo, Q.: A speech enhancement approach using piecewise linear approximation of an explicit model of environmental distortions. In: Interspeech (2008)
27. Hummersone, C., Stokes, T., Brookes, T.S.: On the ideal ratio mask as the goal of computational auditory scene analysis (2014)
28. Yadav, S., Shukla, S.: Analysis of k-fold cross-validation over hold-out validation on colossal datasets for quality classification. In: 2016 IEEE 6th International Conference on Advanced Computing (IACC), pp. 78–83 (2016)
29. Afouras, T., Chung, J.S., Zisserman, A.: LRS3-TED: a large-scale dataset for visual speech recognition. CoRR, abs/1809.00496 (2018)
30. Graetzer, S., et al.: Clarity-2021 challenges: machine learning challenges for advancing hearing aid processing. In: Interspeech (2021)
31. Thiemann, J., Ito, N., Vincent, E.: The diverse environments multi-channel acoustic noise database (DEMAND): a database of multichannel environmental noise recordings. In: Proceedings of Meetings on Acoustics, vol. 19, no. 1, p. 035081 (2013)
32. Reddy, C.K.A., et al.: ICASSP 2021 deep noise suppression challenge. In: ICASSP 2021, pp. 6623–6627 (2021)
33. Chen, H., et al.: The first multimodal information based speech processing (MISP) challenge: data, tasks, baselines and results. In: ICASSP 2022, pp. 9266–9270. IEEE (2022)
34. Rix, A.W., Beerends, J.G., Hollier, M.P., Hekstra, A.P.: Perceptual evaluation of speech quality (PESQ)-a new method for speech quality assessment of telephone networks and codecs. In: Proceedings of the 2001 IEEE International Conference on Acoustics, Speech, and Signal Processing (Cat. No.01CH37221), vol. 2, pp. 749–752 (2001)
35. Taal, C.H., Hendriks, R.C., Heusdens, R., Jensen, J.: An algorithm for intelligibility prediction of time-frequency weighted noisy speech. IEEE Trans. Audio Speech Lang. Process. **19**(7), 2125–2136 (2011)
36. Vincent, E., Gribonval, R., Fevotte, C.: Performance measurement in blind audio source separation. IEEE Trans. Audio Speech Lang. Process. **14**(4), 1462–1469 (2006)

APNet2: High-Quality and High-Efficiency Neural Vocoder with Direct Prediction of Amplitude and Phase Spectra

Hui-Peng Du, Ye-Xin Lu, Yang Ai$^{(\boxtimes)}$, and Zhen-Hua Ling

National Engineering Research Center of Speech and Language Information Processing, University of Science and Technology of China, Hefei, People's Republic of China
{redmist,yxlu0102}@mail.ustc.edu.cn, {yangai,zhling}@ustc.edu.cn

Abstract. In our previous work, we have proposed a neural vocoder called APNet, which directly predicts speech amplitude and phase spectra with a 5 ms frame shift in parallel from the input acoustic features, and then reconstructs the 16 kHz speech waveform using inverse short-time Fourier transform (ISTFT). The APNet vocoder demonstrates the capability to generate synthesized speech of comparable quality to the HiFi-GAN vocoder but with a considerably improved inference speed. However, the performance of the APNet vocoder is constrained by the waveform sampling rate and spectral frame shift, limiting its practicality for high-quality speech synthesis. Therefore, this paper proposes an improved iteration of APNet, named APNet2. The proposed APNet2 vocoder adopts ConvNeXt v2 as the backbone network for amplitude and phase predictions, expecting to enhance the modeling capability. Additionally, we introduce a multi-resolution discriminator (MRD) into the GAN-based losses and optimize the form of certain losses. At a common configuration with a waveform sampling rate of 22.05 kHz and spectral frame shift of 256 points (i.e., approximately 11.6 ms), our proposed APNet2 vocoder outperforms the original APNet and Vocos in terms of synthesized speech quality. The synthesized speech quality of APNet2 is also comparable to that of HiFi-GAN and iSTFTNet, while offering a significantly faster inference speed.

Keywords: Neural vocoder · Amplitude spectrum · Phase spectrum · ConvNeXt v2 · Multi-resolution discriminator

1 Introduction

Neural vocoder technology, which converts speech acoustic features into waveforms, has seen rapid progress in recent years. The vocoder capability signifi-

This work was funded by the Anhui Provincial Natural Science Foundation under Grant 2308085QF200 and the Fundamental Research Funds for the Central Universities under Grant WK2100000033.

cantly affects the performance of several speech generation applications, such as text-to-speech (TTS) synthesis, singing voice synthesis (SVS), bandwidth extension (BWE), speech enhancement (SE), and voice conversion (VC).

The synthesized speech quality and inference efficiency are two major indicators for evaluating vocoders. Although auto-regressive (AR) neural vocoders such as WaveNet [1] and SampleRNN [2] achieved significant improvements in synthesized speech quality compared to traditional signal-processing-based vocoders [3,4], they demanded considerable computational cost and exhibited extremely low generation efficiency due to their auto-regressive inference mode on raw waveforms. Consequently, alternative approaches have been proposed, including knowledge-distilling-based models (e.g., Parallel WaveNet [5] and ClariNet [6]), flow-based models (e.g., WaveGlow [7] and WaveFlow [8]), and glottis-based models (e.g., GlotNet [9] and LPCNet [10]). While these models have substantially improved inference efficiency, their overall computational complexity remains elevated, limiting their applicability in resource-constrained environments like embedded devices. Recently, there has been increasing attention towards waveform generation models that eschew auto-regressive or flow-like structures. An example of such a model is the neural source-filter (NSF) model [11], which integrates neural networks with speech production mechanisms to directly generate speech waveforms using explicit F0 and mel-spectrograms. Furthermore, generative adversarial network (GAN) [12] based vocoders, such as WaveGAN [13], MelGAN [14], and HiFi-GAN [15], employ GANs to achieve both high-quality synthesized speech and improved inference efficiency. These methods directly predict waveforms without relying on complex structures. However, to compensate for the difference in temporal resolution between input acoustic features and output waveforms, typical temporal GAN vocoders need multiple transposed convolutions to upsample the input features to the desired sample rate, incurring substantial computational cost.

To solve this issue, several vocoders (e.g., iSTFTNet [16], Vocos [17], and APNet [18]) turn to predict amplitude and phase spectra and finally use inverse short-time Fourier transform (ISTFT) to reconstruct waveforms. This type of approach effectively avoids the direct prediction of high-resolution waveforms. Compared with HiFi-GAN [15], iSTFTNet [16] employs fewer upsampling layers to estimate amplitude and phase spectra at a large time-domain resolution, which still consumes much computational cost. Furthermore, Vocos [17] leverages ConvNeXt [19] as its backbone network, omitting upsampling layers to directly predict amplitude and phase spectra at the same temporal resolution as the input acoustic features. However, without effective tools for precise phase estimation, iSTFTNet and Vocos only define waveform- and amplitude-related losses without constraining the predicted phase spectra. This may lead to insufficient phase prediction accuracy and black-box problem of phase prediction. In our previous work, we proposed the APNet vocoder [18] as a means to overcome this limitation. The APNet vocoder can also directly predict the amplitude and phase spectra at the original resolution and then reconstruct the waveforms. Differently, to ensure phase prediction accuracy, the APNet vocoder adopts a phase parallel estimation architecture and employs anti-wrapping losses to explicitly model and optimize the phase, respectively. Through experiments at the configuration with a waveform sampling rate of 16 kHz and spectral frame shift of 5 ms, the APNet vocoder achieved comparable synthesized speech quality as the HiFi-GAN vocoder, while achieving a remarkable eightfold increase in inference speed on a CPU.

However, in further investigation, we discover that phase prediction is highly sensitive to spectral frame shift [20]. When frame shift increases, phase continuity deteriorates, and the modeling capabilities of the existing APNet vocoder may prove insufficient. In preliminary experiments, we draw a conclusion that accurately modeling the temporal discontinuity of long-frame-shift phase spectra poses challenges for the APNet vocoder, which can lead to a noticeable reduction in phase prediction accuracy. But, accurately predicting long-frame-shift spectra is crucial in several speech generation tasks. One particular challenge arises when generating short-frame-shift features using attention-based acoustic models [21,22], as aligning texts and features can be difficult and result in improper alignment. Besides, 16 kHz waveform generation may not meet the requirements for high-quality speech synthesis applications. Currently, most neural vocoders take 80-dim mel-spectrograms in the 0–8000 Hz frequency range as input and generate speech waveforms at a 22.05 kHz sampling rate for a fair comparison. This process also implicitly includes the operation of BWE.

Therefore, this paper proposes an improved iteration of APNet called APNet2. The proposed APNet2 vocoder further augments its modeling capabilities, making it adaptable to scenarios with higher waveform sampling rates and extended spectral frame shifts. Similar to APNet, APNet2 is also composed of an amplitude spectrum predictor (ASP) and a phase spectrum predictor (PSP), which predict the speech amplitude and phase spectra from acoustic features in parallel, and then reconstruct the waveforms through ISTFT. Differently, the APNet2 vocoder adopts the ConvNeXt v2 [23] as the backbone network of both ASP and PSP, bolstering its modeling capabilities. Regarding the phase-related loss function, we have updated the anti-wrapping function from a negative cosine function to a linear function to yield more precise phase predictions. In the waveform-related GAN loss, we introduce a multi-period discriminator (MPD) and multi-resolution discriminator (MRD), using a hinge formulation. Under the conditions of a waveform sampling rate of 22.05 kHz and a spectral frame shift of approximately 11.6 ms, our proposed APNet2 vocoder significantly outperforms the original APNet and Vocos, and is comparable to HiFi-GAN and iSTFTNet in terms of synthesized speech quality for both analysis-synthesis and TTS tasks. In terms of efficiency, the inference speed of the proposed APNet2 is obviously faster than that of HiFi-GAN and iSTFTNet.

The rest of this paper is organized as follows. Section 2 briefly reviews some related works including iSTFTNet, Vocos, and APNet. In Sect. 3, we give details of our proposed APNet2 vocoder. The experimental results and analysis are presented in Sect. 4. Finally, we make a conclusion and preview some areas of future research in Sect. 5.

2   Related Work

Neural vocoders have garnered extensive attention in the fields of signal processing and machine learning. The method of using ISTFT to avoid directly predicting time-domain waveforms has gained traction and demonstrated progress [16–18, 24–26]. Figure 1 shows the concise architectures and loss functions of representative models, including iSTFTNet [16], Vocos [17], APNet [18], and the proposed APNet2.

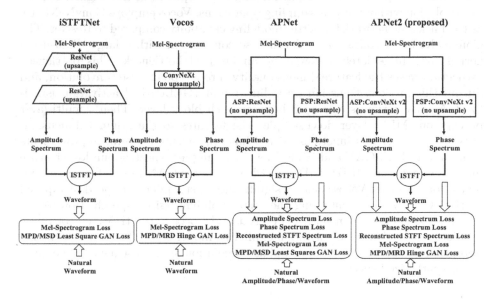

Fig. 1. Overview of the concise architectures and loss functions of iSTFTNet, Vocos, APNet, and our proposed APNet2.

2.1   iSTFTNet

As illustrated in Fig. 1, iSTFTNet [16] initially processes the mel-spectrogram through multiple residual convolutional neural networks (ResNets) with upsampling operations to obtain the amplitude and phase spectra. Subsequently, it reconstructs the waveform through the ISTFT operation. It is evident that, through the upsampling layers, the amplitude and phase spectra predicted by iSTFTNet have higher temporal resolution and lower frequency resolution. As mentioned in the original paper [16], upsampling operations are inevitable to ensure the quality of the synthesized speech. Hence, iSTFTNet does not achieve a truly all-frame-level amplitude and phase prediction, leaving room for improvement in terms of inference efficiency. The loss functions employed by iSTFTNet align with those of HiFi-GAN [15], which includes mel-spectrogram loss, feature matching loss, and MPD/multi-scale discriminator (MSD) based least squares GAN losses. Experimental results show that the iSTFTNet achieves faster inference speed than HiFi-GAN, and its synthesized speech quality is comparable to that of HiFi-GAN.

2.2 Vocos

As illustrated in Fig. 1, Vocos [17] is an all-frame-level neural vocoder, which simultaneously predicts the amplitude and phase spectra at the original temporal resolution without any upsampling operations. Vocos employs ConvNeXt [19] as the backbone network for better modeling capability compared to ResNet. The ConvNeXt block consists of a depth-wise convolution with a larger-than-usual kernel size, immediately followed by an inverted bottleneck. This bottleneck projection raises the feature dimensionality through point-wise convolution, and during this process, Gaussian error linear unit (GELU) [27] activation is utilized. Normalization is employed between each block layer. Then a multi-layer perceptron (MLP) layer downsamples the features to the original dimensionality. In terms of loss functions, Vocos has implemented several improvements compared to HiFi-GAN and iSTFTNet, including mel-spectrogram loss, feature matching loss, and MPD/MRD-based least squares GAN losses. Empirical evidence confirms that Vocos achieves significantly faster inference speeds compared to HiFi-GAN while maintaining the quality of synthesized speech. However, a common issue existing in both iSTFTNet and Vocos is the treatment of phase prediction as a black box without explicitly modeling it. This approach may potentially impact the accuracy of phase prediction and, consequently, the quality of synthesized speech.

2.3 APNet

As illustrated in Fig. 1, APNet [18] consists of an ASP and a PSP. These two components work in parallel to predict the amplitude and phase spectra, which are then employed to reconstruct the waveform through ISTFT. The backbone of both the ASP and PSP is the ResNet without any upsampling operations. Specifically, the ResNet consists of several parallel residual convolutional blocks (ResBlocks), featuring a large number of dilated convolutions. Then, the outputs of each ResBlock are summed, averaged, and finally activated by a leaky rectified linear unit (ReLU) [28] activation. In contrast to ASP, what distinguishes PSP is its emphasis on the characteristics of the wrapped phase and the introduction of a phase parallel estimation architecture at the output end. The parallel estimation architecture is composed of two parallel linear convolutional layers and a phase calculation formula denoted as Φ, simulating the process of computing phase spectra from short-time complex spectra. The formula Φ is a bivariate function defined as follows:

$$\Phi(R, I) = \arctan\left(\frac{I}{R}\right) - \frac{\pi}{2} \cdot Sgn^*(I) \cdot [Sgn^*(R) - 1],   (1)$$

where R and I represent the pseudo-real and imaginary parts output by the two parallel linear convolutional layers, respectively; $Sgn^*(x)$ is a new symbolic function defined in [18]: when $x \geq 0$, $Sgn^*(x) = 1$, otherwise $Sgn^*(x) = -1$.

A series of loss functions are defined in APNet to guide the generation of spectra and waveforms, including: 1) amplitude spectrum loss $\mathcal{L}_A$, which is the L^2 distance of the predicted logarithmic amplitude spectrum and the natural one; 2) phase spectrum loss $\mathcal{L}_P$, which is the sum of instantaneous phase loss, group delay loss, and phase time difference loss, all activated by negative cosine anti-wrapping function, measuring the gap between the predicted phase spectrum and the natural one at various perspectives; 3) reconstructed STFT spectrum loss $\mathcal{L}_S$, which includes the STFT consistency loss between the reconstructed STFT spectrum and the consistent one, and L^1 distance of the real parts and imaginary parts between the reconstructed STFT spectrum and the natural one; 4) final waveforms loss $\mathcal{L}_W$, which is the same as used in HiFi-GAN [15], including mel-spectrogram loss, feature matching loss, and MPD/MSD least squares GAN loss. We still use the names of these functions in APNet2, but only change the definition of some functions which will be introduced in Sect. 3.2.

3   Proposed Method

As illustrated in Fig. 2, the proposed APNet2 vocoder directly predicts speech amplitude spectra and phase spectra at unique raw temporal resolution from input acoustic features (e.g., mel-spectrogram) in parallel. Subsequently, the amplitude and phase spectra are reconstructed to the STFT spectrum, from which the waveform is ultimately recovered via ISTFT.

3.1   Model Structure

The proposed APNet2 vocoder comprises an ASP and a PSP. The ASP and PSP aim to directly predict the logarithmic amplitude spectrum and the wrapped phase spectrum from the input mel-spectrogram, respectively. The structure of ASP is a cascade of an input convolutional layer, a ConvNeXt v2 network [23], and an output convolutional layer, while that of PSP is a cascade of input convolutional layers, a ConvNeXt v2 network, and a phase parallel estimation architecture. The phase parallel estimation architecture, adopted from APNet [18], is specially designed to ensure the direct output of the wrapped phase spectrum.

We employ ConvNeXt v2 as the backbone for both the ASP and PSP in APNet2 rather than the ResNet used in APNet because ConvNeXt v2 has demonstrated strong modeling capabilities in the field of image processing. The ConvNeXt v2 is a cascade of k identical ConvNeXt v2 blocks. As shown in Fig. 2, each ConNeXt v2 block contains a large-kernel-sized depth-wise convolutional layer, a layer normalization operation, a point-wise convolutional layer that elevates feature dimensions, a GELU activation, a global response normalization (GRN) layer, and another point-wise convolutional layer that restores features

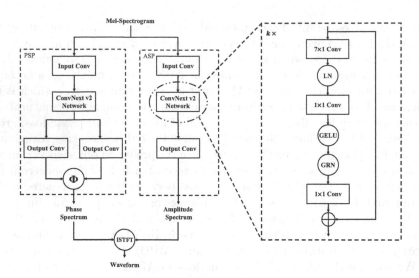

Fig. 2. The architecture of APNet2. Φ and ISTFT represent the phase calculation formula and inverse short-time Fourier transform, respectively. The content in the dotted box is the specific structure of the ConvNeXt v2 block, where LN, GELU, and GRN represent layer normalization, Gaussian error linear unit, and global response normalization, respectively.

to their original dimensionality. Finally, residual connections are employed, with the input added to the output of the last point-wise convolutional layer as the block's output. Compared to ConvNeXt [19] used in Vocos, ConvNeXt v2 introduces a GRN layer positioned after the dimension-expansion MLP and drop layer scale operation, aiming to increase the contrast and selectivity of channels. The GRN layer consists of global feature aggregation, feature normalization, and feature calibration, improving the representation quality by enhancing the feature diversity.

3.2   Training Criteria

In the APNet [18], we incorporate MPD and MSD borrowed from [15]. However, the high parameter count of MSD results in slow training speed. As mentioned in [29], the accuracy discriminative of MSD can reach 100% during training, indicating overfitting of this discriminator. In the proposed APNet2, we replaced MSD with MRD [30], whose key elements are strided 2-D convolutions and leaky ReLU activations [28], as shown in Fig. 3. The MRD is divided into multiple sub-discriminators, each operating at different resolutions. Each sub-discriminator first extracts amplitude spectra at a certain temporal and spectral resolution from input natural/predicted waveforms with certain STFT parameters, and then outputs a discriminant value. The MRD is expected to cover a wide range of scales of speech waveforms as comprehensively as possible. As recommended

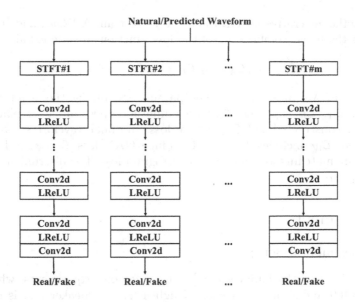

Fig. 3. Architecture of MRD. STFT#m denotes computing the amplitude spectrum using the m-th STFT parameter set and LReLU denotes the leaky ReLU activation.

in [31], we discard the least squares GAN loss used in the APNet, and adopt a hinge GAN loss here as follows:

$$\mathcal{L}_{GAN-G}(\hat{\boldsymbol{x}}) = \frac{1}{L} \sum_l \max(0, 1 - D_l(\hat{\boldsymbol{x}})), \qquad (2)$$

$$\mathcal{L}_{GAN-D}(\boldsymbol{x}, \hat{\boldsymbol{x}}) = \frac{1}{L} \sum_l \left\{ \max(0, 1 - D_l(\boldsymbol{x})) + \max(0, 1 + D_l(\hat{\boldsymbol{x}})) \right\}, \qquad (3)$$

where D_l is the l-th sub-discriminator in MPD and MRD, and L is the total number of sub-discriminators in MPD and MRD. $\boldsymbol{x}$ and $\hat{\boldsymbol{x}}$ are the natural and the synthesized waveforms, respectively.

We retain all the losses in the APNet [18] which are defined on amplitude spectra, phase spectra, reconstructed STFT spectra, and final waveforms. Refer to [32], we make slight modifications to the phase-related losses, i.e., using a linear function f_{AW} as the anti-wrapping function and discarding the negative cosine function used in the APNet. In our preliminary experiments, we confirmed that the linear form is more suitable for phase prediction than the cosine form. The linear function f_{AW} is defined as:

$$f_{AW}(x) = \left| x - 2\pi \cdot round\left(\frac{x}{2\pi}\right) \right|, \qquad (4)$$

and used to activate the instantaneous phase error, group delay error, and phase time difference error between predicted and natural phases for mitigating the error expansion issue caused by phase wrapping.

We use the generative adversarial strategy to train APNet2, and the generator loss is the linear combination of the loss functions mentioned above:

$$\mathcal{L}_G = \lambda_A \mathcal{L}_A + \lambda_P \mathcal{L}_P + \lambda_S \mathcal{L}_S + \lambda_W \mathcal{L}_W, \tag{5}$$

where λ_A, λ_P, λ_S, and λ_W are hyperparameters, taking the same values as in APNet. $\mathcal{L}_A$, $\mathcal{L}_P$, $\mathcal{L}_S$, and $\mathcal{L}_W$ are amplitude spectrum loss, phase spectrum loss, reconstructed STFT spectrum loss, and final waveforms loss, respectively, where $\mathcal{L}_W$ includes MPD/MRD hinge GAN loss $\mathcal{L}_{GAN-G}$ defined as Eq. 2, feature matching loss, and mel-spectrogram loss. The discriminator loss is $\mathcal{L}_D = \mathcal{L}_{GAN-D}$.

4   Experiments

4.1   Experimental Setup

Dataset. We used the LJSpeech [33] dataset for our experiments, which consists of 13,100 audio clips of a single English female speaker and is of about 24 h. The audio sampling rate is 22.05 kHz with a format of 16-bit PCM. We randomly selected 12,000 audio clips for training, 100 for validation, and 500 for testing. Spectral features (e.g., 80-dimensional mel-spectrograms, amplitude spectra, phase spectra) were extracted by STFT with an FFT point number of 1024, frame shift of 256 (i.e., approximately 11.6 ms), and frame length of 1024 (i.e., approximately 46.4 ms). Note that the mel-spectrogram is not a full-band spectrum. It only covers the frequency range of 0 to 8000 Hz to be aligned with the common configuration [15,16]. Therefore, the vocoders generate waveforms from the input mel-spectrogram in this configuration that, in fact, imply the operation of BWE.

Implementation. In the proposed APNet2 vocoder, the number of ConvNeXt v2 blocks k in both ASP and PSP was set to 8. In each ConvNeXt v2 block, the kernel size and channel size of the large-kernel-sized depth-wise convolutional layer were set to 7 and 512, respectively. The channel sizes of the first and the last 1×1 point-wise convolutional layers were set to 512 and 1536, respectively.

We trained our proposed APNet2 vocoder up to 2 million steps, with 1 million steps per generator and discriminator, on a single Nvidia 2080Ti GPU. During training, we randomly cropped the audio clips to 8192 samples and set the batch size to 16. The model is optimized using the AdamW optimizer [34] with $\beta_1 = 0.8$, $\beta_2 = 0.99$, and weight decay of 0.01. The learning rate was set initially to 2×10^{-4} and scheduled to decay with a factor of 0.999 at every epoch.

Baselines. We compared the proposed APNet2 to HiFi-GAN[1], iSTFTNet[2], Vocos[3] and APNet[4]. All these compared vocoders were trained on the same settings as mentioned in APNet2 using their open-source implementations.

Tasks. We applied our proposed APNet2 and baseline vocoders to two tasks in our experiments, i.e., the analysis-synthesis task and the TTS task, which employed natural and predicted mel-spectrograms as vocoders' input, respectively. For the TTS task, we used a Fastspeech2-based acoustic model[5] [35] to predict the mel-spectrograms from texts.

4.2   Evaluation

We used both objective and subjective evaluations to compare the performance of these vocoders. In this study, we employed five objective metrics to assess the quality of synthesized speech, as previously utilized in our work [18]. These metrics include signal-to-noise ratio (SNR), root mean square error (RMSE) of logarithmic amplitude spectra (referred to as LAS-RMSE), mel-cepstrum distortion (MCD), RMSE of F0 (referred to as F0-RMSE), and V/UV error. Additionally, the real-time factor (RTF) was used as an objective metric to evaluate the efficiency of the inference process. RTF is defined as the ratio between the time taken to generate speech waveforms and their total duration. In our implementation, the RTF value was calculated as the ratio between the time consumed to generate all 500 test sentences using a single Nvidia 2080Ti GPU or a single Intel Xeon E5-2620 CPU core and the total duration of the test set.

To assess the subjective quality, we conducted mean opinion score (MOS) tests to compare the naturalness of these vocoders. Each MOS test involved twenty test utterances synthesized by different vocoders, alongside natural utterances. We gathered feedback from a minimum of 25 native English listeners on the Amazon Mechanical Turk crowdsourcing platform[6]. Listeners were asked to rate the naturalness on a scale of 1 to 5, with a score interval of 0.5.

The objective results on the test sets of the LJSpeech dataset for the analysis-synthesis task are listed in Table 1. HiFi-GAN obtained the highest scores for most objective metrics. Our proposed APNet2 was comparable to HiFi-GAN and iSTFTNet, and outperformed the Vocos and APNet. In terms of RTF, the inference speed of APNet2 was second only to the fastest Vocos, and significantly faster than other vocoders on both GPU and CPU. However, Vocos exhibited a noticeable disadvantage in all objective metrics, particularly with a high F0-RMSE, indicating noticeable pronunciation errors. Furthermore, ISTFT-based vocoders (i.e., iSTFTNet, Vocos, APNet, and APNet2) significantly improved

---

[1] https://github.com/jik876/hifi-gan.
[2] https://github.com/rishikksh20/iSTFTNet-pytorch.
[3] https://github.com/charactr-platform/vocos.
[4] https://github.com/yangai520/APNet.
[5] https://github.com/ming024/FastSpeech2.
[6] https://www.mturk.com.

Table 1. Objective results of HiFi-GAN, iSTFTNet, Vocos, APNet, and the proposed APNet2 on the test set of the LJSpeech dataset for the analysis-synthesis task. Here, "$a\times$" represents $a\times$ real time.

	SNR (dB) ↑	LAS-RMSE (dB)↓	MCD (dB) ↓	F0-RMSE (cent) ↓	V/UV error (%) ↓	RTF (GPU) ↓	RTF (CPU) ↓
HiFi-GAN	**3.93**	6.44	**1.62**	45.08	**5.20**	0.0296 (33.75×)	0.297 (3.36×)
iSTFTNet	3.83	6.78	1.71	**43.60**	5.28	0.0057 (175.95×)	0.148 (6.76×)
Vocos	2.30	6.97	2.31	148.33	10.64	**0.0012 (869.90×)**	**0.009 (116.45×)**
APNet	2.65	7.03	2.07	46.68	5.53	0.0028 (358.17×)	0.039 (25.84×)
APNet2	3.84	**6.34**	1.73	44.33	5.31	0.0015 (665.05×)	0.021 (47.73×)

Table 2. Subjective results of HiFi-GAN, iSTFTNet, Vocos, APNet, and the proposed APNet2 on the test set of the LJSpeech dataset. Here, "AS" and "TTS" represent analysis-synthesis and TTS tasks, respectively.

	MOS (AS)	MOS (TTS)
Natural Speech	4.02 ± 0.108	3.93 ± 0.176
HiFi-GAN	**3.88 ± 0.153**	**3.58 ± 0.259**
iSTFTNet	**3.93 ± 0.148**	**3.66 ± 0.232**
Vocos	3.50 ± 0.296	3.25 ± 0.379
APNet	3.53 ± 0.256	3.39 ± 0.329
APNet2	**3.83 ± 0.296**	**3.66 ± 0.230**

inference speed compared to HiFi-GAN. This also validated the effectiveness of predicting low-resolution spectra rather than high-resolution time-domain waveforms for efficiency enhancement.

The subjective MOS test results on the test set of the LJSpeech dataset for both analysis-synthesis and TTS tasks are listed in Table 2. For the analysis-synthesis task, iSTFTNet got the highest average MOS among all vocoders. The average MOS of the proposed APNet2 vocoder was lower than that of iSTFT-Net and HiFi-GAN. To verify the significance of the differences between paired vocoders, we calculated the p-value of a t-test for paired MOS sequences of the test set. Interestingly, the p-value of the results between iSTFTNet and APNet2 was 0.148, and the p-value of the results between HiFi-GAN and APNet2 was 0.515. This indicates that, in terms of synthesized speech quality, there was no significant difference between the proposed APNet2 and two vocoders with high average MOS (i.e., iSTFTNet and HiFi-GAN). Moreover, APNet2 was significantly ($p < 0.01$) better than Vocos and APNet regarding the MOS scores. The conclusions in the TTS task were closely aligned with those in the analysis-synthesis task, and it is gratifying that in the TTS task, APNet2 achieved one of the highest average MOS scores. This indicates that APNet2 exhibited strong robustness when dealing with non-natural acoustic features. Furthermore, by observing the objective and subjective results between APNet and APNet2, it can be concluded that the introduction of ConvNeXt v2, MRD, etc., effectively improved the model's modeling capabilities, making it suitable for waveform generation in high waveform sampling rates and long spectral frame shift scenarios. For a more intuitive experience, please visit our demo page[7].

Table 3. Objective results of different variants of APNet2 and Vocos for the analysis-synthesis task.

	SNR	LAS-RMSE	MCD	F0-RMSE	V/UV error
	(dB) ↑	(dB) ↓	(dB) ↓	(cent) ↓	(%) ↓
Vocos	2.30	6.97	2.31	148.33	10.64
Vocos w/ 100-dim-mel	2.96	5.84	1.80	56.59	6.81
APNet2	**3.84**	6.34	1.73	44.33	5.31
APNet2 w/ 100-dim-mel	3.42	**5.50**	**1.64**	48.69	5.88
APNet2 w/o ConvNeXt v2	1.95	6.90	2.36	76.40	8.18
APNet2 w/o MRD	1.98	7.14	2.48	65.69	8.25
APNet2 w/o HingeGAN	2.61	6.57	2.00	**35.69**	**4.64**

4.3   Analysis and Discussion

We did some analysis experiments and ablation studies to examine the effectiveness of each key component in the APNet2 vocoder. We only make objective evaluations here to compare different vocoder variations.

As shown in Table 1 and Table 2, the performance of Vocos was disappointing. However, in the original paper of Vocos [17], the authors used the 100-dimensional full-band log-mel-spectrogram as input and achieved perfect results. Therefore, we replicated the original configuration in [17] and built Vocos w/ 100-dim-mel. As shown in the results in Table 3, when using 100-dimensional full-band log-mel-spectrograms as input, Vocos exhibited significant improvement in all objective metrics. This suggests that Vocos was highly sensitive to the frequency band range of input features, making it difficult to implicitly achieve BWE. This could be the reason for Vocos' poor performance when using 80-dimensional narrow-band mel-spectrograms as input.

Subsequently, we compared APNet2 with some of its variants. The results are also listed in Table 3. 1) Firstly, we also used 100-dimensional full-band log-mel-spectrogram as APNet2's input (i.e., APNet2 w/ 100-dim-mel). Compared to APNet2, APNet2 w/ 100-dim-mel showed significant improvements in the amplitude-related metrics (i.e., LAS-RMSE and MCD). It is reasonable because 100-dimensional mel-spectrograms provide complete high-frequency amplitude information. 2) Then, we replaced the ConvNeXt v2 with the original ResNet

---

[7] Source codes are available at https://github.com/redmist328/APNet2. Examples of generated speech can be found at https://redmist328.github.io/APNet2_demo.

and built APNet2 w/o ConvNeXt v2. Clearly, with the absence of ConvNeXt v2, all objective metrics show a sharp decline, confirming the superiority of ConvNeXt v2 in modeling capability. 3) Thirdly, in the GAN-based losses, MRD was replaced with MSD (i.e., APNet2 w/o MRD). The performance of APNet2 w/o MRD also significantly deteriorated. This indicates that MRD was better suited for waveform discrimination. 4) Finally, also in the GAN-based losses, we adopted the original least squares form rather than the hinge form (i.e., APNet2 w/o HingeGAN). Although using the least squares significantly improved the accuracy of F0, the waveform and spectral-related metrics were significantly degraded. This suggests that employing the hinge GAN loss was beneficial in improving overall waveform and spectral quality to some extent.

5 Conclusions

In this paper, we proposed a novel APNet2 vocoder, addressing the performance limitations of the original APNet that were constrained by the waveform sampling rate and spectral frame shift. In comparison to APNet, the improvements in APNet2 mainly included using ConvNeXt v2 as the backbone network for amplitude and phase predictions, introducing MRD into the GAN-based losses, and employing the hinge GAN form, etc. Experimental results demonstrated that APNet2 can achieve high-quality and efficient waveform generation at the configuration with a waveform sampling rate of 22.05 kHz and spectral frame shift of approximately 11.6 ms. Moreover, ablation studies verified the effectiveness of the key components in the APNet2 vocoder. Further applying the APNet2 vocoder to other speech generation tasks (e.g., SE and VC) will be the focus of our future work.

References

1. Oord, A.V.D., et al.: WaveNet: a generative model for raw audio. In: Proceedings of the SSW, p. 125 (2016)
2. Mehri, S., et al.: SampleRNN: an unconditional end-to-end neural audio generation model. In: Proceedings of the ICLR (2016)
3. Kawahara, H., Masuda-Katsuse, I., De Cheveigne, A.: Restructuring speech representations using a pitch-adaptive time-frequency smoothing and an instantaneous-frequency-based F0 extraction: possible role of a repetitive structure in sounds. Speech Commun. **27**(3–4), 187–207 (1999)
4. Morise, M., Yokomori, F., Ozawa, K.: WORLD: a vocoder-based high-quality speech synthesis system for real-time applications. IEICE Trans. Inf. Syst. **99**(7), 1877–1884 (2016)
5. Oord, A., et al.: Parallel WaveNet: fast high-fidelity speech synthesis. In: Proceedings of the ICML, pp. 3918–3926 (2018)
6. Ping, W., Peng, K., Chen, J.: ClariNet: parallel wave generation in end-to-end text-to-speech. In: Proceedings of the ICLR (2018)
7. Prenger, R., Valle, R., Catanzaro, B.: WaveGlow: a flow-based generative network for speech synthesis. In: Proceedings of the ICASSP, pp. 3617–3621 (2019)

8. Ping, W., Peng, K., Zhao, K., Song, Z.: WaveFlow: a compact flow-based model for raw audio. In: Proceedings of the ICML, pp. 7706–7716 (2020)
9. Juvela, L., Bollepalli, B., Tsiaras, V., Alku, P.: GlotNet-a raw waveform model for the glottal excitation in statistical parametric speech synthesis. IEEE/ACM Trans. Audio Speech Lang. Process. **27**(6), 1019–1030 (2019)
10. Valin, J.M., Skoglund, J.: LPCNet: improving neural speech synthesis through linear prediction. In: Proceedings of the ICASSP, pp. 5891–5895 (2019)
11. Wang, X., Takaki, S., Yamagishi, J.: Neural source-filter-based waveform model for statistical parametric speech synthesis. In: Proceedings of the ICASSP, pp. 5916–5920 (2019)
12. Goodfellow, I., et al.: Generative adversarial nets. In: Proceedings of the NeurIPS, vol. 27 (2014)
13. Donahue, C., McAuley, J., Puckette, M.: Adversarial audio synthesis. In: Proceedings of the ICLR (2018)
14. Kumar, K., et al.: MelGAN: generative adversarial networks for conditional waveform synthesis. In: Proceedings of the NeurIPS, pp. 14910–14921 (2019)
15. Kong, J., Kim, J., Bae, J.: HiFi-GAN: generative adversarial networks for efficient and high fidelity speech synthesis. In: Proceedings of the NeurIPS, vol. 33, pp. 17022–17033 (2020)
16. Kaneko, T., Tanaka, K., Kameoka, H., Seki, S.: iSTFTNet: fast and lightweight mel-spectrogram vocoder incorporating inverse short-time Fourier transform. In: Proceedings of the ICASSP, pp. 6207–6211 (2022)
17. Siuzdak, H.: Vocos: closing the gap between time-domain and Fourier-based neural vocoders for high-quality audio synthesis. arXiv preprint arXiv:2306.00814 (2023)
18. Ai, Y., Ling, Z.H.: APNet: an all-frame-level neural vocoder incorporating direct prediction of amplitude and phase spectra. IEEE/ACM Trans. Audio Speech Lang. Process. (2023)
19. Liu, Z., Mao, H., Wu, C.Y., Feichtenhofer, C., Darrell, T., Xie, S.: A convnet for the 2020s. In: Proceedings of the CVPR, pp. 11976–11986 (2022)
20. Ai, Y., Lu, Y.X., Ling, Z.H.: Long-frame-shift neural speech phase prediction with spectral continuity enhancement and interpolation error compensation. IEEE Signal Process. Lett. (2023)
21. Shen, J., et al.: Natural TTS synthesis by conditioning WaveNet on mel spectrogram predictions. In: Proceedings of the ICASSP, pp. 4779–4783 (2018)
22. Wang, Y., et al.: Tacotron: towards end-to-end speech synthesis. In: Interspeech 2017 (2017)
23. Woo, S., et al.: ConvNeXt v2: co-designing and scaling convnets with masked autoencoders. In: Proceedings of the CVPR, pp. 16133–16142 (2023)
24. Oyamada, K., Kameoka, H., Kaneko, T., Tanaka, K., Hojo, N., Ando, H.: Generative adversarial network-based approach to signal reconstruction from magnitude spectrogram. In: Proceedings of the EUSIPCO, pp. 2514–2518 (2018)
25. Gritsenko, A., Salimans, T., van den Berg, R., Snoek, J., Kalchbrenner, N.: A spectral energy distance for parallel speech synthesis. In: Proceedings of the NeurIPS, vol. 33, pp. 13062–13072 (2020)
26. Neekhara, P., Donahue, C., Puckette, M., Dubnov, S., McAuley, J.: Expediting TTS synthesis with adversarial vocoding. In: Proceedings of the Interspeech, vol. 2019, pp. 186–190 (2019)
27. Hendrycks, D., Gimpel, K.: Gaussian error linear units (GELUs). In: Proceedings of the ICML, vol. 70, pp. 3441–3450 (2017)
28. Maas, A.L., Hannun, A.Y., Ng, A.Y.: Rectifier nonlinearities improve neural network acoustic models. In: Proceedings of the ICML (2013)

29. Lee, J., Han, S., Cho, H., Jung, W.: PHASEAUG: a differentiable augmentation for speech synthesis to simulate one-to-many mapping. In: Proceedings of the ICASSP, pp. 1–5 (2023)
30. Jang, W., Lim, D., Yoon, J., et al.: UnivNet: a neural vocoder with multi-resolution spectrogram discriminators for high-fidelity waveform generation. In: Proceedings of the Interspeech (2021)
31. Zeghidour, N., Luebs, A., Omran, A., Skoglund, J., Tagliasacchi, M.: SoundStream: an end-to-end neural audio codec. IEEE/ACM Trans. Audio Speech Lang. Process. **30**, 495–507 (2021)
32. Ai, Y., Ling, Z.H.: Neural speech phase prediction based on parallel estimation architecture and anti-wrapping losses. In: Proceedings of the ICASSP, pp. 1–5 (2023)
33. Ito, K., Johnson, L.: The LJ speech dataset (2017). https://keithito.com/LJ-Speech-Dataset
34. Loshchilov, I., Hutter, F.: Decoupled weight decay regularization. In: Proceedings of the ICLR (2018)
35. Ren, Y., et al.: FastSpeech 2: fast and high-quality end-to-end text to speech. In: Proceedings of the ICLR (2020)

Within- and Between-Class Sample Interpolation Based Supervised Metric Learning for Speaker Verification

Jian-Tao Zhang[1], Hao-Yu Song[2], Wu Guo[1], Yan Song[1(✉)], and Li-Rong Dai[1]

[1] National Engineering Research Center of Speech and Language Information Processing, University of Science and Technology of China, Hefei, China
zhangjiantao@mail.ustc.edu.cn, {guowu,songy,lrdai}@ustc.edu.cn
[2] The Australian National University, Canberra, Australia
u7439298@anu.edu.au

Abstract. Metric learning aims to pull together the samples belonging to the same class and push apart those from different classes in embedding space. Existing methods may suffer from inadequate and low-quality sample pairs, resulting unsatisfactory speaker verification (SV) performance. To address this issue, we propose the data augmentation methods in the embedding space to guarantee sufficient and high-quality negative points for metric learning, termed as within-class and between-class points interpolation generation (WBIG). Furthermore, the strategy of hard negative pair mining (HDPM) is also considered in WBIG. It is shown that WBIG is simple and flexible enough to be incorporated into existing metric learning method, such as supervised contrastive loss (SCL). Experiments on CNCeleb and VoxCeleb demonstrate the superiority of WBIG, and achieve relative performance improvement in terms of EER by 9.74% and 9.95% compared to the baseline system, separately.

Keywords: Metric learning · Interpolation generation · Supervised contrastive loss · Hard negative pair mining

1 Introduction

Speaker verification (SV) aims to verify whether a speech utterance was produced by the claimed identity or not. For SV tasks, the tested identities are unseen during training, which is the open-set setting. Therefore, SV tasks are metric learning problems in which utterances are mapped into a discriminative embedding space by a train-well feature extractor, and employ a metric function to measure the similarity between a pair of test utterances.

The prevailing method on speaker verification is to train the network with a multi-class classifier and softmax-based optimization loss (e.g. cross entropy loss [1,2]) in the training set. Although the softmax-based loss can separate speaker embeddings [3,4], they are not well-discriminative since it is not directly designed

© The Author(s), under exclusive license to Springer Nature Singapore Pte Ltd. 2024
J. Jia et al. (Eds.): NCMMSC 2023, CCIS 2006, pp. 81–91, 2024.
https://doi.org/10.1007/978-981-97-0601-3_7

on the embedding space. To compensate this drawback, the PLDA back-ends and its derivative methods [5–7] are often adopted for computing more accurate similarity scores. In [8], angular softmax (A-Softmax) introduces the cosine similarity as logit input to the softmax layer which is superior to the vanilla softmax in SV. Laterly, the variants of angular softmax such as Additive margin variants, AM-Softmax [9,10] and AAM-Softmax [11] adopt cosine similarity margin penalty to increase the inter-class variance. However, the mismatch between classification train and verification test still exists although above variants have achieved good performance [12–14].

Metric learning losses are directly designed on the embedding space and can be powerful alternatives to prevailing classifier-based methods. The open-set speaker verification tasks are typical metric learning problems which aim to learn an embedding space by enforcing the intra-class compactness and inter-class separation. To learn an embedding space, many of the metric learning losses are proposed, which usually take pairs of samples to optimize the loss. Triplet loss [15] and contrastive loss [16–19] are pair-based metric learning losses, which have achieved promising performance in speaker verfication [20,21]. Triplet loss has a common idea: pull the anchor together with the 'positive' sample in embedded space, and separate the anchor from one 'negative' sample which selected from many 'nagative' samples through hard negative pair mining mechanism. In [22], the self-supervised batch contrastive loss is extended to the fully-supervised setting and the supervised contrastive loss (SCL) is proposed, which show consistent outperformance over cross-entropy with large batchsize. Compared with triplet loss, SCL exploits multiple negatives to optimize the model based on the conclusion that the performance of metric learning loss is very dependent on the quality and quantity of 'positive' and 'negative' samples.

Motivatede by triplet loss and SCL [22], we propose a novel augmentation method in the embedding space to generate a large number of hard points, termed as within-class and between-class points interpolation generation (WBIG). WBIG consists of within-class points interpolation generation (WCIG) and between-class points interpolation generation (BCIG), which is shown in Fig. 1. On the basis of making full use of existing negative examples, we can get more sufficient negative points for metric learning. In WCIG, we choose to generate synthetic points by linear interpolation among original points of the same class, to provide more choices for mining hard points. We can confirm that the labels of synthetic points share the high degree certainty with the original points since they are included inside the original class cluster [23]. In BCIG, we generate new synthetic points by linear interpolation among original points from different classes, and the new synthetic points is harder than original points. We are also excited to find that between-class points interpolation generation can bring an implicit margin penalty to increase the inter-class variance, and we will analyze it in Sect. 2.2. Our proposed WBIG can obtain more abundant and high-quality negative points for metric learning. Compared to other hard sample generation methods based on additional sub-network as a generator, our proposed method does not introduce additional parameters and is

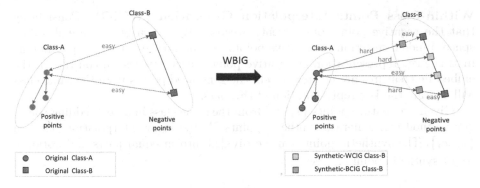

Fig. 1. Illustration of our proposed within-class and between-class points interpolation generation **(WBIG)**for metric learning. **Left:** The standard pair-based metric learning. **Right:** We exploit WBIG to generate more abundant and harder synthetic points. The blue points denote the original point, the yellow points denote the synthetic points generated by WCIG, and the grey points denote the synthetic points generated by BCIG.

simple and flexible enough that can be combined with existing pair-based metric learning losses.

Extensive experiments have been conducted on CNCeleb benchmark datasets, we demonstrate the superiority of our WBIG. Evaluations on CNCeleb achieve a relative performance improvement of about 9.74% over the baseline system in terms of Equal Error Rate (EER). Evaluations on VoxCeleb achieve a relative performance improvement of about 9.95% over the baseline system in terms of Equal Error Rate (EER).

2 Method

2.1 Within-Class and Between-Class Points Interpolation Generation (WBIG)

Inspired by the supervised contrastive loss [22] and triplet loss [15] with hard positive or negative pair mining, the proposed WBIG generates multiple synthetic points by within-class and between-class points interpolation generation in the embedding space for hard negative pairs mining. On the basis of making full use of existing negative examples, we can get more effective negative points for metric learning.

To be specific, we design the interpolation generation for more abundant and harder negative points from two different views. The most natural idea is to generate new negative points from existing negative points which come from one class, and we call it within-class points interpolation generation (WCIG). Another method is to generate new negative points based on the existing positive and negative points, and we call it between-class points interpolation generation (BCIG).

Within-Class Points Interpolation Generation (WCIG). Considering that the negative points of different categories vary greatly in the embedding space, random selection of negative points to generate new negative points may miss the key representation of negative point characteristics. In contrast, if the embedding features are clustered well, the generated points from the same class will contain the key representation of the class.

Given two feature points $\{x_i, x_j\}$ from the same class in an embedding space, our method will generate synthetic points $\widetilde{x}_{ij}^k$ by linear interpolation between $\{x_i, x_j\}$. The synthetic points can be divided into m equal parts and obtain a set of synthetic points $\widetilde{S}_{ij}$ as:

$$\widetilde{x}_{ij}^k = \frac{kx_i + (m+1-k)x_j}{m+1} \tag{1}$$

$$\widetilde{S}_{ij} = \{\widetilde{x}_{ij}^1, \widetilde{x}_{ij}^2, ..., \widetilde{x}_{ij}^m\} \tag{2}$$

where m is the number of synthetic points. Metric learning use L2-normalization to make the feature points be located on the hyper-sphere space with the same norm. Therefore, the generated synthetic points should adopt the L2-normalization to share the same hyper-sphere space with the original points:

$$\overline{x}_{ij}^k = \frac{\widetilde{x}_{ij}^k}{||\widetilde{x}_{ij}^k||^2} \tag{3}$$

$$\overline{S}_{ij} = \{\overline{x}_{ij}^1, \overline{x}_{ij}^2, ..., \overline{x}_{ij}^m\} \tag{4}$$

where $\overline{x}_{ij}^k$ is the synthetic point after L2-normalization, and $\overline{S}_{ij}$ is the set of L2-normalized synthetic points.

By doing so, we can dig out more potential points from the original points and provide more reliable negative points for metric learning.

Between-Class Points Interpolation Generation (BCIG). The above generation method is only focus on points of the same class. In fact, negative points can be generated with the participation of positive points. When we calculating supervised contrastive loss, each positive point will have multiple negative points for joint calculation. We can get new synthetic points based on the existing positive and negative points.

Given two feature points $\{x_i, n_j\}$ from the different class in the embedding space, our method will generate a synthetic point $\widetilde{t}_{ij}$ between two feature points. Similar to the within-class points interpolation generation method, we give different weight coefficients to positive and negative points to linearly synthesize new points.

$$\widetilde{t}_{ij} = \lambda x_i + (1-\lambda)n_j \quad \lambda \in [0, \eta] \tag{5}$$

$$\overline{t}_{ij} = \frac{\widetilde{t}_{ij}}{||\widetilde{t}_{ij}||^2} \tag{6}$$

where λ is the positive point x_i weight coefficient, and η is the set weight upper limit, and $\bar{t}_{ij}$ is the L2-normalized synthetic point.

Whether the synthetic point can be regarded as a positive point? We don't think the synthetic point is suitable for the positive point. If the synthetic point is regarded as a positive point, some examples from different classes will be pulled closer since the synthetic point contains of partial negative point information, which is inconsistent with the idea of pushing apart clusters of samples from different classes. Therefore, it is most appropriate to consider the synthetic point as a negative point.

Considering that the positive sample needs to be pushed far away from the negative sample, the harder the negative sample is, the closer it will be to the positive sample. From this perspective, the larger λ is, the harder negative point is to generate. However, the negative point can not be infinitely difficult. When λ exceeds a certain limit, it may reverse optimization model, which will be discussed in Sect. 3.3. To increase the diversity of generated negative points on the premise of ensuring the quality of generated negative points, we choose the dynamic λ to randomly generate negative points.

2.2 WBIG for Supervised Contrastive Loss

The second step of our method is to adopt the above two methodes of generating synthetic points to improve the supervised contrastive loss [22]:

$$\mathcal{L}_{scl} = \sum_{i \in I} \frac{-1}{|P(i)|} \sum_{p \in P(i)} log \frac{exp(z_i \cdot z_p/\tau)}{\sum_{a \in A(i)} exp(z_i \cdot z_a/\tau)} \tag{7}$$

where $i \in I \equiv \{1...N\}$ in a minibatch, z_i is the L2-normalized embedding of x_i, $A(i) \equiv I \backslash \{i\}$, and $P(i) \equiv \{p \in A(i): y_i = y_p\}$, τ is a scalar temperature parameter. For convenience, we define the denominator as:

$$A_{ia} = \sum_{a \in A(i)} exp(z_i \cdot z_a/\tau) \tag{8}$$

For the **within-class points interpolation generation (WCIG)**, we select the z_i as the anchor point, then the negative point $z_{n \in I}$ can randomly select a point of the same class to obtain a set of synthetic points $S_n = \{z_n^1, z_n^2, ..., z_n^m\}$, and we select the hardest synthetic negative point $s_n = \{z_n^k \mid max(z_n^k \cdot z_i), z_n^k \in S_n\}$. We perform the operations of generating negative points and mine difficult negative points for each negative point, and finally get a set of hard synthetic negative points of $S(i) = \{s_1, s_2, ..., s_n, ..., s_N \mid y_n \neq y_i\}$, and we define:

$$S_{in} = \sum_{s_n \in S(i)} exp(z_i \cdot s_n/\tau) \tag{9}$$

WCIG + Supervised Contrastive Loss can be formulated as:

$$\mathcal{L}_{scl}^{wc} = \sum_{i \in I} \frac{-1}{|P(i)|} \sum_{p \in P(i)} log \frac{exp(z_i \cdot z_p/\tau)}{A_{ia} + S_{in}} \tag{10}$$

For the **between-class points interpolation generation(BCIG)**, we also select the z_i as the anchor point, and can obtain a set of synthetic negative points with each negative point. In the previous analysis, only one negative point needs to be synthesized for each pair of samples, and the difficulty of negative point depends on the weight parameter λ of positive and negative points. Therefore, the mining of hard synthetic negative points is implicitly completed through the dynamic change of λ. Finally, we get a set of hard synthetic negative points of $T(i) = \{t_{i1}, t_{i2}, ..., t_{in}, ..., t_{iN} \mid y_n \neq y_i\}$, and we define:

$$T_{in} = \sum_{t_{in} \in T(i)} exp(z_i \cdot t_{in}/\tau) \tag{11}$$

BCIG + Supervised Contrastive Loss can be formulated as:

$$\mathcal{L}_{scl}^{bc} = \sum_{i \in I} \frac{-1}{|P(i)|} \sum_{p \in P(i)} log \frac{exp(z_i \cdot z_p/\tau)}{A_{ia} + T_{in}} \tag{12}$$

Assuming the each anchor point only has one positive point, and only use the harder synthetic negative points generated by BCIG as the negative points, set scalar temperature parameter $\tau = 1$, we have:

$$
\begin{aligned}
\mathcal{L}_{scl}^{bc} &= -log \frac{exp(z_i \cdot z_p)}{exp(z_i \cdot z_p) + exp[z_i \cdot (\lambda z_i + (1 - \lambda)z_n)]} \\
&= -log \frac{exp(z_i \cdot z_p)}{exp(z_i \cdot z_p) + exp[\lambda + (1 - \lambda)z_i \cdot z_n]} \\
&= -log \frac{exp(z_i \cdot z_p)}{exp(z_i \cdot z_p) + exp\left[(1 - \lambda)\left(\frac{\lambda}{1-\lambda} + z_i \cdot z_n\right)\right]}
\end{aligned}
\tag{13}
$$

Compared with the standard $\mathcal{L}_{scl}$, $\mathcal{L}_{scl}^{bc}$ not only has the role of hard negative points mining, but also has the role of adding implicit margin $\alpha = \frac{\lambda}{1-\lambda}$, and $1 - \lambda$ can be seen as the correction to scalar temperature parameter τ.

WBIG + Supervised Contrastive Loss can be formulated as:

$$\mathcal{L}_{scl}^{wb} = \sum_{i \in I} \frac{-1}{|P(i)|} \sum_{p \in P(i)} log \frac{exp(z_i \cdot z_p/\tau)}{A_{ia} + S_{in} + T_{in}} \tag{14}$$

Our propose within-class & between-class points interpolation generation (WBIG) combines the advantages of the above two methods, and more abundant and high-quality negative points can be obtained for SCL.

3 Experiments

3.1 Experimental Settings

Our proposed method is experimentally evaluated on CNCeleb [24] and Voxceleb [25,26]. The CNCeleb contains more than 130k utterances from 3000 Chinese celebrities and covers 11 diverse genres including movie, interview, and so on. We adopt 2,768 speakers for training and 200 speakers for testing. For evaluating on Voxceleb, we use VoxCeleb2-dev as training set, containing 1,092,009 utterances from 5,994 speakers, and adopt official VoxCeleb-O as evaluation set.

Input Features: The Kaldi toolkit [27] is used for the feature extraction process. In our experiments, 41-dimensional filter bank (FBank) acoustic features are transformed from 25 ms windows with a 10 ms shift between frames. Energy-based voice activity detection (VAD) is used to remove silent segments. The training features are randomly truncated into short slices ranging in length from 2 to 4 s.

Baseline Configuration: The default baseline feature extractor is ResNet-18 with feature maps sizes [256, 400, 1, 128] ([B, T, F, C]). And we adopt attentive bilinear pooling (ABP) [28] to obtain speaker embeddings of dimension 256. The batchsize is 256, with each mini-batch consisting of 128 speakers for pair-based metric learning. The networks are optimized using stochastic gradient descent (SGD), with momentum of 0.9 and weight decay of 5e−4. We set the m in Eq. (1) as four by default. The range of positive point weight coefficient λ defaults to [0, 0.03] in Eq. (5).

3.2 Main Results

Table 1. EER results (%) of the comparison systems on CNCeleb evaluation with different losses. The backbone is ResNet-18. λ is the positive point weight coefficient of between-class points interpolation generation in Eq. (5).

System	Loss	range of λ	EER (%)
Baseline	CE	–	9.95
Baseline	AM	–	9.81
Baseline	AAM	–	9.79
Baseline	SCL	–	10.05
WCIG	SCL	–	9.40
BCIG	SCL	[0, 0.03]	9.18
WBIG	SCL	[0, 0.03]	**8.98**

To measure the performance of WBIG in metric learning, we introduce softmax-based classification loss such as CE, AM-Softmax and AAM-Softmax loss for comparison. We adopt supervised contrastive loss (SCL) as a representative of contrastive learning, which has achieved promising results.

To show the superiority of our method based on metric learning, we conduct a study for comparing different systems with different loss on CNCeleb evaluation. The results are shown in Table 1. In the baseline system, we can find that the usage of AM-Softmax and AAM-Softmax loss can achieve higher performance than CE loss in terms of EER. The variants of angular softmax adopt cosine similarity margin penalty to increase the inter-class variance, which is benefit for speaker verification. Besides, the performance of direct use of SCL loss is similar to above softmax-based classification losses. In the system of WCIG, we utilize the method of within-class points interpolation generation to generate more synthetic points for SCL loss, and achieve a significant improvement compared with baseline. In the system of BCIG, we utilize the method of between-class points interpolation generation with the range of positive point weight coefficient λ set as $[0, 0.03]$ to generate more synthetic points for SCL loss, and get improvement over the baseline SCL loss with 8.65% relative reduction in terms of EER. This indicates that the potential implicit margin penalty introduced by BCIG when providing new synthetic points will continue to increase the inter-class variance and improve the performance. **WBIG** combines advantages of the above two systems, and more abundant and high-quality negative points can be obtained. As we expected, the performance of **WBIG** gets further improvement over the baseline SCL loss with 10.19% relative reduction in terms of EER owing to obtaining a better discriminative embedding.

Table 2. EER results (%) of WBIG and the baseline with different backbones on CNCeleb evaluation.

Backbone	System	Loss	EER (%)
ResNet-18	Baseline	CE	9.95
	WBIG	SCL	**8.98**
ResNet-34	Baseline	CE	9.76
	WBIG	SCL	**8.89**

Table 3. EER results (%) of WBIG and the baseline with different backbones on VoxCeleb-O evaluation.

Backbone	System	Loss	EER (%)
ResNet-18	Baseline	CE	2.01
	WBIG	SCL	**1.81**
ResNet-34	Baseline	CE	1.77
	WBIG	SCL	**1.62**

Table 2 shows the comparative results with different backbone on CNCeleb evaluation. The performance of **WBIG** gets further improvement over the base-line CE loss with 9.74% and 8.98% relative reduction in terms of EER on back-bone of ResNet-18 and ResNet-34 separately. It is worth noting that the deeper and wider ResNet-34 can achieve better baseline result, and obtain consistent performance improvement. The above results show that our proposed method is plug-and-play for models with different parameter sizes.

In Table 3, we validate the generalizability of our approach using the Vox-Celeb dataset. We also use two types of backbones, ResNet-18 and ResNet-34, to compare performance with the baseline. The performance of **WBIG** gets further improvement over the baseline CE loss with 9.95% and 8.47% relative reduction in terms of EER on backbone of ResNet-18 and ResNet-34 separately. The above results validate the generalizability of our method.

3.3 Ablation Experiments

The range of positive point weight coefficients λ is the sole hyper-parameter of our proposed method. As illustrated in Fig. 2, we conduct an study with the backbone ResNet-18 by differentiating the range of positive point weight coef-ficient λ on WBIG. The peak of the performance is the positive point weight coefficient λ near the range of [0, 0.03], before and after which the performance starts decreasing. We infer that if the weight coefficient λ is too small, the syn-thetic points will be too close to the original points, and can not get much gain for model optimization. Besides, if the weight coefficient λ is too large, it will bring too large margin penalty to the inter-class variance. Therefore, in consid-eration of performance, the range of positive point weight coefficient λ defaults to [0, 0.03] in this paper.

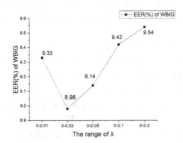

Fig. 2. EER(%) performance of WBIG by the range of positive point weight coefficients λ in Eq. (5). The backbone of each model is Res-Net 18.

4 Conclusion

The performance of metric learning loss is very dependent on the quality and quantity of negative samples. In this paper, we propose within-class and between-class points interpolation generation (**WBIG**) to generate more abundant and high-quality negative points in the embedding space for metric learning. Within-class points interpolation generation can provide more choices for mining hard points. Between-class points interpolation generation can bring an implicit margin penalty to increase the inter-class variance while generating hard negative points. Besides, **WBIG** is simple without introducing additional parameters that can be combined with existing pair-based metric learning losses. Experiments on CNCeleb and VoxCeleb demonstrate the capability of the proposed WBIG method.

Acknowledgement. This work was supported by 2022ZD0160600.

References

1. Variani, E., Lei, X., McDermott, E., Moreno, I.L., Gonzalez-Dominguez, J.: Deep neural networks for small footprint text-dependent speaker verification. In: 2014 IEEE International Conference on Acoustics, Speech and Signal Processing (ICASSP), pp. 4052–4056. IEEE (2014)
2. Snyder, D., Garcia-Romero, D., Sell, G., Povey, D., Khudanpur, S.: X-vectors: robust DNN embeddings for speaker recognition. In: 2018 IEEE International Conference on Acoustics, Speech and Signal Processing (ICASSP), pp. 5329–5333. IEEE (2018)
3. Ravanelli, M., Bengio, Y.: Speaker recognition from raw waveform with SincNet. In: 2018 IEEE Spoken Language Technology Workshop (SLT), pp. 1021–1028. IEEE (2018)
4. Okabe, K., Koshinaka, T., Shinoda, K.: Attentive statistics pooling for deep speaker embedding. In: Proceedings of the Interspeech 2018, pp. 2252–2256 (2018)
5. Villalba, J., Lleida, E.: Bayesian adaptation of PLDA based speaker recognition to domains with scarce development data. In: Odyssey 2012-The Speaker and Language Recognition Workshop (2012)
6. Prince, S.J.D., Elder, J.H.: Probabilistic linear discriminant analysis for inferences about identity. In: 2007 IEEE 11th International Conference on Computer Vision, pp. 1–8. IEEE (2007)
7. Li, L., Zhang, Y., Kang, J., Zheng, T.F., Wang, D.: Squeezing value of cross-domain labels: a decoupled scoring approach for speaker verification. In: ICASSP 2021–2021 IEEE International Conference on Acoustics, Speech and Signal Processing (ICASSP), pp. 5829–5833. IEEE (2021)
8. Liu, W., Wen, Y., Yu, Z., Li, M., Raj, B., Song, L.: SphereFace: deep hypersphere embedding for face recognition. In: Proceedings of the IEEE Conference on Computer Vision and Pattern Recognition, pp. 212–220 (2017)
9. Wang, F., Cheng, J., Liu, W., Liu, H.: Additive margin softmax for face verification. IEEE Signal Process. Lett. **25**(7), 926–930 (2018)
10. Wang, H., et al.: CosFace: large margin cosine loss for deep face recognition. In: Proceedings of the IEEE Conference on Computer Vision and Pattern Recognition, pp. 5265–5274 (2018)
11. Deng, J., Guo, J., Xue, N., Zafeiriou, S.: ArcFace: additive angular margin loss for deep face recognition. In: Proceedings of the IEEE/CVF Conference on Computer Vision and Pattern Recognition, pp. 4690–4699 (2019)
12. Liu, Y., He, L., Liu, J.: Large margin softmax loss for speaker verification. In: Proceedings of the Interspeech 2019, pp. 2873–2877 (2019)

13. Garcia-Romero, D., Snyder, D., Sell, G., McCree, A., Povey, D., Khudanpur, S.: X-vector DNN refinement with full-length recordings for speaker recognition. In: Interspeech, pp. 1493–1496 (2019)
14. Xiang, X., Wang, S., Huang, H., Qian, Y., Yu, K.: Margin matters: towards more discriminative deep neural network embeddings for speaker recognition. In: 2019 Asia-Pacific Signal and Information Processing Association Annual Summit and Conference (APSIPA ASC), pp. 1652–1656. IEEE (2019)
15. Schroff, F., Kalenichenko, D., Philbin, J.: FaceNet: a unified embedding for face recognition and clustering. In: Proceedings of the IEEE Conference on Computer Vision and Pattern Recognition, pp. 815–823 (2015)
16. Chopra, S., Hadsell, R., LeCun, Y.: Learning a similarity metric discriminatively, with application to face verification. In: 2005 IEEE Computer Society Conference on Computer Vision and Pattern Recognition (CVPR 2005), vol. 1, pp. 539–546. IEEE (2005)
17. Sohn, K.: Improved deep metric learning with multi-class n-pair loss objective. In: Advances in Neural Information Processing Systems, vol. 29 (2016)
18. Tian, Y., Sun, C., Poole, B., Krishnan, D., Schmid, C., Isola, P.: What makes for good views for contrastive learning? In: Advances in Neural Information Processing Systems, vol. 33, pp. 6827–6839 (2020)
19. Zhang, X., Jin, M., Cheng, R., Li, R., Han, E., Stolcke, A.: Contrastive-mixup learning for improved speaker verification. In: ICASSP 2022–2022 IEEE International Conference on Acoustics, Speech and Signal Processing (ICASSP), pp. 7652–7656. IEEE (2022)
20. Zhang, C., Koishida, K., Hansen, J.H.L.: Text-independent speaker verification based on triplet convolutional neural network embeddings. IEEE/ACM Trans. Audio Speech Lang. Process. **26**(9), 1633–1644 (2018)
21. Chung, J.S., et al.: In defence of metric learning for speaker recognition. In: Proceedings of the Interspeech 2020, pp. 2977–2981 (2020)
22. Khosla, P., et al.: Supervised contrastive learning. In: Advances in Neural Information Processing Systems, vol. 33, pp. 18661–18673 (2020)
23. Ko, B., Gu, G.: Embedding expansion: augmentation in embedding space for deep metric learning. In: Proceedings of the IEEE/CVF Conference on Computer Vision and Pattern Recognition, pp. 7255–7264 (2020)
24. Li, L., et al.: CN-Celeb: multi-genre speaker recognition. Speech Commun. **137**, 77–91 (2022)
25. Chung, J., Nagrani, A., Zisserman, A.: Voxceleb2: deep speaker recognition. Interspeech 2018 (2018)
26. Nagrani, A., Chung, J.S., Zisserman, A.: Voxceleb: a large-scale speaker identification dataset. In: Proceedings of the Interspeech, pp. 2616–2620 (2017)
27. Povey, D., et al.: The kaldi speech recognition toolkit. In: IEEE 2011 Workshop on Automatic Speech Recognition and Understanding, number CONF. IEEE Signal Processing Society (2011)
28. Liu, Y., Song, Y., Jiang, Y., McLoughlin, I., Liu, L., Dai, L.-R.: An effective speaker recognition method based on joint identification and verification supervisions. In: Interspeech, pp. 3007–3011 (2020)

Joint Speech and Noise Estimation Using SNR-Adaptive Target Learning for Deep-Learning-Based Speech Enhancement

Xiaoran Li[1], Zilu Guo[1], Jun Du[1(✉)], Chin-Hui Lee[2], Yu Gao[3], and Wenbin Zhang[3]

[1] University of Science and Technology of China, Hefei, Anhui, China
lixiaoran@mail.ustc.edu.cn, jundu@ustc.edu.cn
[2] Georgia Institute of Technology, Atlanta, GA, USA
[3] AI Innovation Center, Midea Group Co., Ltd., Shanghai, China

Abstract. In this paper, we propose an SNR-adaptive target learning strategy and apply it to a joint speech and noise estimation network to address the mismatch between speech enhancement (SE) and automatic speech recognition (ASR) modules. The progressive learning (PL) methods have revealed the importance of retaining residual noise in the training targets of the enhancement model to alleviate this mismatch. Inspired by this, we adopt an SNR-adaptive target learning strategy to optimize the SNR targets for the SE model, thereby achieving adaptive denoising of the enhancement model in a data-driven manner and further improving its performance on the backend ASR task. Next, we extend the SNR-adaptive target learning strategy to a joint speech and noise estimation network and validate the adaptability of the target learning strategy with the noise prediction branch. We demonstrate the effectiveness of our proposed method on a public benchmark, achieving a significant relative word error rate (WER) reduction of approximately 37% compared to the WER results obtained from unprocessed noisy speech.

Keywords: speech recognition · speech enhancement · adaptive noise reduction

1 Introduction

Single-channel speech enhancement [1,2] refers to the suppression of background noise from a mono microphone's noisy signal. It plays a crucial role in enhancing speech perceptual quality and improving backend speech-based applications such as automatic speech recognition (ASR) [3]. Over the years, significant efforts have been devoted to addressing this problem, continually advancing the performance of speech enhancement and enabling the expansion of speech-related applications in diverse and complex acoustic environments [4].

Before the emergence of machine learning, several traditional algorithms were developed, such as spectral subtraction and subspace-based methods. However,

© The Author(s), under exclusive license to Springer Nature Singapore Pte Ltd. 2024
J. Jia et al. (Eds.): NCMMSC 2023, CCIS 2006, pp. 92–101, 2024.
https://doi.org/10.1007/978-981-97-0601-3_8

these traditional algorithms are often built upon certain assumptions, which limit their applicability in complex acoustic environments. The advent of deep-learning-based techniques has enabled speech enhancement to handle a wider range of diverse and unstable noises. Early work utilizing deep neural networks (DNNs) improved the generality of SE models by employing deeper structures [5]. Additionally, the application of Long Short-Term Memory (LSTM) in speech enhancement demonstrated better performance in low signal-to-noise ratio (SNR) environments, thanks to their ability to capture long-term dependencies in speech [6,7]. Another notable technique is the progressive learning (PL) method, which decomposes the direct mapping from noisy to clean speech into multiple stages. This approach progressively increases the SNR by guiding hidden layers to learn intermediate targets with reduced noise residuals [8,9]. The PL method has shown promising results in achieving improved perceptual quality.

Despite advancements in speech enhancement performance, directly feeding the enhanced results into the ASR system often fails to produce the expected improvements in word error rate (WER) due to a mismatch between the speech enhancement and ASR modules [10]. This mismatch primarily arises from differences in the training data used for the SE and ASR modules [11]. One possible solution to address this mismatch is joint training of the two modules [12]. However, joint training can be time-consuming, and may not be feasible if the ASR backend is accessible only through an application programming interface (API) [13]. Additionally, it is worth noting that the ASR model is more sensitive to artificial speech distortion rather than original noise residuals [14]. Consequently, the intermediate layer output of PL models has been reported to benefit ASR tasks [11,15], as it significantly reduces speech distortion despite introducing additional noise residuals. In previous PL methods, the SNR improvement targets, known as SNR targets, are fixed for intermediate layers. However, the optimal SNR targets can vary depending on factors such as the type of noise and the input prior SNRs [16].

This paper aims to enhance the ASR results based on previous PL methods. The main contributions of this work are as follows:

1. Proposal of an SNR-adaptive target learning strategy: In this paper, we introduce a novel strategy to optimize the SNR targets of the SE model in the training process. The effectiveness of this training strategy is demonstrated through experiments conducted on CHiME4 ASR tasks.
2. Expansion of the strategy to a joint speech and noise estimation network: We extend the SNR-adaptive strategy to a joint speech and noise estimation network, which further enhances the recognition results of the enhanced speech.

2   System Description

2.1   Backgrounds

In this paper, the observed mixture audio signal can be decomposed as $X(d, l) = S(d, l) + N(d, l)$, where $S(d, l)$, and $N(d, l)$ denote the target speech spectrum and the additive noise spectrum. d and l represent the time and frequency indices, respectively. And all the following experiments are conducted in the time-frequency domain.

Progressive Learning (PL) for Speech Enhancement. Despite notable advancements in enhancement models, accurately predicting clean speech in real-world scenarios remains a challenging task. To tackle this issue, PL aims to break down the direct mapping from noisy to clean targets into a multi-learning task that progressively improves the SNR in the hidden layers [8,9].

$$\mathcal{L} = \|Y(d,l) - \hat{Y}(d,l)\|_1 \tag{1}$$

$$Y(d,l) = S(d,l) \tag{2}$$

The training objective and loss function commonly used in the enhancement model are denoted by Eq. 1 and Eq. 2. In these equations, $Y(d,l)$ represents the training objective, corresponding to the input's target speech, denoted as $S(d,l)$. $\hat{Y}(d,l)$ represents the estimated speech output by the model. In the context of progressive learning, the optimization objective, illustrated by Eq. 3 and 4, involves a multi-level joint loss across the intermediate layers of the enhancement model. Here, m denotes the layer index, while η_m represents the weighted coefficients for each layer's loss. The optimization objective Y_m encompasses not only the input target speech $S(d,l)$ but also residual noise that gradually decreases as the layers progress, which is controlled by the scalar k_m. The output of the intermediate m-th layer in the progressive model is denoted as $\hat{Y}_m$. In [8], the first, second, and final layers of the model are designed to achieve fixed target SNR gains of 10 dB, 20 dB, and infinite dB (representing a clean target), respectively. For more detailed information, please refer to the references [8,9].

$$\mathcal{L}' = \sum_m \eta_m * \|Y_m(d,l) - \hat{Y}_m(d,l)\|_1 \tag{3}$$

$$Y_m(d,l) = S(d,l) + k_m * N(d,l) \tag{4}$$

The final output of the SE model with progressive learning strategy is reported to exhibit better perceptual quality, while the output of intermediate layers is found to be effective in enhancing backend ASR's performance [11,15].

2.2   SNR-Adaptive Target Learning Strategy

As mentioned in Subsect. 2.1, the utilization of intermediate layers in PL models has demonstrated significant improvements in ASR performance. In previous PL methods, researchers commonly selected the output from a single intermediate layer or fused multiple intermediate layers for ASR tasks. However, the optimal SNR improvement targets for input speech are not fixed. For example, speech corrupted with stationary noise can be easily separated, making a high SNR

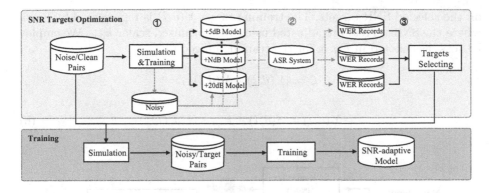

Fig. 1. The framework of the SNR-adaptive target learning strategy

improvement target preferable for backend tasks. Conversely, when dealing with unstable noise characterized by impulsive characteristics, which poses challenges for the SE model, a high SNR gain in the enhanced speech often introduces excessive distortions.

The PL model has demonstrated the feasibility of adjusting the SNR improvement degrees of the enhancement model by regulating the SNR improvement targets within the training set. Therefore, it is possible to enhance the adaptivity of SE models by replacing fixed SNR improvement targets with targets that are optimized for backend ASR tasks. To address this, a three-step adjustment method for SNR improvement targets in the training process is proposed, namely the SNR-adaptive target learning strategy.

As depicted in Fig. 1, the proposed training strategy comprises two main parts: the SNR targets optimization process and the training process. In the SNR targets optimization process, we follow three steps to select relatively optimal SNR targets for the training corpus.

Firstly, we simulate a range of different SNR targets using the same training corpus. For instance, in this study, we utilize four distinct targets: +5 dB, +10 dB, +15 dB, and +20 dB, to train a set of enhancement models with varying SNR gains. In the second step, these models are employed to enhance the noisy speech samples from the training set, and we measure the corresponding word error rates (WERs) at different SNR gains using an ASR system. Subsequently, we record these WERs for further analysis. In the third step, we select the optimal SNR targets based on the obtained WERs. To prevent overfitting of the training samples, we set a threshold to differentiate between preferences for high and low gains. For instance, in this paper, if the average WERs of 5 dB and 10 dB gains differ from the average WERs of 15 dB and 20 dB gains by more than 0.02, we classify the optimal SNR targets as either high gain (17.5 dB) or low gain (7.5 dB), respectively. If the difference is not significant, the SNR target is set to the intermediate gain level (12.5 dB). Moreover, we introduce a jitter of ±2.5 dB during the simulation to facilitate faster model convergence. After the SNR targets optimization process, the training pairs are re-simulated based on the selected SNR targets. The training target $Y(d,l)$ is represented by Eq. 6, where the SNR targets are adjusted using the optimized scalar k_{opt}. We employ L1 loss, as denoted in Eq. 5, for the training procedure.

$$\mathcal{L} = \|Y(d,l) - \hat{Y}(d,l)\|_1 \tag{5}$$

$$Y(d,l) = S(d,l) + k_{opt} * N(d,l) \tag{6}$$

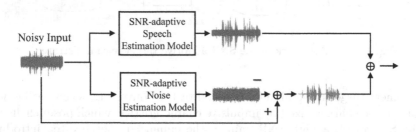

Fig. 2. The framework of the joint estimation network using the proposed strategy

2.3 Joint Speech and Noise Estimation Using SNR-Adaptive Target Learning

Currently, most models optimized for SNR targets are limited to speech estimation. Zheng et al. [17] have demonstrated the complementary role of noise estimation in speech prediction models. Therefore, in this study, we extend the SNR strategy to a joint speech-noise prediction network. The network structure is illustrated in Fig. 2. It consists of two branches: a speech estimation network and a noise estimation network. These branches are trained separately using the SNR-adaptive target learning strategy with L1 loss. Similar to speech estimation using SNR-adaptive target learning, the target formulation for the noise estimation network, with a given "SNR improvement" (where noise is considered as signal and speech is considered as noise for SNR calculation), is represented in Eq. 7. k_n is used to control the speech residuals of the noise estimation network. When k_n is set to 0, the network is turned into a pure noise estimation network.

$$Y_n(d,l) = k_n * S(d,l) + N(d,l) \tag{7}$$

$$\hat{Y}(d,l) = (1 - \lambda) * \hat{Y}_s(d,l) + \lambda * (X(d,l) - \hat{Y}_n(d,l)) \tag{8}$$

In the SNR-adaptive training process for noise estimation networks, the initial step involves the base noise estimation models producing noise estimates along with speech residuals. These estimates are then subtracted from the noisy input $X(d,l)$ to obtain an estimation of the clean speech. In the subsequent step, the target selection is based on the WERs of the clean speech estimation.

Regarding the joint speech and noise estimation network, as depicted in Fig. 2 and described by Eq. 8, the noise estimation network's output $\hat{Y}_n(d, l)$ is also subtracted from the noisy input $X(d, l)$ during the inference stage. Subsequently, the output of this subtraction, along with the output of the speech estimation network $\hat{Y}_s(d, l)$, are combined or fused together. The parameter λ is employed to adjust the relative contribution of the outputs from the speech estimation network and the noise estimation network.

3 Experiments and Results Analysis

3.1 Data Corpus

CHiME-4 single channel source is chosen as test datasets to demonstrate our algorithms and strategies, which are recorded in real scenarios including four conditions: cafe (CAF), street junction (STR), public transport (BUS), and pedestrian area (PED) [18]. Background noise is provided as a simulation source, which consists of around 8 h of a single-channel noise record. Clean speech is derived from the WSJ0 corpus, which consists of 7138 utterances (around 12 h of reading-style speech) from 83 speakers. 21000 training tuples are sampled from the sources and are simulated at a range of -5 to 10 dB SNR as noisy input.

3.2 Implementation Details

The CRN structure is chosen as the speech enhancement model and the settings of the model are similar to the original design proposed in [19]. Input time-domain signal is transformed to STFT spectrum and the predicted spectrum is restored back to the speech signal using iSTFT. All the ablation models are trained for 30 epochs. The learning rate is initialized as 0.001, and the Adam optimizer is used. As for the backend, the official CHiME4 backend ASR system is adopted without acoustic model retraining. The acoustic model is a DNN-HMM discriminatively trained with the sMBR criterion [20] and the language models are 5-gram with Kneser-Ney smoothing for the first-pass decoding and the simple RNN-based language model for restoring [21].

3.3 Experiments

SNR-Adaptive Target Learning Strategy. Based on the SNR-adaptive algorithm outlined in Sect. 2.2, we conducted comparative experiments on the CHiME4 real test sets using the CRN model. The experimental results are presented in Table 1. In these experiments, we employed four fixed SNR gain enhancement models for the optimization process of SNR targets: +5 dB, +10 dB, +15 dB, and +20 dB. These models are labeled as "SNR_+ndB" in the table. The adaptive models trained using the SNR-adaptive algorithm are denoted as "SNR_adapt". For comparison purposes, we included WERs of the original noisy

Table 1. WER (%) comparison of the speech estimation models with different fixed SNR targets and SNR-adaptive targets

Enhancement	BUS	CAF	PED	STR	AVG
Unprocessed	36.55	24.73	19.94	14.16	23.84
CRN_baseline	38.04	27.90	20.63	15.20	25.44
SNR_+5 dB	27.74	16.74	12.63	11.30	17.10
SNR_+10 dB	27.52	17.05	**12.21**	10.45	16.80
SNR_+15 dB	27.87	17.22	12.97	10.14	17.05
SNR_+20 dB	29.50	18.85	14.44	10.57	18.34
SNR_adapt	**26.17**	**16.42**	12.52	**9.47**	**16.14**

speech without enhancement ("Unprocessed") and the enhanced speech generated by the CRN model, which directly predicts clean speech without utilizing SNR targets ("CRN_baseline").

The comparison of WERs reveals that the "SNR_+10 dB" fixed SNR improvement target model is the optimal choice among the four base models. It significantly alleviates the mismatch problem, demonstrating an improvement of approximately 34% compared to the "CRN_baseline" model. Regarding the SNR-adaptive target learning strategy, the "SNR_adapt" model further enhances performance compared to the base models. It outperforms the results of the four base models in the "BUS", "CAF", and "STR" scenarios. Ultimately, it achieves an average WER of 16.14%. However, it should be noted that the "SNR_adapt" model exhibits a slight decline in performance in the "PED" scenario. This suggests that the adaptability algorithm may not perfectly predict the optimal distribution of SNR targets. The relatively small size of the current simulation dataset and potential deviations when transitioning to real test sets could contribute to this observation.

Table 2. WER (%) comparison of the joint speech and noise estimation networks with and without SNR-adaptive target learning strategy

Enhancement	BUS	CAF	PED	STR	AVG
Unprocessed	36.55	24.73	19.94	14.16	23.84
OMLSA [22]	33.91	26.62	21.88	13.17	23.89
Conv-TasNet [23]	37.63	28.05	20.07	14.37	25.03
PL-ANSE [11]	25.50	16.80	12.45	**9.13**	15.97
SNR_+10dB_joint	25.78	15.99	13.25	9.32	16.08
SNR_adapt_joint	**24.14**	**14.59**	**11.49**	9.41	**14.90**
-speech_est_branch	26.17	16.42	12.52	9.47	16.14
-noise_est_branch	35.65	27.25	20.63	13.49	24.25

Expansion on Joint Speech and Noise Estimation Networks. Table 2 provides a comprehensive comparison of our expanded joint speech and noise estimation model employing the SNR-adapt strategy, with other baseline models on the CHiME4 real test sets. The "OMLSA" column represents the WER results of the traditional SE algorithm OMLSA on CHiME4 [22]. The "PL-ANSE" column denotes a representative PL model from previous research, trained under the same CRN settings as our model [11]. In addition to the CRN model, which serves as the baseline for our strategy (as shown in Table 1), the results of Conv-Tasnet [23], which targets clean speech as the objective, are also included in Table 2. The "SNR_+10 dB_joint" column represents the extension of the optimal model from the fixed SNR gain base models to the joint estimation structure. On the other hand, the "SNR_adapt_joint" column indicates our final joint estimation model trained based on the SNR-adapt strategy. The "speech_est_branch" and "noise_est_branch" columns correspond to the WER results of the speech prediction from the speech estimation branch and the noise prediction branch of the "SNR_adapt_joint".

From the observations in Table 2, it is evident that traditional algorithms and enhancement models targeting clean speech as the objective struggle to achieve significant improvements in terms of WER on real datasets like CHiME4. When compared to the "Unprocessed" results, "OMLSA", "Conv-Tasnet", and the CRN baseline (as shown in Table 1) all exhibit some degree of degradation. In contrast, "SNR_adapt_joint" demonstrates significant enhancements over both its speech prediction branch and noise prediction branch. Specifically, compared to the speech prediction branch trained using the SNR-adapt strategy, the joint estimation model achieves an improvement of 7.6% in WER. This highlights that the joint estimation structure can further enhance the recognition results compared to speech estimation alone. Furthermore, when comparing "SNR_adapt_joint" to "SNR_+10 dB_joint", it becomes apparent that the SNR-adapt strategy yields notable improvements even for fixed SNR gain dual-stream prediction models. This indicates that the SNR-adapt training strategy remains effective when transitioning from the speech prediction network to the joint estimation network. Comparing our results with PL-ANSE further emphasizes the advantages of our strategy and structure over the previously proposed progressive learning framework.

4   Conclusion

This paper introduces a joint speech and noise estimation network that utilizes SNR-adaptive target learning, achieving an improved balance between noise reduction and speech distortion. The final ASR recognition results exhibit a significant reduction in WER, approximately 37%, compared to unprocessed speech. The conducted experiments demonstrate the favorable impact of the adaptive strategy and joint model structure on ASR tasks. However, the current method incurs high computational costs and relies on a specific ASR API during the search process, which may restrict the application of the enhancement model to other backend tasks. Future efforts will be dedicated to reducing the complexity of the SNR target search and enhancing the model's flexibility for transfer learning across various downstream tasks.

References

1. Weiss, M.R., Aschkenasy, E., Parsons, T.W.: Study and development of the INTEL technique for improving speech intelligibility (1975)
2. Luo, Y., Bao, G., Xu, Y., Ye, Z.: Supervised monaural speech enhancement using complementary joint sparse representations. IEEE Signal Process. Lett. **23**(2), 237–241 (2016)
3. Kleinschmidt, M., Tchorz, J., Kollmeier, B.: Combining speech enhancement and auditory feature extraction for robust speech recognition. Speech Commun. **34**, 75–91 (2001)
4. Wang, Y., Narayanan, A., Wang, D.L.: On training targets for supervised speech separation. IEEE/ACM Trans. Audio Speech Lang. Process. **22**(12), 1849–1858 (2014)
5. Tamura, S., Waibel, A.: Noise reduction using connectionist models. In: ICASSP-88, International Conference on Acoustics, Speech, and Signal Processing, vol. 1, pp. 553–556 (1988)
6. Sun, L., Du, J., Dai, L.-R., Lee, C.-H.: Multiple-target deep learning for LSTM-RNN based speech enhancement. In: 2017 Hands-free Speech Communications and Microphone Arrays (HSCMA), pp. 136–140 (2017)
7. Liu, M., Wang, Y., Wang, J., Wang, J., Xie, X.: Speech enhancement method based on LSTM neural network for speech recognition. In: 2018 14th IEEE International Conference on Signal Processing (ICSP), pp. 245–249 (2018)
8. Gao, T., Du, J., Dai, L.-R., Lee, C.-H.: SNR-based progressive learning of deep neural network for speech enhancement. In: Interspeech, pp. 3713–3717 (2016)
9. Gao, T., Du, J., Dai, L.-R., Lee, C.-H.: Densely connected progressive learning for LSTM-based speech enhancement. In: 2018 IEEE International Conference on Acoustics, Speech and Signal Processing (ICASSP), pp. 5054–5058 (2018)
10. Wang, P., Tan, K., Wang, D.L.: Bridging the gap between monaural speech enhancement and recognition with distortion-independent acoustic modeling. IEEE/ACM Trans. Audio Speech Lang. Process. **28**, 39–48 (2020)
11. Nian, Z., Tu, Y.-H., Du, J., Lee, C.-H.: A progressive learning approach to adaptive noise and speech estimation for speech enhancement and noisy speech recognition. In: ICASSP 2021 - 2021 IEEE International Conference on Acoustics, Speech and Signal Processing (ICASSP), pp. 6913–6917 (2021)
12. Menne, T., Schlüter, R., Ney, H.: Investigation into joint optimization of single channel speech enhancement and acoustic modeling for robust ASR. In: ICASSP 2019 - 2019 IEEE International Conference on Acoustics, Speech and Signal Processing (ICASSP), pp. 6660–6664 (2019)
13. Sato, H., Ochiai, T., Delcroix, M., Kinoshita, K., Kamo, N., Moriya, T.: Learning to enhance or not: neural network-based switching of enhanced and observed signals for overlapping speech recognition. In: ICASSP 2022 - 2022 IEEE International Conference on Acoustics, Speech and Signal Processing (ICASSP), pp. 6287–6291 (2022)
14. Iwamoto, K., et al.: How bad are artifacts?: analyzing the impact of speech enhancement errors on ASR. In: Interspeech (2022)

15. Nian, Z., Du, J., Yeung, Y.T., Wang, R.: A time domain progressive learning approach with snr constriction for single-channel speech enhancement and recognition. In: ICASSP 2022 - 2022 IEEE International Conference on Acoustics, Speech and Signal Processing (ICASSP), pp. 6277–6281 (2022)

16. Koizumi, Y., Karita, S., Narayanan, A., Panchapagesan, S., Bacchiani, M.A.U.: SNRI target training for joint speech enhancement and recognition. In: Proceedings of the Interspeech (2022)

17. Zheng, C., Peng, X., Zhang, Y., Srinivasan, S., Lu, Y.: Interactive speech and noise modeling for speech enhancement. In: Proceedings of the AAAI Conference on Artificial Intelligence, pp. 14549–14557 (2021)

18. Vincent, E., Watanabe, S., Nugraha, A.A., Barker, J., Marxer, R.: An analysis of environment, microphone and data simulation mismatches in robust speech recognition. Comput. Speech Lang. **46**, 535–557 (2017)

19. Tan, K., Wang, D.L.: A convolutional recurrent neural network for real-time speech enhancement. In: Proceedings of the Interspeech 2018, pp. 3229–3233 (2018)

20. Saon, G., Soltau, H.: A comparison of two optimization techniques for sequence discriminative training of deep neural networks. In: 2014 IEEE International Conference on Acoustics, Speech and Signal Processing (ICASSP), pp. 5567–5571 (2014)

21. Kneser, R., Ney, H.: Improved backing-off for m-gram language modeling. In: 1995 International Conference on Acoustics, Speech, and Signal Processing, vol. 1, pp. 181–184 (1995)

22. Cohen, I.: Noise spectrum estimation in adverse environments: improved minima controlled recursive averaging. IEEE Trans. Speech Audio Process. **11**(5), 466–475 (2003)

23. Luo, Y., Mesgarani, N.: Conv-TasNet: surpassing ideal time-frequency magnitude masking for speech separation. IEEE/ACM Trans. Audio Speech Lang. Process. **27**(8), 1256–1266 (2019)

Data Augmentation by Finite Element Analysis for Enhanced Machine Anomalous Sound Detection

Zhixian Zhang⬤, Yucong Zhang, and Ming Li(✉)⬤

Suzhou Municipal Key Laboratory of Multimodal Intelligent Systems,
Duke Kunshan University, Kunshan, China
ming.li369@dukekunshan.edu.cn

Abstract. Current data augmentation methods for machine anomalous sound detection (MASD) suffer from insufficient data generated by real world machines. Open datasets such as audioset are not tailored for machine sounds, and fake sounds created by generative models are not trustworthy. In this paper, we explore a novel data augmentation method in MASD using machine sounds simulated by finite element analysis (FEA). We use Ansys, a software capable for acoustic simulation based on FEA, to generate machine sounds for further training. The physical properties of the machine, such as geometry and material, and the material of the medium is modified to acquire data from multiple domains. The experimental results on DCASE 2023 Task 2 dataset indicates a better performance from models trained using augmented data.

Keywords: Data augmentation · Finite element analysis · Machine anomalous sound detection

1 Introduction

Machine anomalous sound detection (MASD) is the task to identify whether a sound clip is generated by a machine working normally or anomalously. It is commonly used in automatic machine condition monitoring due to its capability to reduce labor work and detect subtle changes in machine working condition that could be otherwise missed by manually monitoring. However, these advantages have the premise that the detection system has seen enough anomalous data. Due to the low probability of occurrence and high variance of the anomalous sounds, the training process often involves only normal sounds, and the task is often considered as an unsupervised learning problem [4,5,15].

To deal with the problems mentioned above, both generative methods and self-supervised methods are proposed in recent years. Without the anomalous data, generative methods allow the researchers to directly model the distribution of the normal data, and any data that does not fit the distribution is regarded as anomaly. Autoencoders (AE) [7,9], interpolation deep neural networks (IDNN) [20] and generative adversarial networks (GAN) based

© The Author(s), under exclusive license to Springer Nature Singapore Pte Ltd. 2024
J. Jia et al. (Eds.): NCMMSC 2023, CCIS 2006, pp. 102–110, 2024.
https://doi.org/10.1007/978-981-97-0601-3_9

approaches [8,13] are popular generative models for MASD. However, generated methods are hard to extract effective features [17]. The other way to model the normal data is to build a feature extractor by self-supervised methods [2,11,18,25]. By categorizing various machine sounds into different machine types and IDs, self-supervised methods can learn more effective and more compact representations than the generative methods. The anomalies can then be identified in the feature space using distance metrics, such as cosine distance.

However, the lack of data is still a limitation when training a feature extractor in the self-supervised method. One way to overcome this issue is to use large open-source dataset to train the model. In [19], the researchers explore the effectiveness of using various pre-trained models that trained from human speech [1,3,12,22]. Despite the good performance in the field of automatric speech recognition (ASR), those pre-trained models are not trained by machine sounds and might not be suitable for MASD. [21] uses a pre-trained model PANNs [16], which is trained on AudioSet [6]. Although AudioSet is not a dataset for speech, it still contains sounds from various kinds of categories and is not tailored for machine sounds. Another way to address the lack of data is to use simulated data. Recently, the AI-generated content (AIGC) is a hot topic, so researchers who study MASD try to apply the AIGC techniques to generate machine sounds [13,21] or embeddings [24] for data augmentation. However, the AI-generated sounds are not explainable and trustworthy, since it is generated using deep learning techniques.

To tackle the issues above, we propose a novel data augmentation method by finite element analysis in this paper. FEA is a numerical approach that represents geometries as a system of linear equations built up by meshed nodes and links. The solution is formula-based and has actual physical meaning. FEA is widely applied in engineering to approximate solutions for various problems. The acoustic simulation based on FEA is a reliable data augmentation tool. To the best of our knowledge, we are the first to explore the FEA acoustic simulation for data augmentation in MASD.

2   Methods

We utilize FEA to conduct modal analysis aiming to identify its natural vibrational modes. Additionally, we perform harmonic frequency response analysis to ascertain the localized stresses occurring within the speaker. We use Ansys 2023 R1, a software capable for performing acoustic analysis using FEA, to simulate real world sound. We build a speaker model as the geometry to generate sound. To make the speaker vibrates, we simulate a harmonic force exert on the diaphragm of the speaker. The vibration of the diaphragm will generate sound by various kinds of mediums in the defined enclosure. A far-field microphone is set in front of the speaker to record sound signals. The generated sound is trained together with the machine sounds from DCASE 2023 dataset. The acoustic simulation of the speaker consists of geometry design, model analysis, harmonic response analysis and acoustic simulation. The overall pipeline is shown in the Fig. 1.

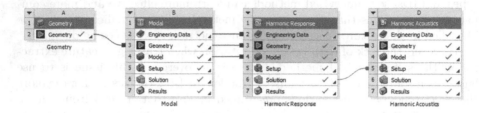

Fig. 1. Overall pipeline of the simulation

2.1 Geometric Model

The geometry where we perform acoustic simulation is a simple speaker model. The design is done using SpaceClaim in Ansys. The speaker consists of a diaphragm and body holding the diaphragm. An enclosure is defined outside of the speaker to simulate the medium that the sound transmits. Figure 2 shows the geometry of the speaker and the enclosure.

Fig. 2. Imported geometry in Ansys   **Fig. 3.** Mesh of the speaker

Fig. 4. Example of deformation of the speaker

2.2   Model Analysis

The model analysis is performed to get the mesh of the model, creating the finite element model shown in Fig. 3. The natural frequencies of resonation which depends on the material assignment is also calculated to provide basic dynamic information of the model. The boundary condition of the speaker is fixed to the bottom of the body. The mesh of the model is created with the size of 5 mm for the body and 3 mm for the diaphragm. The speaker is therefore represented by the finite element model consists of nodes and links generated by the mesh.

2.3   Harmonic Analysis

In harmonic analysis, we first import a constant force of magnitude 1N on the center of the diaphragm on the speaker. Harmonic response results in a sinusoidal excitation at each frequency. So this analysis can simulate vibration of the diaphragm propelled by the electromagenetic voice coil in real world speakers. An example of deformation response of the speaker is shown in Fig. 4. When calculating frequency response, the frequency range 0–20000 Hz is divided into 50 parts with an attenuation ratio of 10% is assumed. The solution contains the deformation in every frequency bins.

2.4   Harmonic Acoustics

After we get the vibration frequencies and velocity results in harmonic analysis, we can import the computed information to acoustic simulation. The frequency information in harmonic analysis will be synchronized to harmonic acoustics. The acoustic simulation will suppress the solid part of the model, and calculate the response in fluid domain. The material of the medium in the enclosure can be any fluids. In this paper, we use air and water as fluid options to further increase the sensitivity of the classifier. The mesh size is calculated based on the wavelength of the sound and the maximum frequency we want to achieve. Based on the sound speed c in the medium, the wavelength $\lambda = c/f$, where f is the maximum frequency we want to get. The size of the mesh is $\lambda/6$. After we have the mesh size for each fluid, we can create mesh of the fluid in the enclosure. We then set the boundary condition to radiation boundary, which means that when the energy hits the boundary of the enclosure it gets radiated away instead of being reflected back. That means our acoustic analysis is accurate and is focused solely on the vibration of the speaker. We set a far-field microphone in front of the speaker to collect sound information. After solving the acoustic scenario described above, we get information of sound pressure level. We then export the result in the microphone to wav file to get the augmented data.

2.5   Backbone Model

We adopt and re-implement the similar model described in [23] for MASD, where researchers show the effectiveness of this model on previous DCASE 2022 challenge. We use this model as our backbone model for two reasons. First, it is a

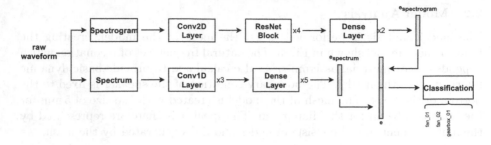

Fig. 5. Model Overview

popular and effective model. State-of-the-art performance is reported on both DCASE 2022 [23] and DCASE 2023 challenge [14]. Second, the model has two encoding pathways, which is investigated already by [18] and proved to be effective for MASD. Hence, in this paper, we use this dual-path model to test the effectiveness of our data augmentation technique.

Table 1. Structure of the encoder. n indicates the number of layers or blocks, c is the number of output channels, k is the kernel size and s is the stride. h and w are the output height and width of the ResNet Blocks.

Operator	n	c	k	s
Conv2D 7 × 7	1	32	(7, 7)	(2, 2)
MaxPooling	–	–	(3, 3)	(2, 2)
ResNet block	4	(64, 128, 128, 128)	(3, 3)	(2, 2)
MaxPooling	–	–	(h, w)	(h, w)

As is shown in Fig. 5, the input waveform is encoded by two pathways. In the spectrum pathway, the machine sound is transformed into utterance-level spectrum, which aims at finding patterns over the whole utterance. The spectrum is processed by three 1D convolutional layers and five dense layers. For the spectrogram pathway, the machine sound is first converted to spectrogram by short-term fourier transform (STFT), and then fed to a 2D convolutional layer and 4 ResNet [10] blocks shown in Table 1.

3   Experiments and Results

3.1   Acoustic Simulation

To simulate the condition of anomalous machines, we manually create holes of different sizes and locations on the diaphragm. Since the diaphragm is the key component of sound generation on the speaker, the changes on the diaphragm

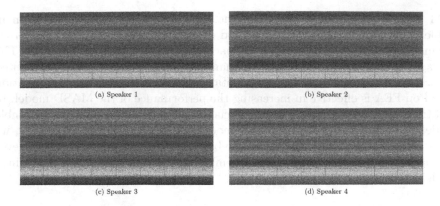

(a) Speaker 1

(b) Speaker 2

(c) Speaker 3

(d) Speaker 4

Fig. 6. Spectrogram of generated sounds on 4 speakers with different holes

can be well represented by the sound the speaker generates. In the experiment, we have four speaker geometry designs and each of them contains different holes on the diaphragm to create different types of sounds. The comparison of spectrogram of sound generated by four geometry designs is shown in Fig. 6.

For material assignment, normally we have wood on the speaker body and polypropylene for the diaphragm of the speaker. To make the classifier detect more subtle differences between sounds generated by the speaker, we expanded the material set for body: {wood, steel, plastic} and diaphragm: {steel, polypropylene}. In acoustic analysis, the set for material of the medium in the enclosure is {air, water}. We created geometry model on every combination of the materials. The far-field microphone is located 30 mm away from the center of the diaphragm of the speaker.

For every speaker geometry design, we assign all possible material combinations in the material sets mentioned above. After solving the result in the microphone and get the sound pressure level at each frequency, we use inverse fourier transform to generate the sounds with a fixed sample rate of 16kHz. As a result, we get 48 different classes of simulated sounds. Within each class, we get 100 audio clips of 10 s long each.

3.2 Dataset

The experiments are conducted on the development part of DCASE 2023 Task 2 dataset [4], containing audio clips from seven distinct machine types. Each machine type has about 1,000 audio clips. Each audio clip lasts 10 s with a sampling rate of 16 kHz. Only normal data in the training dataset is used for training, and the results are evaluated on the test dataset.

3.3 Result

We compare the AUC results of training the MASD model with and without the augmented data. The result comparing model performance using only DCASE

2023 dataset for training or adding the simulated data together is shown in Table 2. The data augmentation method of adding simulated data can improve the model performance on most of the machine types in the development set. The average AUC of all machines is also enhanced by the introduction of augmented data. This means that the data augmentation method using acoustic simulation based on FEA is effective in increasing the performance of the MASD model. In addition, we show the results of the official baseline [7] and top-ranked ensemble model [19] using multiple pre-trained models for comparison. From Table 2, we can see that our system outperforms the baseline for a large margin and achieves comparable performance with the ensemble model that uses large pre-trained models.

Table 2. Average AUC % and pAUC % of every machine type on DCASE 2023 development dataset

Machine type		Simulated Data		Official Baseline [7]	Z. Lv et al. [19]
		No	Yes		
Valve	AUC	**87.73**	75.00	53.74	73.66
	pAUC	**62.15**	59.47	51.28	53.68
Gearbox	AUC	**78.81**	73.79	71.58	82.28
	pAUC	**63.05**	55.32	54.84	62.47
Fan	AUC	68.63	**70.52**	61.89	65.97
	pAUC	**58.63**	58.10	58.42	56.32
Bearing	AUC	57.90	**69.78**	59.77	78.80
	pAUC	49.89	**51.05**	50.68	62.26
ToyTrain	AUC	51.29	**57.66**	48.73	64.82
	pAUC	48.05	**50.36**	48.05	49.32
ToyCar	AUC	52.40	58.10	**59.20**	65.47
	pAUC	50.74	**52.79**	49.18	49.47
Slide Rail	AUC	86.11	**92.28**	79.25	94.74
	pAUC	65.57	**78.63**	56.18	76.68
Average	AUC	68.98	**71.02**	62.02	75.11
	pAUC	56.86	**57.96**	52.66	58.60

4 Conclusion

This paper proposes a novel data augmentation method of simulating machine sounds based on finite element analysis (FEA). Subtle changes in geometry design and different combinations of materials enable the MASD model to be more sensitive to sound variation under different machine types and different domains. The performance of MASD model trained with simulated data outperforms the one that is trained without out simulated data. It can be proved that the acoustic simulation based on FEA is effective in enhancing the performance of the MASD model. In the future, we might change the basic structure of the geometry to add more varieties to the machine types or even build geometry that is tailored for certain machine type, so that the simulated data can be more effective.

Acknowledgement. This research is funded in part by Science and Technology Program of Suzhou City (SYC2022051) and National Natural Science Foundation of China (62171207). Many thanks for the computational resource provided by the Advanced Computing East China Sub-Center.

References

1. Baevski, A., Zhou, H., rahman Mohamed, A., Auli, M.: Wav2vec 2.0: a framework for self-supervised learning of speech representations. ArXiv abs/2006.11477 (2020)
2. Chen, H., Song, Y., Dai, L., Mcloughlin, I., Liu, L.: Self-supervised representation learning for unsupervised anomalous sound detection under domain shift. In: Proceedings of the ICASSP 2022, pp. 471–475 (2022)
3. Chen, S., et al.: WavLM: large-scale self-supervised pre-training for full stack speech processing. IEEE J-STSP **16**, 1505–1518 (2021)
4. Dohi, K., et al.: Description and discussion on DCASE 2023 challenge task 2: first-shot unsupervised anomalous sound detection for machine condition monitoring. ArXiv abs/2305.07828 (2023)
5. Dohi, K., et al.: Description and discussion on DCASE 2022 challenge task 2: unsupervised anomalous sound detection for machine condition monitoring applying domain generalization techniques. In: Proceedings of DCASE 2022 Workshop (2022)
6. Gemmeke, J.F., et al.: Audio set: an ontology and human-labeled dataset for audio events. In: Proceedings of ICASSP 2017, pp. 776–780 (2017)
7. Harada, N., Niizumi, D., Ohishi, Y., Takeuchi, D., Yasuda, M.: First-shot anomaly sound detection for machine condition monitoring: a domain generalization baseline. ArXiv abs/2303.00455 (2023)
8. Hatanaka, S., Nishi, H.: Efficient GAN-based unsupervised anomaly sound detection for refrigeration units. In: Proceedings of ISIE, pp. 1–7. IEEE (2021)
9. Hayashi, T., Yoshimura, T., Adachi, Y.: Conformer-based id-aware autoencoder for unsupervised anomalous sound detection. Technical report, DCASE 2020 Challenge (2020)
10. He, K., Zhang, X., Ren, S., Sun, J.: Deep residual learning for image recognition. In: Proceedings of CVPR 2016, pp. 770–778 (2015)
11. Hojjati, H., Armanfard, N.: Self-supervised acoustic anomaly detection via contrastive learning. In: Proceedings of ICASSP 2022, pp. 3253–3257 (2021)
12. Hsu, W.N., Bolte, B., Tsai, Y.H.H., Lakhotia, K., Salakhutdinov, R., rahman Mohamed, A.: HuBERT: self-supervised speech representation learning by masked prediction of hidden units. IEEE/ACM TASLP **29**, 3451–3460 (2021)
13. Jiang, A., Zhang, W.Q., Deng, Y., Fan, P., Liu, J.: Unsupervised anomaly detection and localization of machine audio: a GAN-based approach. In: Proceedings of ICASSP, pp. 1–5. IEEE (2023)
14. Jie, J.: Anomalous sound detection based on self-supervised learning. Technical report, DCASE 2023 Challenge (2023)

15. Koizumi, Y., et al.: Description and discussion on DCASE2020 challenge task2: unsupervised anomalous sound detection for machine condition monitoring. In: Proceedings of DCASE 2020 Workshop (2020)
16. Kong, Q., Cao, Y., Iqbal, T., Wang, Y., Wang, W., Plumbley, M.D.: PANNs: large-scale pretrained audio neural networks for audio pattern recognition. IEEE/ACM TASLP **28**, 2880–2894 (2019)
17. Kuroyanagi, I., Hayashi, T., Takeda, K., Toda, T.: Improvement of serial approach to anomalous sound detection by incorporating two binary cross-entropies for outlier exposure. In: Proceedings of EUSIPCO 2022, pp. 294–298 (2022)
18. Liu, Y., Guan, J., Zhu, Q., Wang, W.: Anomalous sound detection using spectral-temporal information fusion. In: Proceedings of ICASSP 2022, pp. 816–820 (2022)
19. Lv, Z., Han, B., Chen, Z., Qian, Y., Ding, J., Liu, J.: Unsupervised anomalous detection based on unsupervised pretrained models. Technical report, DCASE 2023 Challenge (2023)
20. Suefusa, K., Nishida, T., Purohit, H., Tanabe, R., Endo, T., Kawaguchi, Y.: Anomalous sound detection based on interpolation deep neural network. In: Proceedings of ICASSP 2020, pp. 271–275. IEEE (2020)
21. Tian, J., et al.: First-shot anomalous sound detection with GMM clustering and finetuned attribute classification using audio pretrained model. Technical report, DCASE 2023 Challenge (2023)
22. Wang, C., et al.: UniSpeech: unified speech representation learning with labeled and unlabeled data. ArXiv abs/2101.07597 (2021)
23. Wilkinghoff, K.: Design choices for learning embeddings from auxiliary tasks for domain generalization in anomalous sound detection. In: Proceedings of ICASSP 2023, pp. 1–5 (2023)
24. Zeng, X., Song, Y., McLoughlin, I., Liu, L., Dai, L.: Robust prototype learning for anomalous sound detection. In: Proceedings of INTERSPEECH 2023 (2023)
25. Zhang, Y., Hongbin, S., Wan, Y., Li, M.: Outlier-aware inlier modeling and multi-scale scoring for anomalous sound detection via multitask learning. In: Proceedings of INTERSPEECH 2023, pp. 5381–5385 (2023)

A Fast Sampling Method in Diffusion-Based Dance Generation Models

Puyuan Guo[iD], Yichen Han[iD], Yingming Gao[iD], and Ya Li[(✉)][iD]

Beijing University of Posts and Telecommunications, Beijing, China
{guopy,adelacvgaoiro,yingming.gao,yli01}@bupt.edu.cn

Abstract. Recently, diffusion models have attracted much attention and have also been used in many fields, including dance movement generation. However, the slow generation speed of previous sampling methods makes the diffusion-based dance generation models limited in many application scenarios. In this paper, we use a more advanced algorithm to speed up the generation of a diffusion model based system to generate long dance movements. The algorithm does not require retraining the model. Instead, it only needs to map the noise schedule of the existing model to a new time step sequence and then construct a second-order solver for the data prediction diffusion ODEs based on the estimated higher-order differentials using multi-step methods. In order to apply this algorithm to generate long sequences, we employ a technique similar to inpainting, continuously updating the correlations between short sequences during the iteration process, and eventually concatenating multiple short sequences to form a longer one. Experimental results show that our improved sampling method not only makes the generation speed faster, but also maintains the quality of the dance movements.

Keywords: Dance Generation · Diffusion Models · Sampling Methods

1 Introduction

Dance, as a form of artistic expression in human culture, plays a significant role in constructing a more realistic digital world. Dance generation technologies have wide-ranging applications in various fields such as film production and game development. With the advancement of generative models, there have been diverse solutions for dance generation tasks. For example, there are many models based on Generative Adversarial Networks (GAN) [1]. However, because of its use of adversarial training, GANs are slow and difficult to converge, which means the model parameters tend to oscillate. In addition, there is also the problem of mode collapse which means all generated results of GANs look the same [2,3]. Some other generative methods have various shortcomings. For instance, Variational Autoencoders (VAE) [4] often produce ambiguous results with low quality. Flow-based generative models [5] require the construction of reversible

© The Author(s), under exclusive license to Springer Nature Singapore Pte Ltd. 2024
J. Jia et al. (Eds.): NCMMSC 2023, CCIS 2006, pp. 111–124, 2024.
https://doi.org/10.1007/978-981-97-0601-3_10

mapping functions, and their fitting ability is limited, resulting in poor model performance. The emergence of diffusion models [6] compensates for the above shortcomings, which can not only facilitate training, but also generate diverse results. Diffusion models are generative models based on the phenomenon of physical diffusion process. Its application in the field of image generation [7,8] has attracted wide attention, and some works have tried to apply diffusion models to other fields, such as singing voice synthesis [9], motion generation [10,11], and achieved excellent results.

Despite these advantages of diffusion models, their biggest drawback is that the generation process requires iterative sampling, thus seriously slowing down their generation speed. There exist some methods for improvement, such as using resampling [12,13] to reduce the number of iterations, or using distillation [14,15] to obtain the prior knowledge of the diffusion models. In this paper, we adopt a more advanced resampling algorithm DPM-Solver++ [16] to improve the generation speed of diffusion models and apply it to a dance motion generation system [17] based on diffusion models. This system uses the sampling algorithm of the original diffusion model [6] and the DDIM model [13] for generation, but it still takes a lot of time, especially when generating long dance movement sequences. Therefore, for this discrete-time diffusion model trained on $n = 0, 1, \cdots, T - 1$, we convert the discrete steps to continuous time steps by $t_n = (n + 1)/T$, and then we solve the corresponding diffusion ordinary differential equation from time t_{T-1} to t_0. For long sequence generation, we employ a technique which is similar to inpainting in text-to-image and used in the dance generation system. During each iteration process, we use the second half of a short sequence as the first half of the next time-correlated one. This allows us to control the correlations between the different short sequences. Finally, we concatenate these to form a longer sequence that is temporally continuous. We also perform some experiments where the algorithm we use is able to maintain the quality of the generated results to some extent and speed up the generation compared to some other algorithms.

2   Related Works

2.1   Denoising Diffusion Probabilistic Models

Diffusion models are a family of probabilistic generative models that progressively inject noises into the data in a forward progress, and then learn to remove noises iteratively in a reverse progress in order to restore the original data distribution [18]. In this way, the models can learn to reduce from a purely noisy sample to get a data sample for the purpose of generation. Since the dance generation system in this paper uses a denoising diffusion probabilistic model (DDPM) [6], this kind of models are described in detail. In addition, there are other forms of diffusion models, such as score matching with Langevin dynamics models (SMLD) [19], and improved denoising diffusion probabilistic models (improved DDPM) [20].

A denoising diffusion probabilistic model makes use of two Markov chains: a forward chain that preturbs data to noise and a reverse chain that converts noise back to data. Formally, given a data distribution $\mathbf{x}_0 \sim q(\mathbf{x}_0)$, the forward Markov progress generates a sequence of random variables $\mathbf{x}_1, \mathbf{x}_2, \cdots, \mathbf{x}_T$ by progressively injecting Gaussian noises with different σ as follows:

$$q(\mathbf{x}_t|\mathbf{x}_{t-1}) = \mathcal{N}(\mathbf{x}_t; \sqrt{1 - \beta_t}\mathbf{x}_{t-1}, \beta_t\mathbf{I}), \tag{1}$$

where $\beta_t \in (0,1)$ is a constant sequence that increases monotonically with the increase of t, which is a predefined series of values. In the reverse Markov process, we get a sample from the standard normal distribution, and denoise it to restore the original data distribution. Formally, the sample is denoted as $p(\mathbf{x}_T) = \mathcal{N}(\mathbf{x}_T; \mathbf{0}, \mathbf{I})$, and the denoising operation is represented as $p_\theta(\mathbf{x}_{t-1}|\mathbf{x}_t)$, which takes the form of

$$p_\theta(\mathbf{x}_{t-1}|\mathbf{x}_t) = \mathcal{N}(\mathbf{x}_{t-1}; \mu_\theta(\mathbf{x}_t, t), \Sigma_\theta(\mathbf{x}_t, t)), \tag{2}$$

where θ denotes model parameters, and the mean μ_θ and variance Σ_θ are parameterized by deep neural networks. According to Ho et al. [6], the final training objective of DDPM is as follows by minimizing Kullback-Leibler (KL) divergence of $q(\mathbf{x}_0, \mathbf{x}_1, \cdots, \mathbf{x}_T)$ and $p_\theta(\mathbf{x}_0, \mathbf{x}_1, \cdots, \mathbf{x}_T)$:

$$\mathcal{L}_{simple}(\theta) = E_{t,\mathbf{x}_0,\epsilon}[\|\epsilon - \epsilon_\theta(\sqrt{\bar{\alpha}_t}\mathbf{x}_0 + \sqrt{1 - \bar{\alpha}_t}\epsilon, t)\|^2], \tag{3}$$

where ϵ denotes the noises added into data in the forward process, ϵ_θ denotes the model which learns to predict the added noises.

2.2   Dance Generation Diffusion Model

The dance movement generation model experimented in this paper is based on the work of Tseng et al. [17]. They implement music-to-dance by using a diffusion-based model which conditions on Jukebox features [21] in a classifier-free manner, similarly to text-to-image [8,22]. The model structure is shown in Fig. 1. Formally, a sample $\mathbf{x}_0 \sim q(\mathbf{x}_0)$ is drawn from the data distribution, and a series of noised samples $\{\mathbf{x}_t\}_{t=0}^T$ are obtained by following the DDPM forward process.

In their setting with paired music conditioning $\mathbf{c}$, they reverse the forward diffusion process by learning to estimate $\mathbf{x}_\theta(\mathbf{x}_t, t, \mathbf{c}) \approx \mathbf{x}_0$ with model parameters θ for all t, which is different from the reverse process of above DDPMs. The model $\mathbf{x}_\theta$ is called data prediction diffusion model. Therefore, its "simple" objective is also different from Eq. (3):

$$\mathcal{L}_{simple} = E_{\mathbf{x}_0,t}[\|\mathbf{x}_0 - \mathbf{x}_\theta(\mathbf{x}_t, t, \mathbf{c})\|_2^2]. \tag{4}$$

In order to facilitate the extraction of music features and the training process, the model slices the dance movements and the corresponding music into several 5-s segments. However, this makes it impossible to use the model to directly

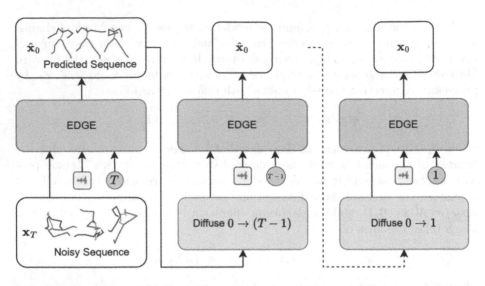

Fig. 1. The dance generation model takes a nosiy sequence $\mathbf{x}_T \sim \mathcal{N}(\mathbf{0}, \mathbf{I})$ and produces the estimated final sequence $\hat{\mathbf{x}}_0$, noising it back to $\mathbf{x}_{T-1}$ and repeating until $t = 0$ [17].

generate dance movements longer than 5 s. To solve this problem, they use a inpainting technique which is proposed in the text-to-image field [23] to utilize the generation of multiple 5-s segments and combine them to form longer dance movements.

As shown in Fig. 2, several 5-s noisy sequences are randomly generated to form a batch. Then the model is utilized for denoising generation to obtain a batch of dance sequences. Next, each fragment in this batch is re-composed into a new batch in the way that the first 2.5 s of each sequence are constrained to match the last 2.5 s of previous one. Then the new batch is diffused according to t. Finally, a batch with higher quality is obtained by repeating the above operation and iterating it according to Fig. 1, and this batch can be unfolded to get a longer sequence of dance movements.

2.3 Resampling Techniques

From Fig. 1, it can be seen that the generation of diffusion models needs to be denoised step by step from the moment of T until $t = 1$. Moreover, the value of T is generally very large, and assuming that $T = 1000$, it means that the loop in the algorithm of generation process needs to be executed in 1000 steps, which is a very time-consuming process. Therefore, how to reduce the number of denoising operations and ensure the quality of the generated results becomes an important issue in optimizing the diffusion models.

In recent years, Song et al. [13] have proposed DDIM and a fast generation algorithm called resample, which allows the generation of diffusion models 10 to 50 times faster than the original generation algorithm. The basic principle of

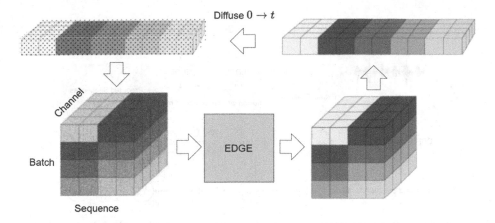

Fig. 2. An example of long-term generation. In this example, the model constrains the first 2.5 s of each sequence to match the last 2.5 s of the previous one to generate a 12.5-s clip, as represented by the temporal regions of distinct clips in the batch that share the same color [17].

this generation algorithm is that the model is denoised starting from T moments, and it can skip some moments after that and directly generate samples from, for example, $T - 20$ moments. By skipping intermediate moments in this way, the number of denoising operations can be reduced from 1000 to 200, and the quality of the generated results will not lose too much. After that, Lu et al. [12] proposed a faster generation algorithm DPM-Solver by solving the ordinary differential equation of the diffusion models. Then the latest and best generation algorithm is DPM-Solver++ [16], which is improved from the above algorithm, and can be used for the data prediction diffusion models $\mathbf{x}_\theta$.

3   Proposed Methods

The overall generation process is shown in Fig. 3 (up), and the time consumed mainly consists of the extraction of music features and the generation process of the diffusion model. We can use cached features to save the time on music feature extraction. Therefore, the main way to speed up the overall generation process is to improve the sampling speed of the diffusion model. Our proposed method is used in the generation process of the diffusion model. As shown in Fig. 3 (below), compared with the sampling algorithm of DDPM [6], we use resample technique to reduce the denoising operations of the model, thereby greatly saving the time in generation process. The basic principles and specific process of our proposed method are discussed in detail below.

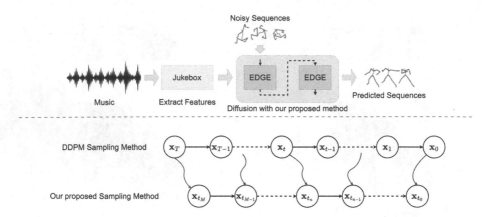

Fig. 3. Overall generation process (up). Our proposed method is used in the generation process of diffusion models. **Comparison with DDPM [6] (below).** Our method uses less sampling steps than DDPM. The blue arrows only indicate the same time step (subscript), not the same sample between the both ends. (Color figure online)

The generation process of a diffusion model can be translated into the problem of solving the corresponding ordinary differential equation (ODE) [19]:

$$\frac{\mathrm{d}\mathbf{x}_t}{\mathrm{d}t} = \left(f(t) + \frac{g^2(t)}{2\sigma_t^2}\right)\mathbf{x}_t - \frac{\alpha_t g^2(t)}{2\sigma_t^2}\mathbf{x}_\theta(\mathbf{x}_t, t), \tag{5}$$

where the coefficients $f(t) = \frac{\mathrm{d}\log\alpha_t}{\mathrm{d}t}, g^2(t) = \frac{\mathrm{d}\sigma_t^2}{\mathrm{d}t} - 2\frac{\mathrm{d}\log\alpha_t}{\mathrm{d}t}\sigma_t^2$ [4]. We follow the notations in Lu et al. [12]. Given a sequence $\{t_i\}_{i=0}^M$ decreasing from $t_0 = T$ to $t_M = 0$ and an initial value $\mathbf{x}_{t_0} \sim \mathcal{N}(\mathbf{0}, \tilde{\sigma}^2\mathbf{I})$, the solver of the ODE aims to iteratively compute a sequence $\{\tilde{\mathbf{x}}_{t_i}\}_{i=0}^M$ to approximate the exact solution at each time t_i, and the final value $\tilde{\mathbf{x}}_{t_M}$ is the approximated sample by the diffusion ODE. Denote $h_i := \lambda_{t_i} - \lambda_{t_{i-1}}$ for $i = 1, \cdots, M$. By solving Eq. (5), we can obtain that for any moment $s > 0$, given an initial value $\mathbf{x}_s$, the solution of the above ordinary differential equation at $t \in [0, s]$ takes the following form:

$$\mathbf{x}_t = \frac{\sigma_t}{\sigma_s}\mathbf{x}_s + \sigma_t \int_{\lambda_s}^{\lambda_t} e^\lambda \hat{\mathbf{x}}_\theta(\hat{\mathbf{x}}_\lambda, \lambda)\mathrm{d}\lambda. \tag{6}$$

Substituting $s = t_{i-1}$ and $t = t_i$ into Eq. (6) and applying the Taylor expansion formula of order $k - 1$ for $\mathbf{x}_\theta$ at $\lambda_{t_{i-1}}$, we can yield an approximate solution:

$$\tilde{\mathbf{x}}_{t_i} = \frac{\sigma_{t_i}}{\sigma_{t_{i-1}}}\tilde{\mathbf{x}}_{t_{i-1}} + \sigma_{t_i}\sum_{n=0}^{k-1}\mathbf{x}_\theta^{(n)}(\hat{\mathbf{x}}_{\lambda_{t_{i-1}}}, \lambda_{t_{i-1}})\int_{\lambda_{t_{i-1}}}^{\lambda_{t_i}} e^\lambda\frac{(\lambda - \lambda_{t_{i-1}})^n}{n!}\mathrm{d}\lambda + o(h_i^{k+1}),$$
$$\tag{7}$$

where the integral term can be computed analytically by the method of integration by parts, $o(h_i^{k+1})$ can be neglected, and ultimately only $\mathbf{x}_\theta^{(n)}$ needs to be estimated [16]. According to Atkinson et al. [24], we can use multistep methods to approximate the derivatives for $n \geq 1$. That is, given the previous values $\{\tilde{\mathbf{x}}_{t_j}\}_{j=0}^{i-1}$ at time t_{i-1}, multistep methods just reuse the previous values to approximate the high-order derivatives.

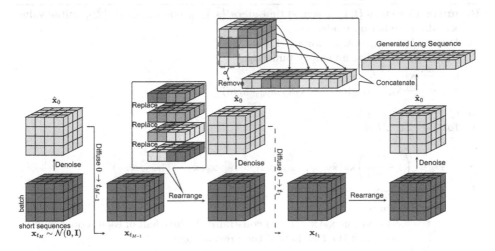

Fig. 4. Our proposed sampling method. The deep gray color in the figure represents the samples obtained after adding noise or sampling from noise, while the light gray color represents the samples obtained after denoising. The other colors in the dialog boxes are used to illustrate the connection between short sequences. In fact, the samples in "Rearrange" refer to the ones after adding noise, while the samples in "Concatenate" refer to the ones after denoising. (Color figure online)

Based on the process of solving the diffusion ODE for continuous time variables in Eq. (5), our proposed sampling method is shown in Fig. 4. Compared with the sampling method of DDPM [6], our method uses less steps so that it can save more time in the generation process. Firstly, we convert the discrete steps $n = 0, 1, \cdots, T - 1$ into continuous time steps using the formula: $t_n = (n+1)/T$. In the dance movement generation system mentioned above, $T = 1000$, so we obtain the mapped sequence $\{0.001, 0.002, \cdots, 1\}$. Then, according to the number of sampling steps M, we extract from the discrete time steps and the remapped continuous time steps at equal distances to form new sequences $\{t_i\}_{i=0}^{M}$ and $\{s_i\}_{i=0}^{M}$, where $t_0 = 1000, s_0 = 1$. It is important to note that these two sequences decrease as i increases, unlike t_n. Similar to the general diffusion model generation process, we sample from a standard normal distribution to obtain a Gaussian noise as the initial value $\mathbf{x}_T$, and then we use the multi-step approximation method for Eq. (7) to iteratively denoise it. After the diffusing operation in every iteration, we "rearrange" the noisy sequences $\mathbf{x}_{t_n}$ to constrain the first half of each sequence to match the last half of the previous one. Finally, we "concatenate" the predicted short sequences to a long sequence according to the illustration shown in Fig. 4. Combined with the long-term generation algorithm of the dance movement generation model, our improved sampling algorithm is shown in Algorithm 1.

Algorithm 1. Long-term DPM++ Sampling.

Require: time steps $\{t_i\}_{i=0}^M$, scaled time steps $\{s_i\}_{i=0}^M$, conditions $\{c_i\}_{i=0}^M$, initial value $\mathbf{x}_T$, data prediction model $\mathbf{x}_\theta$.

1: Denote $h_i := \lambda_{s_i} - \lambda_{s_{i-1}}$ for $i = 1, \cdots, M$.
2: $\tilde{\mathbf{x}}_{s_0} \leftarrow \mathbf{x}_T$. Initialize an empty buffer Q.
3: $Q \overset{\text{buffer}}{\longleftarrow} \mathbf{x}_\theta(\tilde{\mathbf{x}}_{s_0}, t_0, c_0)$
4: $\tilde{\mathbf{x}}_{s_1} \leftarrow \frac{\sigma_{s_1}}{\sigma_{s_0}}\tilde{\mathbf{x}}_{s_0} - \alpha_{s_1}(e^{-h_1} - 1)\mathbf{x}_\theta(\tilde{\mathbf{x}}_{s_0}, t_0, c_0)$
5: $Q \overset{\text{buffer}}{\longleftarrow} \mathbf{x}_\theta(\tilde{\mathbf{x}}_{s_1}, t_1, c_1)$
6: **for** $i \leftarrow 2$ to M **do**
7:     $r_i \leftarrow \frac{h_{i-1}}{h_i}$
8:     $D_i \leftarrow \left(1 + \frac{1}{2r_i}\right)\mathbf{x}_\theta(\tilde{\mathbf{x}}_{s_{i-1}}, t_{i-1}, c_{i-1}) - \frac{1}{2r_i}\mathbf{x}_\theta(\tilde{\mathbf{x}}_{s_{i-2}}, t_{i-2}, c_{i-2})$
9:     $\tilde{\mathbf{x}}_{s_i} \leftarrow \frac{\sigma_{s_i}}{\sigma_{s_{i-1}}}\tilde{\mathbf{x}}_{s_{i-1}} - \alpha_{s_i}(e^{-h_i} - 1)D_i$
10:    **if** $t_i > 0$ **then**
11:        the model output $\mathbf{x}_\theta(\tilde{\mathbf{x}}_{s_i}, t_i, c_i)$ constrains the first half of each sequence in its batch to match the last half of the previous one.
12:    **end if**
13:    **if** $i < M$ **then**
14:        $Q \overset{\text{buffer}}{\longleftarrow} \mathbf{x}_\theta(\tilde{\mathbf{x}}_{s_i}, t_i, c_i)$
15:    **end if**
16: **end for**
17: **return** $\tilde{\mathbf{x}}_{t_M}$

---

4   Experiments and Results

4.1   Implementation Details

We conduct our experiments using the model parameters of the above dance movement generation model trained on the AIST++ dataset [25], which is one of the large datasets commonly used in the dance generation field. It contains a total of 60 music tracks from 10 different genres, and a total of 18,694.6 s of dance movement data, with approximately 1,800 s of movement duration for each genre. The model is trained on a single graphics card NVIDIA GeForce RTX 3090 using mixed-precision mode, with batch size of 128. It is trained for a total of 2000 epochs in about 3 days. We use this trained model to conduct comparative experiments with DDPM [6], DDIM [13] and our proposed method. We test their generative effects on both in-set musics and out-of-set musics.

In-Set Musics: We randomly select 25 musics from the musics in training set and use them to generate dance movements. The first 30 s of each music are selected as the condition for generating dance movements, and we obtain a total of 750 s of dance movements. The in-set musics are all instrumental and rhythmic, which contain no vocal parts.

Out-of-Set Musics: We select five popular musics from YouTube as the out-of-set musics:

- Doja Cat - Woman
- Luis Fonsi - Despacito ft. Daddy Yankee
- ITZY - LOCO
- Saweetie -My Type
- Rihanna - Only Girl (In The World)

Every music selects its first 150 s as the condition for generating dance movements, resulting in a total of 750 s of dance movements. The out-of-set musics contain vocal parts, which are not processed when used to generate movements.

4.2 Quality and Speed

Different from the commonly used Frechet Inception Distance (FID) [8,22], we use the Physical Foot Contact score (PFC) proposed by Tseng et al. [17] to evaluate the quality of generated dance movements. FID requires a large number of training samples and generated results in order to summarize the real distribution and generated distribution. It calculates the similarity between these two distributions to evaluate the quality of generated results. However, for dance generation tasks, its data is still scarce, making it difficult to effectively summarize the data distribution. Therefore, applying FID metric to evaluate the quality of generated dance movements is somewhat inadequate. In contrast, PFC is a physically-inspired metric that requires no explicit physical modeling, which is suitable for this dance movement generation model. This metric arises from two simple, related observations:

1. On the horizontal (xy) plane, any center of mass (COM) acceleration must be due to static contact between the feet and the ground. Therefore, either at least one foot is stationary on the ground or the COM is not accelerating.
2. On the vertical (z) axis, any positive COM acceleration must be due to static foot contact.

Therefore, we can represent adherence to these conditions as an average over time of the below expression, scaled to normalize acceleration:

$$s^i = \|\bar{\mathbf{a}}^i_{COM}\| \cdot \|\mathbf{v}^i_{LeftFoot}\| \cdot \|\mathbf{v}^i_{RightFoot}\|, \tag{8}$$

$$PFC = \frac{1}{N \cdot \max_{1 \le j \le N} \|\bar{\mathbf{a}}_{COM}^j\|} \sum_{i=1}^{N} s^i,$$ (9)

where

$$\bar{\mathbf{a}}_{COM}^i = \begin{pmatrix} a_{COM,x}^i \\ a_{COM,y}^i \\ \max(a_{COM,z}^i, 0) \end{pmatrix}$$

and the superscript i denotes the frame index.

Furthermore, we use Beat Alignment score (Beat Align.) [26] to evaluate how much the generated dance movements correlate to the corresponding music. This metric measures the similarity between the kinematic beats of dance movements and music beats. We employ the librosa toolkit [27] to extract the music beats and calculate the kinematic beats using the local minima of the kinetic velocity. Therefore, the formula for calculating the Beat Alignment score is as follows:

$$\text{BeatAlign.} = \frac{1}{m} \sum_{i=1}^{m} \exp\left(-\frac{\min_{\forall t_j^y \in B^y} \|t_i^x - t_j^y\|^2}{2\sigma^2}\right),$$ (10)

where $B^x = \{t_i^x\}$ is the kinematic beats, $B^y = \{t_j^y\}$ is the music beats and σ is a normalization parameter, usually set to 3. We use the time taken to generate dance movements of the same duration to evaluate the generation speed, and more time means slower generation.

The results of our experiments are shown in Table 1. We can see that the quality of the generated results of DDPM [6] is the best, but at the same time, it also requires the longest time, whether for in-set musics or out-of-set musics. On the other hand, the results of DDIM [13] and our proposed algorithm are slightly inferior to DDPM in terms of PFC scores. However, after visualizing the generated dance movement, there is a not significant difference in sensory perception among them, as shown in Fig. 5. While in terms of time consumption, the generation speed of our improved algorithm far exceeds the other two algorithms. From the perspective of the Beat Alignment metric, there is not much difference among the three algorithms, as the generated dance movements are well synchronized with the rhythm of their corresponding musics.

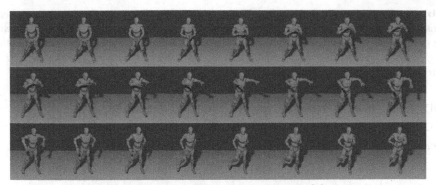

(a) The results generated by DDPM [6].

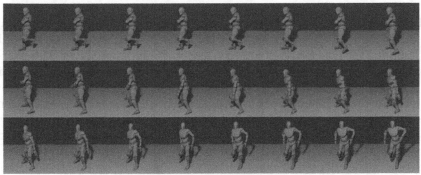

(b) The results generated by DDIM [13].

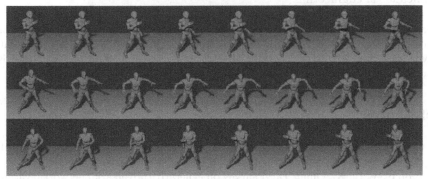

(c) The results generated by our improved algorithm.

Fig. 5. The generated results of the three algorithms using the same piece of music are compared by extracting 24 frames of images from the same time period. The images are arranged in chronological order, with each frame arranged from left to right and top to bottom. Due to the inherent diversity of the results generated by diffusion models, the dance movements of the three algorithm are not identical.

Table 1. Experimental Results. The column "Steps" means the number of iterations in the algorithm. The column "Time" is measured in seconds, and "<1" means the generation time is less than 1 s. ↓ means lower is better. ↑ means higher is better.

Methods	Steps	In-set Musics			Out-of-set Musics		
		PFC ↓	Beat Align. ↑	Time ↓	PFC ↓	Beat Align. ↑	Time ↓
DDPM [6]	1000	**1.7098**	0.2578	30	**0.3983**	0.2374	129
DDIM [13]	200	1.8354	0.2372	6	0.5475	0.2306	25
Ours	20	1.8583	**0.2889**	<1	0.6697	**0.2580**	**2**

5   Conclusions

In this paper, we propose a fast sampling algorithm for a diffusion-based dance generation system. This algorithm not only achieves faster sampling speed, but also supports the generation of long sequences. We improve the DPM-Solver++ algorithm by incorporating a long sequence generation method similar to inpainting, and apply it to the dance movement generation system. Our experiments demonstrate that the improved algorithm greatly speeds up the process of generating long-time sequences, and to some extent ensures that the quality of the generated results does not lose too much. Meanwhile, our work also illustrates that for other generation frameworks based on diffusion models, the DPM-Solver++ algorithm can be adapted in a targeted way to improve the sampling algorithm of these frameworks and accelerate the generation speed, which is of some practical significance in engineering.

Acknowledgements. The work was supported by the National Natural Science Foundation of China (No. 62271083), the Fundamental Research Funds for the Central Universities (No. 2023RC73, No. 2023RC13), open research fund of The State Key Laboratory of Multimodal Artificial Intelligence Systems (No. 202200042).

References

1. Goodfellow, I., et al.: Generative adversarial nets. In: Advances in Neural Information Processing Systems, vol. 27 (2014)
2. Jabbar, A., Li, X., Omar, B.: A survey on generative adversarial networks: variants, applications, and training. ACM Comput. Surv. (CSUR) **54**(8), 1–49 (2021)
3. Salimans, T., Goodfellow, I., Zaremba, W., Cheung, V., Radford, A., Chen, X.: Improved techniques for training GANs. In: Advances in Neural Information Processing Systems, vol. 29 (2016)
4. Kingma, D., Salimans, T., Poole, B., Ho, J.: Variational diffusion models. In: Advances in Neural Information Processing Systems, vol. 34, pp. 21696–21707 (2021)
5. Rezende, D., Mohamed, S.: Variational inference with normalizing flows. In: International Conference on Machine Learning, pp. 1530–1538. PMLR (2015)

6. Ho, J., Jain, A., Abbeel, P.: Denoising diffusion probabilistic models. In: Advances in Neural Information Processing Systems, vol. 33, pp. 6840–6851 (2020)
7. Rombach, R., Blattmann, A., Lorenz, D., Esser, P., Ommer, B.: High-resolution image synthesis with latent diffusion models. In: Proceedings of the IEEE/CVF Conference on Computer Vision and Pattern Recognition, pp. 10684–10695 (2022)
8. Saharia, C., et al.: Photorealistic text-to-image diffusion models with deep language understanding. In: Advances in Neural Information Processing Systems, vol. 35, pp. 36479–36494 (2022)
9. Liu, J., Li, C., Ren, Y., Chen, F., Zhao, Z.: DiffSinger: singing voice synthesis via shallow diffusion mechanism. In: Proceedings of the AAAI Conference on Artificial Intelligence, vol. 36, pp. 11020–11028 (2022)
10. Tevet, G., Raab, S., Gordon, B., Shafir, Y., Cohen-Or, D., Bermano, A.H.: Human motion diffusion model. arXiv preprint arXiv:2209.14916 (2022)
11. Zhang, M., et al.: MotionDiffuse: text-driven human motion generation with diffusion model. arXiv preprint arXiv:2208.15001 (2022)
12. Lu, C., Zhou, Y., Bao, F., Chen, J., Li, C., Zhu, J.: DPM-solver: a fast ode solver for diffusion probabilistic model sampling in around 10 steps. In: Advances in Neural Information Processing Systems, vol. 35, pp. 5775–5787 (2022)
13. Song, J., Meng, C., Ermon, S.: Denoising diffusion implicit models. arXiv preprint arXiv:2010.02502 (2020)
14. Luhman, E., Luhman, T.: Knowledge distillation in iterative generative models for improved sampling speed. arXiv preprint arXiv:2101.02388 (2021)
15. Salimans, T., Ho, J.: Progressive distillation for fast sampling of diffusion models. arXiv preprint arXiv:2202.00512 (2022)
16. Lu, C., Zhou, Y., Bao, F., Chen, J., Li, C., Zhu, J.: DPM-solver++: fast solver for guided sampling of diffusion probabilistic models. arXiv preprint arXiv:2211.01095 (2022)
17. Tseng, J., Castellon, R., Liu, K.: EDGE: editable dance generation from music. In: Proceedings of the IEEE/CVF Conference on Computer Vision and Pattern Recognition, pp. 448–458 (2023)
18. Yang, L., et al.: Diffusion models: a comprehensive survey of methods and applications. arXiv preprint arXiv:2209.00796 (2022)
19. Song, Y., Sohl-Dickstein, J., Kingma, D.P., Kumar, A., Ermon, S., Poole, B.: Score-based generative modeling through stochastic differential equations. arXiv preprint arXiv:2011.13456 (2020)
20. Nichol, A.Q., Dhariwal, P.: Improved denoising diffusion probabilistic models. In: International Conference on Machine Learning, pp. 8162–8171. PMLR (2021)
21. Castellon, R., Donahue, C., Liang, P.: Codified audio language modeling learns useful representations for music information retrieval. arXiv preprint arXiv:2107.05677 (2021)
22. Ramesh, A., Dhariwal, P., Nichol, A., Chu, C., Chen, M.: Hierarchical text-conditional image generation with clip latents. arXiv preprint arXiv:2204.06125 1(2), 3 (2022)
23. Wang, S., et al.: Imagen editor and editbench: advancing and evaluating text-guided image inpainting. In: Proceedings of the IEEE/CVF Conference on Computer Vision and Pattern Recognition, pp. 18359–18369 (2023)
24. Atkinson, K., Han, W., Stewart, D.E.: Numerical Solution of Ordinary Differential Equations. Wiley, Hoboken (2011)

25. Li, R., Yang, S., Ross, D.A., Kanazawa, A.: Learn to dance with AIST++: music conditioned 3D dance generation (2021)
26. Li, R., Yang, S., Ross, D.A., Kanazawa, A.: AI choreographer: music conditioned 3D dance generation with AIST++. In: Proceedings of the IEEE/CVF International Conference on Computer Vision, pp. 13401–13412 (2021)
27. McFee, B., et al.: librosa: audio and music signal analysis in python. In: Proceedings of the 14th Python in Science Conference, vol. 8, pp. 18–25 (2015)

End-to-End Streaming Customizable Keyword Spotting Based on Text-Adaptive Neural Search

Baochen Yang[1,2], Jiaqi Guo[3], Haoyu Li[1,2], Yu Xi[1,2], Qing Zhuo[4], and Kai Yu[1,2(✉)]

[1] X-LANCE Lab, Department of Computer Science and Engineering, MoE Key Lab of Artificial Intelligence, AI Institute, Shanghai Jiao Tong University, Shanghai, China
{baochen1202,haoyu.li.cs,yuxi.cs,kai.yu}@sjtu.edu.cn
[2] State Key Laboratory of Media Convergence Production Technology and Systems, Beijing, China
[3] AISpeech Ltd., Suzhou, China
jiaqi.guo@aispeech.com
[4] Department of Automation, Tsinghua University, Beijing, China
zhuoqing@tsinghua.edu.cn

Abstract. Streaming keyword spotting (KWS) is an important technique for voice assistant wake-up. While KWS with a preset fixed keyword has been well studied, test-time customizable keyword spotting in streaming mode remains a great challenge due to the lack of pre-collected keyword-specific training data and the requirement of streaming detection output. In this paper, we propose a novel end-to-end text-adaptive neural search architecture with a multi-label trigger mechanism to allow any pre-trained ASR acoustic model to be effectively used for fast streaming customizable keyword spotting. Evaluation results on various datasets show that our approach significantly outperforms both traditional post-processing baseline and the neural search baseline, meanwhile achieving a 44x search speedup compared to the traditional post-processing method.

Keywords: Keyword spotting · Streaming · Customizable keyword · Text-adaptive search

1 Introduction

The deployment of intelligent appliances and voice assistants has been increasingly widespread. The KWS module, also known as wake-up word detection or voice trigger, serves as an indispensable component for smart devices to operate in always-listen mode and promptly respond to user at anytime. Therefore, the accuracy and latency of the KWS system are vital for ensuring positive user experience.

J. Jia et al. (Eds.): NCMMSC 2023, CCIS 2006, pp. 125–137, 2024.
https://doi.org/10.1007/978-981-97-0601-3_11

Conventional KWS studies focus on the detection of a *preset fixed keyword* in the continuous speech stream, typically composed of an acoustic model and a post-processing module. Early approaches retain the use of the hidden Markov model (HMM), characterizing the acoustic representations from the deep neural networks as keyword and filler [8]. Later, HMM-free systems are proposed to directly predict the keyword or sub-word token sequence [5]. The architecture of the acoustic model varies, including convolutional neural network [22], recurrent neural network (RNN) [2], the attention mechanism [3] and other composite networks [18]. Moreover, to mitigate the dependency on alignments, some approaches have leveraged connectionist temporal classifier [23] systems and RNN-Transducer systems [15].

Though the aforementioned works have significantly improved performance in preset keyword scenarios, they are mostly not applicable for supporting users to use arbitrary customized keyword at test-time [1]. The requirements of extensive keyword-specific training data to build the system further limit the generalization of these methods. To address the above issues, various approaches [20] have been developed for test-time *customizable keyword* spotting. These methods usually perform adaptive detection of keywords by incorporating textual information into the model [4] or utilizing query-by-example methods to match keywords [9,14]. While these methods have achieved promising results, most of them operate in the offline or non-streaming mode, which are unsuitable for on-the-fly detection requirements. Although some methods have been designed to work in streaming mode [10,17], they still require additional overhead of search and detection process which result in further latency.

In this paper, we propose a fully end-to-end *streaming customizable KWS* system to address the issues of model complexity and detection latency. Our approach leverages text-adaptive neural search to detect arbitrary keyword in continuous speech stream. We conduct comparison against a traditional search method [12] as well as a neural network search approach [4] to demonstrate the effectiveness. The core contributions can be summarized as follows:

- The proposed end-to-end framework significantly reduces the model complexity and moderates the adaptation detection latency. The elimination of post-processing modules ensures the consistency of the training and evaluation process and enables joint optimization of the model.
- We introduce a multi-label mechanism in the trigger module, firstly regarding the KWS task as a cascading multi-label trigger task. The objective function directly optimizes the probabilities of all tokens jointly, enforcing the model to consider all information of the arbitrary keyword.
- The proposed neural search process is decoupled from the acoustic model, which is solely responsible for providing acoustic representation. It implies that the search process is independent of any specific acoustic model and easily extended to different acoustic models.

The rest of the paper is organized as follows. In Sect. 2, we explain in detail the architecture and training process of our method. Then we describe the implementation detail in Sect. 3 and the experiment results in Sect. 4. In Sect. 5, we finally conclude our work in this paper.

2 End-to-End Streaming Framework

In this section, we detail the architecture and training process of our end-to-end streaming customizable keyword spotting framework. In Sect. 2.1, we construct the training dataset by sampling the Audio-(Text Set) pairs from common recognition dataset. In Sect. 2.2, the sampled Audio and Text Set are encoded by a pretrained AM and a BiLSTM respectively. In Sect. 2.2, we propose the keyword-constrained attention with an additional attention loss to fuse the encoded Audio and Text Set while guide the attention scores to be more focused on the target keyword. In Sect. 2.3, we propose the cascading trigger module with additional multi-label loss to utilize the cumulative pattern of keyword spotting and determine the final wake-up. The overview of our framework is shown in Fig. 1.

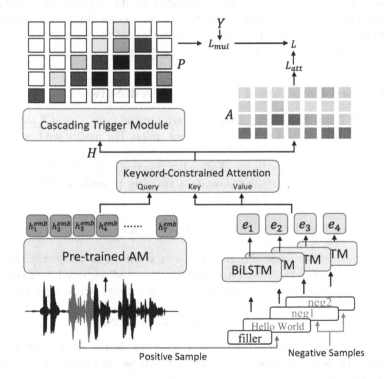

Fig. 1. Overview of our end-to-end streaming customizable keyword spotting framework. The sampled Audio-(Text Set) pairs are encoded by pre-trained AM and BiLSTM respectively. The keyword-constrained attention fuses acoustic and textual information and guides the attention scores to be more focused. The cascading trigger module determines the final wake-up from a multi-label classification view. Note that the negative samples in gray are only used in the training phase. (Color figure online)

2.1 Training Sample Construction

In the absence of a training dataset specifically designed for customizable keyword spotting tasks, we construct training samples from common recognition dataset by sampling the Audio-(Text Set) pairs. For each utterance $\mathbf{X}$, we construct a text set of size N: $\mathcal{S}, |\mathcal{S}| = N$. $\mathcal{S}$ contains three types of text: 1) a zero vector $\mathbf{0}$ serving as the filler, 2) a positive keyword phrase to be spotted, 3) $N-2$ negative inputs that helps training. For each utterance, we sample two consecutive words from the transcript corresponding to the audio as the positive keyword phrase input $\mathbf{w}^{(+)}$. To enhance the model's discriminative ability and robustness to false alarms, additional $N-2$ negative phrases $\{\mathbf{w}_1^{(-)}, \mathbf{w}_2^{(-)}, \ldots, \mathbf{w}_{N-2}^{(-)}\}$ are incorporated during the training process. Inspired by contrastive learning [7] and discriminative training [19], we employ two sampling strategies to create these negative samples: 1) Randomly sample phrases from other utterances in the same mini-batch. 2) Perform rule-based substitution on the phoneme sequences of positive keyword phrase. These rules are based on a manually constructed confusion set where each phoneme is mapped to its confused phonemes. The maximum length of every sampled phoneme sequences is limit to 20 empirically. In summary, we construct an Audio-(Text Set) pair for each training sample: $(\mathbf{x}, \mathcal{S})$, where $\mathcal{S} = \{\mathbf{0}, \mathbf{w}^{(+)}, \mathbf{w}_1^{(-)}, \mathbf{w}_2^{(-)}, \ldots, \mathbf{w}_{N-2}^{(-)}\}$.

2.2 Keyword-Constrained Attention Based Network

Audio and Text Encoding. Given a training sample pair $(\mathbf{X}, \mathcal{S})$, we first encode the audio and text inputs into latent space. For audio, we first pre-train a frame-level AM via frame-level force-aligned labels, utilizing the cross-entropy loss as the optimization criterion. Following pre-training, all parameters of the AM are frozen throughout the subsequent training procedure. Then we embed the audio into a sequence of vectors:

$$\mathbf{H}^{emb} = [\mathbf{h}_1^{emb}, \ldots, \mathbf{h}_T^{emb}] = AM(\mathbf{X}), \tag{1}$$

where $\mathbf{h}_t^{emb} \in \mathbb{R}^h$ denotes the encoded t-th frame. For the text inputs, the phoneme sequences of the words are encoded by a bidirectional long short-term memory (BiLSTM) and the last hidden states are regarded as aggregated word embeddings:

$$\mathbf{e}_i = BiLSTM(\mathbf{w}_i), \text{for } \mathbf{w}_i \in \mathcal{S}, \tag{2}$$

We use $\mathbf{E} = [\mathbf{e}_1, \mathbf{e}_2, \ldots, \mathbf{e}_N]$ to represent the embeddings of text inputs where $\mathbf{e}_1$ denotes the filler, $\mathbf{e}_2$ denotes the positive phrase and other vectors denotes negative samples.

Keyword-Constrained Attention. The encoded audio and text representations are fused through cross attention mechanism [13]. We follow the attention implementation in [16] and use encoded audio as query, encoded text sets as both key and value in the cross attention calculation:

$$\mathbf{Q} = \mathbf{W}_q\mathbf{H}^{emb}, \mathbf{K} = \mathbf{W}_k\mathbf{E}, \ \mathbf{V} = \mathbf{W}_v\mathbf{E}, \tag{3}$$

$$\mathbf{A} = softmax(\frac{\mathbf{Q}\mathbf{K}^T}{\sqrt{d_k}}), \tag{4}$$

$$\mathbf{H} = \mathbf{A}\mathbf{V} + \mathbf{H}^{emb}, \tag{5}$$

where $\mathbf{W}_q, \mathbf{W}_k, \mathbf{W}_v \in \mathbb{R}^{h \times h}$ are used to perform projection. $\mathbf{A} \in \mathbb{R}^{T \times N}$ is the attention score matrix and $\mathbf{A}_t \in \mathbb{R}^N$ indicates the t-th frame's attention score on the N text inputs in the text set. $\mathbf{H} \in \mathbb{R}^{T \times h}$ is the final attention output and $\mathbf{H}_t \in \mathbb{R}^h$ is the fused representation of the t-th frame. To guide the distribution of attention scores to become more focused, an additional attention loss $\mathcal{L}_{att}$ is employed on the attention scores of the last attention layer. The $\mathcal{L}_{att}$ can be calculated as:

$$\mathcal{L}_{att} = -(\sum_{t \notin \mathcal{P}} \log \mathbf{A}_{t,1} + \sum_{t \in \mathcal{P}} \log \mathbf{A}_{t,2}), \tag{6}$$

where $\mathcal{P}$ represents frames corresponding to the positive keyword. Note that we use w_1 for the filler input and w_2 for the positive keyword, the loss $\mathcal{L}_{att}$ encourages the attention to be activated on the positive item if the frame belongs to the keyword, otherwise on the filler.

2.3   Cascading Trigger Module

The trigger module is responsible for making the final decision on whether the expected keyword is spotted. Some previous end-to-end methods [1,4] directly output the likelihood score of the keyword without any post-processing searching procedure. These methods regard the problem as a binary classification task, design totally end-to-end frameworks, and reduce the complexity of the system. To retain the advantages of the end-to-end systems while better modeling the discriminative ability of keywords, we further reformulate the problem as a cascading multi-label trigger task. For the t-th frame, we assign a label $\mathbf{Y}_t \in [0, 1]^{l_{max}+1}$ (l_{max} denotes the maximum length of the keyword phoneme sequence mentioned in Sect. 2.1) to it:

The keyword is totally non-triggered at time step t,

$$j \geq 0 : \mathbf{Y}_{t,j} = 0; \ \mathbf{Y}_{t,0} = 1, \tag{7}$$

The first i phonemes have been triggered at time step t,

$$1 \leq j \leq i : \mathbf{Y}_{t,j} = 1; \ \text{Otherwise}: \mathbf{Y}_{t,j} = 0, \tag{8}$$

Based on the above definition, only when all the phonemes of keyword have been triggered (i.e. $1 \leq j \leq D : \mathbf{Y}_{t,j} = 1$; Otherwise: $\mathbf{Y}_{t,j} = 0, D$ is length of the phoneme sequence), the keyword is considered triggered and causes a wake-up. An example of the label generation process is shown in Fig. 2.

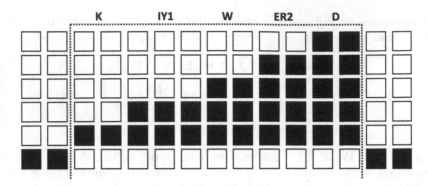

Fig. 2. The example multi-label for keyword "keyword". We first assign labels of 0 to all frames of the sequence that are not related to the keyword. Next, we assign labels of 1 to the frames corresponding to the first phoneme of the keyword ("K" in this example). Then we assign labels of 1 and 2 to the frames corresponding to the second phoneme "IY1" and so on, until all frames corresponding to the keyword are labeled.

Assuming the distribution between frames and labels is distribution-independent, we solve the aforementioned cascading trigger problem by multiple binary classifiers. The loss $\mathcal{L}_{mul}$ of the multi-label classification can be calculated as:

$$\mathcal{L}_{mul} = -\sum_{t=1}^{T} \sum_{i=0}^{l_{max}} \left(\mathbf{Y}_{t,i} \log \mathbf{P}_{t,i} + (1 - \mathbf{Y}_{t,i}) \log(1 - \mathbf{P}_{t,i}) \right), \qquad (9)$$

where $\mathbf{P}_{t,i} \in (0,1)$ denotes the predicted probability that the i-th phoneme has been triggered at t-th frame. The likelihood score of the keyword for each frame is calculated by taking the mean score of all the labels. The advantages of multi-label mechanism will be further discussed in Sect. 4.3.

The overall end-to-end model is trained jointly using loss $\mathcal{L}$ in Eq. 10. The α serves as a hyper-parameter to balance the relative significance of each part.

$$\mathcal{L} = \mathcal{L}_{mul} + \alpha \, \mathcal{L}_{att} . \qquad (10)$$

3   Experiment Configuration

Our proposed text-adaptive search method is compared against a traditional search method **automatic-gain** [12] and a neural search network **detection-filters** [4]. All the models were trained on the same **LibriKWS** dataset. To provide a better comparison of performance with existing methods, we also include a publicly available and commonly used dataset: **Hey-Snips** [6].

3.1   Dataset

LibriKWS: To our best knowledge, current customizable keyword spotting task lacks a widely used and recognized standard dataset. Thus, we build a keyword spotting version of the LibriSpeech [11] dataset to train and evaluate our proposed method. The keywords for testing were selected based on their frequency rank in the testing dataset, and the least frequent keyword selected appears only three times in the testing set. Due to the textual domain shift between the training dataset and the testing dataset, some **unseen or less frequent words** in training set were selected, such as "Bartley", "brahman", "Wilfrid" etc. Specifically, we select 15 keywords with at least six phonemes per word from the test-clean and test-other datasets respectively. And the remained audios irrelevant to any keyword are assigned to the false alarm dataset. The duration of the false alarm dataset is about 3 h for both test-clean and test-other datasets.

Notably, the match of the keyword is based on the phoneme sequence, meaning that any word with a keyword as a prefix, suffix, or sub-word is considered as a positive sample. For instance, "accompany" will also be regarded as a positive sample of the keyword "company".

Hey-snips: Hey-snips is a crowd-sourced dataset for the *hey snips* wake word. The unbalanced and non-i.i.d acoustic environment can effectively reveal the robustness to false alarms. We kept the train set and dev set untouched and only used the test set for evaluation. All evaluated methods perform a zero-shot detection of *hey-snips* to compare their effectiveness of customizable keyword spotting.

3.2 Baseline Models Setup

The **Automatic-gain** [12] follows the pipeline proposed by [5], composed of an AM and a post-processing algorithm. The frame-level AM consists of 5 layers of deep fast sequential memory network (DFSMN) [21], and the post-processing algorithm is a dynamic programming algorithm, which can effectively reduce false alarm rate during decoding compared to the original posterior handling algorithm in [5]. The hidden size and the projection size in the DFSMN are 256 and 128, respectively. The total trainable parameters of the DFSMN are about 1440K.

The **Detection-filters** [4] introduced a keyword spotting neural network, utilizing a keyword encoder predicts the weights of the detection filters to detect the keyword from the output of the acoustic model at each time step. We re-implement this method and include it in our baseline. We replaced the LSTM acoustic model with the same DFSMN mentioned above and kept the other settings consistent with the descriptions in the original paper.

3.3 Proposed Model Setup

Our end-to-end model uses the architecture depicted in Fig. 1. The same DFSMN in the baseline models is used as the pre-trained AM in our method. The BiLSTM consists of 3 layers of network with a hidden size of 128. And we apply 3 layers of cross-attention between the acoustic representations and textual information with the 128 dimensions of hidden size. The trigger module contains a 2-layer fully connected network. The size of the text set fed to the system N mentioned in Sect. 2.1 during the training phase is set to 4, including one filler, one positive phrase, and two negative phrase sampled from two sample strategy respectively. The α in Equation (10) will be further analysed in Sect. 4.1.

During the testing phase, the model takes a text set consisting of a keyword and a filler without negative samples as input. As the attention score on the keyword item of the keyword frames is typically high under the constrained attention, only frames with an attention score above 0.8 (empirically set) are considered potential keyword-activated frames. The mean score of all the labels corresponding to the last phoneme is then used as the confidence score. In summary, the wake-up event is triggered when the confidence scores of those potential frames exceed the wake-up threshold.

4   Results and Analysis

Table 1 shows the comparison results with the baseline models on LibriKWS dataset. We report the micro-averaged recall (higher is better) of all the keywords.

Table 1. Micro-averaged recall of keywords on LibriKWS dataset, FAR stands for false alarm rate.

Model	FAR	LibriKWS	
		test-clean	test-other
Automatic-gain [12]	1 FA/hour	0.934	0.739
Detection-filters [4]	1 FA/hour	0.753	0.444
Detection-filters [4]	4 FA/hour	0.910	0.683
Ours	1 FA/hour	**0.966**	**0.758**

The ability to detect unseen keywords is a significant advantage of customizable keyword spotting. Thus, we conduct zero-shot evaluation on the Hey-snips dataset and present the results in Table 2. Our model significantly outperforms baseline models on both datasets. The Detection-filters [4] is prone to the false alarms, especially when the keyword is short in length.

To further investigate the effectiveness of our sample strategies and introduced multi-label mechanism, we conduct a series of comparative experiments on LibriKWS dataset.

4.1   Relative Importance of the Modules

To balance the relative importance of the modules, we conduct experiments on the hyper-parameter α in Eq. 10. Results are shown in Fig. 3. A small value of α (≤ 0.2) diminishes the keyword-constrained attention mechanism, weakening the model's ability to **discriminate confusing words** and resulting in poor performance in complex acoustic environments. Conversely, a large value of α (≥ 1.5)

Table 2. Zero-shot performance of keyword hey-snips on Hey-snips dataset, FAR stands for false alarm rate.

Model	FAR	Hey-snips
Automatic-gain [12]	0.3 FA/hour	0.739
Detection-filters [4]	0.3 FA/hour	0.702
Ours	0.3 FA/hour	**0.856**

reduces the influence of the multi-label mechanism, weakening the model's ability to **detect the keyword** and leading to poor performance. In the balanced intermediate settings, the model consistently achieves good performance, validating the effectiveness of the two mechanisms in our model.

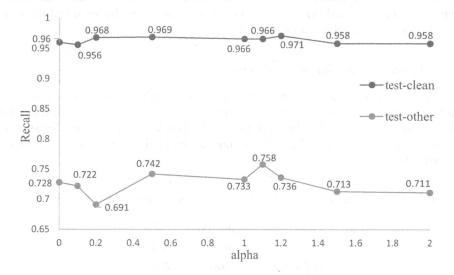

Fig. 3. Impact of hyper-parameter α at 1 FA per hour.

4.2 Impact of Negative Samples

To study the impact of negative samples during the training process, we evaluate the model performance with different negative sample construction strategies. Hybrid sample refers to using both random samples and more confusing samples. All_random and All_hard refer to using only randomly sampled negative inputs and manually constructed confusing negative inputs, respectively. And No_neg denotes using only positive keyword and filler during training.

As shown in Table 3, the Hybrid sample obtains the best performance on both test datasets. The results of All_random and No_neg suggest that random negative samples can effectively improve the detection ability of the model in

Table 3. The evaluation results of different sample strategies at 1 FA per hour.

Sample strategy	LibriKWS	
	test-clean	test-other
Hybrid sample	**0.966**	**0.758**
All_random	0.961	0.689
All_hard	0.919	0.697
No_neg	0.948	0.676

continuous audio. It should be due to greater coverage of the textual space and better fits to the testing scenarios of the keyword spotting task. On this basis, the poor performance on the test-other dataset indicates that the model struggles in a complex environment, leading to limited recall with the same false alarm rate. The usage of additional confusing negative samples during discrimination training leads to further gains in performance.

4.3   Impact of Multi-label Mechanism

In this section, we analyse the impact of the multi-label mechanism in our cascading trigger module. To do this, we remove the multi-label mechanism and directly use the score of either the first or the last label as the confidence score for detecting the keyword. In this case, our method will degenerate into a binary classification method.

Table 4. The evaluation results of different label methods at 1 FA per hour.

Label method	LibriKWS	
	test-clean	test-other
Multi-label	**0.966**	**0.758**
First label	0.806	0.621
Last label	0.950	0.697

The results in Table 4 clearly show that our multi-label mechanism is crucial for better performance. When using only the first label, the model is susceptible to false alarms caused by similar-pronounced words, such as common prefixes, and suffixes. While using only the last label is similar to the approach proposed in [1] and leads to a significant drop in performance, especially under complex acoustic circumstances. The potential reason is the lack of task-specific training data. Our multi-label mechanism is more suitable for the trigger process of keywords with different number of phonemes and leads to significantly better performance.

4.4 Comparison of the Model Efficiency

We conducted a comparison of the average processing time between our proposed end-to-end model and the post-processing method [12] under identical conditions. Our end-to-end model based on text-adaptive neural search directly outputs confidence scores without any post-processing, while Automatic-gain [12] requires additional time to run a dynamic programming algorithm. Despite having more parameters, our end-to-end model achieves a **44x** relative speedup.

5   Conclusions

This paper presents a fully end-to-end streaming customizable keyword spotting system that performs a text-adaptive neural search during run-time. We utilize a keyword-constrained cross-attention mechanism to incorporate textual information with audio representations. Meanwhile, we introduce a novel multi-label mechanism to achieve better modeling of the arbitrary keyword. Noting that the neural search process is decoupled from the acoustic model and can be transferred to other acoustic models. Evaluation results demonstrate the superior robustness of our model compared to the baseline models against the false alarms. Furthermore, our methods achieves about 44x relative speedup than conventional post-processing method. In future work, we will focus on extending the model to multiple keywords scenarios and enhancing the robustness in the complex acoustic environment.

Acknowledgements. This research was funded by Scientific and Technological Innovation 2030 under Grant 2021ZD0110900 and the Key Research and Development Program of Jiangsu Province under Grant BE2022059.

References

1. Alvarez, R., Park, H.: End-to-end streaming keyword spotting. In: Proceedings of the IEEE ICASSP, pp. 6336–6340 (2019). https://doi.org/10.1109/ICASSP.2019.8683557
2. Baljekar, P., Lehman, J.F., Singh, R.: Online word-spotting in continuous speech with recurrent neural networks. In: Proceedings of the IEEE SLT, pp. 536–541 (2014). https://doi.org/10.1109/SLT.2014.7078631
3. Berg, A., O'Connor, M., Cruz, M.T.: Keyword transformer: a self-attention model for keyword spotting. In: Proceedings of the ISCA Interspeech, pp. 4249–4253 (2021). https://doi.org/10.21437/Interspeech.2021-1286
4. Bluche, T., Gisselbrecht, T.: Predicting detection filters for small footprint open-vocabulary keyword spotting. In: Proceedings of the ISCA Interspeech, pp. 2552–2556 (2020). https://doi.org/10.21437/Interspeech.2020-1186
5. Chen, G., Parada, C., Heigold, G.: Small-footprint keyword spotting using deep neural networks. In: Proceedings of the IEEE ICASSP, pp. 4087–4091 (2014). https://doi.org/10.1109/ICASSP.2014.6854370

6. Coucke, A., Chlieh, M., Gisselbrecht, T., Leroy, D., Poumeyrol, M., Lavril, T.: Efficient keyword spotting using dilated convolutions and gating. In: Proceedings of the IEEE ICASSP, pp. 6351–6355 (2019). https://doi.org/10.1109/ICASSP.2019. 8683474

7. Gao, T., Yao, X., Chen, D.: SimCSE: simple contrastive learning of sentence embeddings. In: Proceedings of the EMNLP, pp. 6894–6910 (2021). https://doi.org/10. 18653/v1/2021.emnlp-main.552

8. Ge, F., Yan, Y.: Deep neural network based wake-up-word speech recognition with two-stage detection. In: Proceedings of the IEEE ICASSP, pp. 2761–2765 (2017). https://doi.org/10.1109/ICASSP.2017.7952659

9. Huang, J., Gharbieh, W., Wan, Q., Shim, H.S., Lee, H.C.: QbyE-MLPMixer: query-by-example open-vocabulary keyword spotting using MLPMixer. In: Proceedings of the ISCA Interspeech, pp. 5200–5204 (2022). https://doi.org/10.21437/Interspeech

10. Liu, Z., Li, T., Zhang, P.: RNN-T based open-vocabulary keyword spotting in mandarin with multi-level detection. In: Proceedings of the IEEE ICASSP, pp. 5649–5653 (2021). https://doi.org/10.1109/ICASSP39728.2021.9413588

11. Panayotov, V., Chen, G., Povey, D., Khudanpur, S.: Librispeech: An ASR corpus based on public domain audio books. In: Proceedings of the IEEE ICASSP, pp. 5206–5210 (2015). https://doi.org/10.1109/ICASSP.2015.7178964

12. Prabhavalkar, R., Alvarez, R., Parada, C., Nakkiran, P., Sainath, T.N.: Automatic gain control and multi-style training for robust small-footprint keyword spotting with deep neural networks. In: Proceedings of the IEEE ICASSP, pp. 4704–4708 (2015). https://doi.org/10.1109/ICASSP.2015.7178863

13. Pundak, G., Sainath, T.N., Prabhavalkar, R., Kannan, A., Zhao, D.: Deep context: end-to-end contextual speech recognition. In: Proceedings of the IEEE SLT, pp. 418–425 (2018)

14. Reuter, P.M., Rollwage, C., Meyer, B.T.: Multilingual query-by-example keyword spotting with metric learning and phoneme-to-embedding mapping. In: Proceedings of the IEEE ICASSP, pp. 1–5. IEEE (2023)

15. Tian, Y., Yao, H., Cai, M., Liu, Y., Ma, Z.: Improving RNN transducer modeling for small-footprint keyword spotting. In: Proceedings of the IEEE ICASSP, pp. 5624–5628 (2021). https://doi.org/10.1109/ICASSP39728.2021.9414339

16. Vaswani, A., et al.: Attention is all you need. In: Proceedings of the NeurIPS, pp. 5998–6008 (2017)

17. Xi, Y., Tan, T., Zhang, W., Yang, B., Yu, K.: Text adaptive detection for customizable keyword spotting. In: Proceedings of the IEEE ICASSP, pp. 6652–6656 (2022). https://doi.org/10.1109/ICASSP43922.2022.9746647

18. Yan, H., He, Q., Xie, W.: CRNN-CTC based mandarin keywords spotting. In: Proceedings of the IEEE ICASSP, pp. 7489–7493 (2020). https://doi.org/10.1109/ ICASSP40776.2020.9054618

19. Yang, Z., Lv, H., Wang, X., Zhang, A., Xie, L.: Minimizing sequential confusion error in speech command recognition. In: Proceedings of the ISCA Interspeech, pp. 3193–3197 (2022)

20. Yang, Z., et al.: CaTT-KWS: a multi-stage customized keyword spotting framework based on cascaded transducer-transformer. In: Proceedings of the ISCA Interspeech, pp. 1681–1685 (2022). https://doi.org/10.21437/Interspeech

21. Zhang, S., Lei, M., Yan, Z., Dai, L.: Deep-FSMN for large vocabulary continuous speech recognition. In: Proceedings of the IEEE ICASSP, pp. 5869–5873 (2018). https://doi.org/10.1109/ICASSP.2018.8461404

22. Zhou, H., Hu, W., Yeung, Y.T., Chen, X.: Energy-friendly keyword spotting system using add-based convolution. In: Proceedings of the ISCA Interspeech, pp. 4234–4238 (2021). https://doi.org/10.21437/Interspeech
23. Zhuang, Y., Chang, X., Qian, Y., Yu, K.: Unrestricted vocabulary keyword spotting using LSTM-CTC. In: Morgan, N. (ed.) Proceedings of the ISCA Interspeech, pp. 938–942 (2016). https://doi.org/10.21437/Interspeech

The Production of Successive Addition Boundary Tone in Mandarin Preschoolers

Aijun Li[1,2(✉)] [iD], Jun Gao[1] [iD], and Zhiwei Wang[1] [iD]

[1] Institute of Linguistics, Chinese Academy of Social Sciences, Beijing, China
{liaj,gao-jun}@cass.org.cn, 2020110015@stu.bisu.edu.cn
[2] Corpus and Computational Linguistics Center, CASS, Beijing, China

Abstract. In Mandarin Chinese, a falling tone or a rising tone can be added to the final lexical tone of an utterance, called successive addition boundary tone (SuABT). It can be used to convey pragmatic or paralinguistic information during conversation. Our primary concern is to understand how Mandarin-speaking children master the use of successive addition boundary tones (SuABT) during their interactive language development. In pursuit of this objective, the current study investigated the phonetic characteristics of falling SuABT produced by 25 preschoolers, ranging in age from 2;6 to 6;8, as well as their mothers who engaged in child-directed speech (CDS). Results showed that children as early as 2 years old were able to use falling SuABT to encode the phonological functions of tone and intonation simultaneously, however, the performance of SuABT production was tone-dependent. For children between 2–3 years old, errors were predominantly observed in the phonetic control of the lexical-tone part. Specifically, the pitch targets of tone 2 and 4 were undershot for children's production speech (CPS) compared with those found in CDS. The pitch range of the SuABT was smaller in CPS, while their duration was longer. The production of falling SuABT exhibited significant developmental changes during the 3–4 age range, stabilized during 4–5 years, and approached the level of CDS after 6 years old. There was also evidence of prosodic entrainment between CDS and CPS.

Keywords: Tone production · Mandarin-speaking preschooler · Child- directed speech · Successive addition boundary tone · Prosodic entrainment

1 Introduction

Boundary tones are pitch patterns employed at the end of spoken phrases or sentences to convey pragmatic and paralinguistic information. In tone languages, the phonetic realization of boundary tones reflects a mixed relationship of tone and intonation. Chao [1] proposed that in the case of Mandarin, this relationship can be manifested as simultaneous addition and successive addition of the tonal and intonational features at the boundary. Simultaneous addition boundary tone (SiABT) is achieved as the algebraic sum of tonal and intonational features. It alters the phonetic realization without changing the phonological features of the original tone. Lin and Li [2] identified three patterns of SiABT based on acoustic analysis of Mandarin tones in declarative and interrogative sentences. Successive addition boundary tone (SuABT), on the other hand, is achieved by adding a falling or rising tail after the lexical tone without changing the lexical meaning.

To date, many studies have focused on the SiABT by examining the production of monosyllabic and disyllabic tones in Mandarin-speaking children under declarative intonation, as in Xu et al. [4], Wong [5, 6], Wang [7] and Wang et al. [8]. Based on the comparison of the pitch range, pitch contour, and duration between monosyllabic words in preschool child speech and adult-directed speech, it can be observed that the children's production of SiABT shows a developmental tendency towards adult-like patterns as they grow older. Additionally, during the interaction between children and their parents, there is an entrainment in the pitch patterns between children's production and adult-directed speech.

On the other hand, studies on the production of SuABT have been raw rare, except our prior work on the relationship between the function and form of SuABT in emotional intonations and provided its phonetic and phonological descriptions [3] and Mueller-Liu's re-discussion on it [13]. This idea is inspired by our observation of the longitudinal corpus CASS_Child_Word, in which both children and their parents frequently used SuABT to convey various moods and modalities. For instance, parents use SuABT to emphasize certain words while teaching their children vocabulary, as exemplified in Fig. 1.

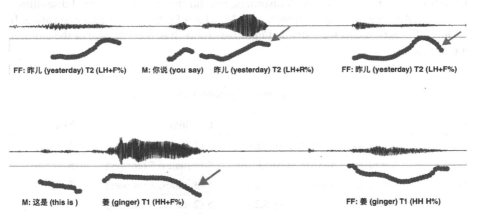

FF: 昨儿 (yesterday) T2 (LH+F%)   M: 你说 (you say)   昨儿 (yesterday) T2 (LH+R%)   FF: 昨儿 (yesterday) T2 (LH+F%)

M: 这是 (this is)   姜 (ginger) T1 (HH+F%)   FF: 姜 (ginger) T1 (HH H%)

Fig. 1. The upper panel is the waveform of a conversation between FF and her mother when FF was 13 months old. They both used SuABT (arrows). The bottom panel is the waveform of a conversation when FF was 16 months old. Her mother used falling SuABT (arrow), but FF did not. Here we denote the rising and falling SuABT as T + R% and T + F%, respectively, where T stands for the lexical tones, including high tone T1 (HH), rising tone T2 (LH), low dipping tone T3 (LL(H)) and falling T4 (HL).

In the upper panel of Fig. 1, we see a 13-month-old baby girl named FF first produced a word in T2 with a segmental pronunciation error. Her mother (M) corrected her with a rising SuABT, and then FF repeated the word in correct pronunciation with a falling SuABT. In another conversation between FF and her mother at 16 months (bottom panel of Fig. 1), her mother used a falling SuABT to emphasize a high-level-tone word "

姜" (ginger). However, in her response, FF just used the high-level tone without using SuABT at all, demonstrating that she is not only able to produce the complex SuABT but can also discriminate the tonal and pragmatic information conveyed by SuABT at a very young age.

Despite the above inspiring observation, it remains to be clarified in terms of how children acquire tone categories from complex intonation patterns that convey both lexical and post-lexical information simultaneously. To answer the question, this study started with the production of falling SuABT by Mandarin-speaking preschool children during vocabulary learning. It aims to explore the developmental phonetic pattern of SuABT in terms of pitch and duration by comparing the patterns between children's production speech (CPS) without guidance and child-directed speech (CDS) that are guided by their parents. This uncharted territory holds the potential to offer a unique perspective regarding the developmental traces of SuABT acquisition.

2 Methods

2.1 Participants

The study comprised 25 mother-child dyads, with children aged from 2;6 to 6;8 (i.e., between two and a half years old and six years and eight months old). None of the children had a history of hearing or speech disorders, nor did they have intellectual disabilities. Moreover, their parents predominantly used Mandarin for their daily communication. In Table 1, the participants were divided into five groups based on the age of the children at one-year intervals.

Table 1. Demographic information of participants.

Groups	Age of children	Children	Mothers
G1	2–3 years (2;6–2;8, Mean: 2;6)	5 (3 males, 2 females)	5
G2	3–4 years (3;4–3;10, Mean: 3;7)	5 (0 males, 5 females)	5
G3	4–5 years (4;1–4;11, Mean: 4;6)	5 (3 males, 2 females)	5
G4	5–6 years (5;1–5;4, Mean: 5;2)	5 (2 males, 3 females)	5
G5	6–7 years (6;1–6;8, Mean: 6;5)	5 (2 males, 3 females)	5

2.2 Materials and Procedure

The study employed 12 monosyllabic words that were well-known to the participants and were organized into 6 minimal pairs of tones in Mandarin, as illustrated in Table 2. These words were accompanied by 12 corresponding pictures and were recorded by a female speaker in a CDS register, both with and without the falling SuABT. These recorded materials served as both visual and auditory prompts during the recording session.

Table 2. List of monosyllabic stimuli.

Target Tone Pairs	Chinese Word	English Meaning	IPA	Pinyin
T1-T2	鸭-牙	duck-tooth	/ja1/-/ja2/	ya1-ya2
T1-T3	三-伞	three-umbrella	/san1/-/san3/	san1-san3
T1-T4	书-树	book-tree	/ʂu1/-/ʂu4/	shu1-shu4
T2-T3	鼻-笔	nose-pen	/pi2/-/pi3/	bi2-bi3
T2-T4	狼-浪	wolf-wave	/laŋ2/-/laŋ4/	lang2-lang4
T3-T4	土-兔	soil-rabbit	/tʰu3/-/tʰu4/	tu3-tu4

During the recording session, the child participants were engaged in three tasks. Task 1 involved a picture-naming exercise, where each child was instructed to name the pictures displayed on the screen. If a child was unable to name a picture correctly, the experimenter provided a hint to the mother to guide the child. For each target word, the child was asked to spontaneously pronounce it once, either with or without the SuABT. Task 2 was an imitation task, where the child was tasked with naming the picture prompted by the pre-recorded sound without the SuABT. Following the child's pronunciation, the mother repeated the word to validate the child's pronunciation. If there were multiple repetitions, only the clearest production was chosen for analysis. Task 3 was also an imitation task, with the child producing the falling SuABT. The procedure mirrored Task 2, except that the audio prompts were delivered with a falling SuABT. This approach simulated the interaction between the child and the mother. Since the children were not pressured to produce the SuABT, they might not always successfully produce the SuABT.

All the sounds were recorded using a laptop computer, equipped with an ex- ternal USB sound blaster card and a microphone, and were sampled at a rate of 44.1 kHz with a 16-bit resolution. A total of 1,587 tokens were collected, which included 780 tokens with SuABT (423 produced by children and 357 by mothers). For phonetic annotation and analysis, Praat (https://www.fon.hum.uva.nl/praat/) was used. F0 values were initially extracted by Praat and subsequently verified manually. Additionally, the pitch and duration of the final (vowel) of the target word were extracted. The pitch of each final was uniformly sampled at 10 points and converted into semitones using Formula 1, in which "f0" represented the frequency in Hz, and "f0r" denoted the reference frequency set to 75 Hz. To mitigate the influence of individual characteristics, z-scores were applied when examining the average pitch contours. All results were based on the semitone-transformed F0 values.

$$ST = 12 * log_2(f0/f0r) \tag{1}$$

3 Results

3.1 Pitch Development Patterns

This section examined the patterns of syllables with falling SuABT in children's production speech (CPS) and child-directed speech (CDS). Figure 2 illustrates the average pitch contours of the falling SuABT on T1-T4 in CPS and CDS across different age groups (G1-G5).

Comparing the CPS (left column) with CDS (right column) in Fig. 2, it's notable that the SuABT of the target syllables was well-produced as T + F% in all CPS groups. However, the pitch realization of the lexical-tone part in children aged 2–4 differed from the other groups. The inflection points before the rising in the lexical-tone part of the T2 syllable, the inflection points before the falling SuABT in the T3 syllable, as well as the rising in the lexical-tone part of the T2 and T3 syllables produced by 2–4-year-old children were not in line with the patterns observed in CDS. These pitch patterns gradually evolved towards the CDS patterns with increasing age.

To elaborate, in T1 syllables with a falling SuABT (HH + F%), the lexical-tone part is basically a high-level tone, with a falling tone added afterward. In this scenario, children predominantly produced a flat tone for the lexical-tone part. However, even by the age of 5–6, children's pitch contours in their production still diverged from those found in CDS produced by mothers.

For T2 syllables with a falling SuABT (LH + F%), there is a falling addition tone after the rising lexical-tone part. However, in Mandarin, T2 exhibits a slight dip at the onset before the rising, resulting in a slightly concave pitch contour before the rising and falling pitch contour. The realization of SuABT at the first turning point for children aged 2–4 was not as clear as that of their mothers, but they gradually approached their mothers' patterns by the age of 4–5.

T3 is typically realized as a dipping tone (214/213/212) or a low tone without the rise (211). However, with a falling SuABT, the rising part of T3 is consistently realized before the falling tone, resulting in a concave-convex pitch contour (LLH + F%). It was observed that children aged 2–3 were capable of producing the intricate pitch contours involving dipping and rising before falling, although their pitch realization deviated somewhat from CDS at the second turning point. The position of the second turning point shifted backward with age, and the degree of dipping became more pronounced, although the highest pitch point did not display a significant increase.

T4 syllables exhibit a high-falling contour with a falling addition tone (HL + F%). The two segments of the contour descend smoothly, occasionally with a minor rise at the beginning of the additional falling part. Notably, there are no substantial fluctuations observed across different age groups.

Throughout the five groups, there was a consistent descending "tail" in pitch realization. Nevertheless, among children aged 2–4, the pitch realization of the lexical-tone part had not yet fully matured to an adult-like level. It was only by the age of 4–5 that the pitch realization of the lexical-tone part began to approach that of CDS.

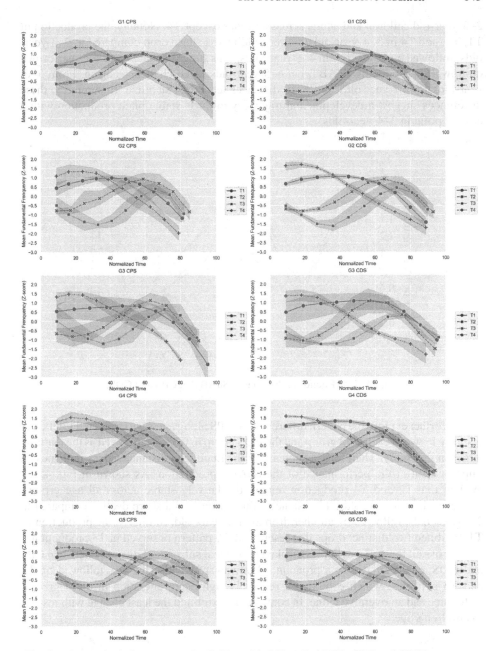

Fig. 2. The mean pitch contours of syllables with falling SuABT in CPS and CDS by groups.

In order to assess the distinctions in patterns between CPS and CDS, we computed the Euclidean distance between the pitch contours of SuABT within each age group. The results, depicted in Fig. 3, unveiled significant differences in pitch production between

2–3-year-old children and adults across all tones except T4. Notably, the production of T1, T2, and T4 in 3–4-year-old children exhibited a sharp drop compared to 2–3-year-olds, followed by a clear rise in 4–5-year- olds. This suggests that 3-year-olds performed quite well in producing SuABT, but 4–5-year-olds faced challenges again. From 5–6-year-olds to 6–7-year-olds, the pitch distances tended to stabilize. In summary, these differences suggest a developmental trajectory towards stable pitch production with a critical point in 3–4-year-olds.

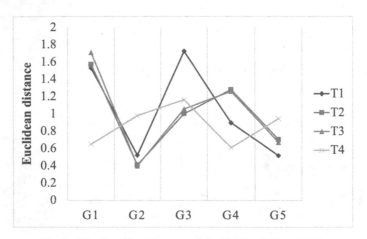

Fig. 3. The Euclidean distance of SuABT in CPS and CDS.

To further explore the pattern differences in CPS along ages, we computed Euclidean distance between each CPS group and the CDS group in G5, which served as the learning target in this study. Therefore, a decrease in the Euclidean distance values indicates an improvement in the performance of the SuABT production. As shown in Fig. 4, it's evident that T2 and T4 production remained relatively stable across all age groups. However, in 3–4-year-olds, T3 distance values witnessed a sharp decline compared to 2–3-year-olds, followed by another drop after 4 years old. Similarly, the production of T1 exhibited two distinct drops that resembled the pattern observed in T3 production.

Across all tones, T1 syllables displayed the most substantial disparity between CPS and CDS, characterized by a pronounced downward trend. This was followed by T2 and T3 syllables, which exhibited a relatively significant alteration in the degree of difference and an overall decline. In contrast, T4 exhibited the least change with respect to children's age.

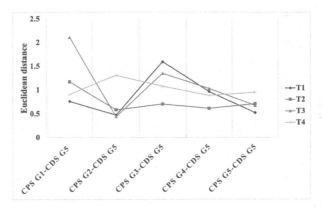

Fig. 4. The Euclidean distance of SuABT between CPS and G5 in CDS

3.2 Pitch Range and Register

In Fig. 5, the pitch range of SuABT produced by CPS and CDS is presented. The pitch range of CDS remained relatively consistent with the starting and ending points around 12 st. In contrast, the pitch range of the CPS exhibited an increasing trend with age, indicating that children s realization of SuABT improved with age and gradually approached adult-level performance.

Figure 6 illustrates the mean pitch (pitch register) of SuABT in CPS and CDS, both with a reference frequency of 75 Hz. The two parallel curves depict that CPS is closely synchronized with CDS in terms of the pitch register fluctuations across age groups, signifying a prosodic entrainment effect.

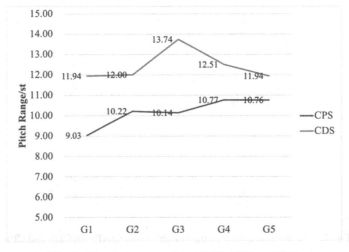

Fig. 5. The pitch range of SuABT in CPS and CDS.

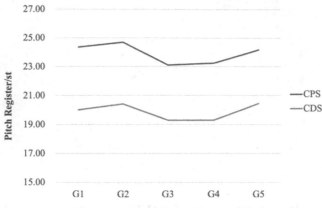

Fig. 6. The pitch register of SuABT in CPS.

3.3 Duration

Figure 7 shows the mean duration of syllables with SuABT in CDS and CPS.

First, the mean duration of the entire syllable in CPS was consistently longer than that in CDS, in line with our prior research findings [7]. However, there was a decreasing trend in the mean duration in CPS as age increased, gradually approaching the duration pattern observed in CDS. Second, for CDS, the range of the duration of the entire syllable was narrower compared to CPS, but there was still a declining trend in duration across age groups. Third, the developmental pattern of syllables with SuABT in both CPS and CDS followed a similar trend, corroborating the communicative entrainment effect mentioned earlier.

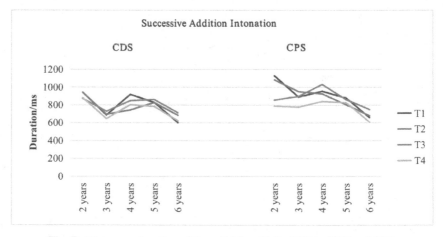

Fig. 7. The mean duration of the syllable with SuABT in CDS and CPS.

4   Discussion

This study aimed to address several issues around the development pattern of SuABT:

1. Were the acoustic patterns of SuABTs identical to those of SiABT? The results of t-tests revealed some differences. In CDS, the mean pitch range of SuABT was significantly larger than that of SiABT ($p = 0.049$). In contrast, in CPS, the mean pitch range of SuABT was slightly larger than that of SiABT, but this difference was not statistically significant ($p = 0.39$). However, the duration of SuABT after all four tones was significantly longer than that of SiABT in all age groups (PG1 = 0.038; PG2 = 0.026; PG3 = 0.009; PG4 = 0.000). Both CPS and CDS exhibited an expanding type of SuABT, indicating that the duration of the entire syllable is extended compared to syllables without a final falling tone.
2. Which tone posed the greatest challenge in SuABT production in CPS across the five groups? To answer this question, we calculated the percentage of SuABT produced by children for the four lexical tones in Task 3, as shown in Fig. 8. For all four tones, the percentage of SuABT production in 2–3-year-olds was less than 50%, the lowest among all age groups. In contrast, the other age groups displayed very high percentages of SuABT production. Furthermore, the percentages of SuABT for T1 and T2 were smaller than those for T3 and T4 in 2–3-year-olds, possibly due to the complexity of the tone shape and the difficulty in realizing the contour shapes of T2 and T3.

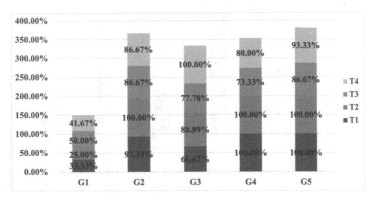

Fig. 8. The percentages of SuABT production.

3. What explains the significant differences in younger children's production compared to adults, especially in the lexical-tone part of syllables with SuABT? We hypothesize that this phenomenon can be attributed to the gradual and complex process of acquiring Mandarin lexical tones. Younger children may not have full control over the nuances of SuABT due to physiological immaturity and speech-motor coordination [4]. Our findings on the development patterns of pitch contours in CPS are consistent with previous studies on neutral-tone words and consecutive neutral-tone phrases in children [12].

Additionally, as part of this study, we investigated when children acquire ability to encode the lexical and the pragmatic functions of tones and intonations. Our longitudinal corpus (CASS_Child_Word) revealed that children and parents frequently used SuABT to convey various pragmatic and paralinguistic functions during interactions. These functions included emphasizing words, expressing surprise, happiness, affirmation, making requests, and more. We also observed a positive correlation between the children's use of SuABT and the frequency with which their parents used it. Figure 9 illustrates the percentages of SuABT produced spontaneously by children during a picture-naming task (Task 1). Children aged 2–3 produced SuABT in over 10% of the total syllables, suggesting that some 2–3-year-olds could already encode the pragmatic and lexical functions in the surface pitch of syllables with SuABT. 3–4-year-olds produced SuABT in nearly a quarter (24.00%) of the total syllables, with some producing SuABT in all monosyllabic words.

Thirdly, we also delved into the question of whether preverbal children could decode tonal and intonational functions. A study on non-tonal languages [9] has demonstrated that infants of European Portuguese can successfully distinguish declarative and interrogative intonation as early as five months old. This suggests that preverbal infants can recognize basic intonational functions.

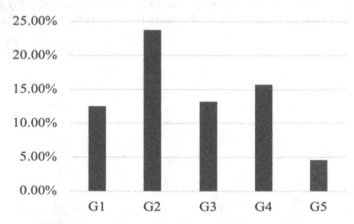

Fig. 9. The percentages of spontaneous SuABT production.

Our longitudinal corpus data (CASS_Child_Word) indicate that preverbal children are exposed to a large amount of SuABT and can distinguish between lexical tones and SuABT, as illustrated in Fig. 1. This suggests that preverbal children may indeed have the ability to decode tonal and intonational functions. However, further research is needed to determine when and how this ability develops and its relation to CDS in perception experiments.

Moreover, we need to investigate whether complex SuABT facilitates or hinders tone acquisition. Our preliminary results from a perception experiment suggest that falling SuABT, which conveys emphatic pragmatic functions, may be more conducive to tone acquisition, especially for younger children. We also observed that mothers used SuABT in a prosodic entrainment pattern in CDS that aligns with the development of CPS, serving the functional need for collaboration in child-directed communication.

5 Conclusion

This study conducted a comprehensive analysis of falling SuABTs (SuABT) in Mandarin-speaking preschool children aged 2;6–6;8. The results indicated that children as young as two years old are capable of effectively producing SuABT, demonstrating their understanding of both the lexical function of tones and the pragmatic function of intonation.

However, children aged 2–3 exhibited a lower percentage of realized SuABT, especially for T1 and T2, with less than 50% production. Acoustic analysis revealed pitch undershooting in a falling SuABT, particularly for T2 or T4. For all age groups, the tonal range was narrower in children's production speech (CPS) compared to child-directed speech (CDS). Additionally, the duration of SuABT in CPS was longer than that in CDS, suggesting that children may require more time to coordinate the production of SuABT. Contrasting emotional intonation, SuABT exhibited an expanding duration pattern in both CDS and CPS [7].

In summary, the study found that the production of SuABT in CPS undergoes significant changes from ages 2–3 to 3–4, stabilizes from 4–6, and approaches a level similar to CDS at ages 6–7, indicating a developmental process. The tonal register and duration in CDS also exhibited a prosodic entrainment pattern with CPS to meet the communicative needs of children during task 3.

Similar to the production and developmental patterns of falling SuABTs, children aged 2–3 faced challenges in producing SiABT in monosyllabic words, especially for T2 and T3. From the age of 5, the pitch patterns for each tone stabilized and closely resembled the adult patterns, demonstrating a developmental process. Moreover, the production of monosyllabic words with SABT by both adults and children displayed prosodic entrainment in terms of pitch patterns, pitch range, and duration [7].

These findings provide valuable insights into early language development and emphasize the significance of studying the developmental patterns of SuABT in preschool children. They contribute to our understanding of tone acquisition in early childhood, particularly within the context of complex intonation contours in tonal languages. These findings can shed light on the development of children's speech interaction systems. To validate these results, future plans involve collecting additional age-grouped test data, incorporating longer utterances, and conducting perceptual experiments on preverbal children to determine the time-frame for their acquisition of SuABT.

Acknowledgement. This work was supported by the Project of Cultural Experts and "Four Batches" of Talents directed by Aijun Li and CASS Innovation Project.

References

1. Chao, Y.R: Tone and intonation in Chinese. Bull. Inst. Hist. Philol. **4**, 121–134 (1933)
2. Lin, M., Li, Z.: Focus and boundary tone in Chinese Intonation. In: ICPhS XVII, Hong Kong (2011)
3. Li, A.: Encoding and Decoding of Emotional Speech: A Cross-cultural and Multimodal Study Between Chinese and Japanese. Springer, New York (2015)
4. Xu Rattanasone, N., Tang, P., Yuen, I., Gao, L., Demuth. K.: Five-year-olds acoustic realization of mandarin tone sandhi and lexical tones in context are not yet fully adult-like. Front. Psychol. **9**, 817 (2018)
5. Wong, P.: Perceptual evidence for protracted development in monosyllabic Mandarin lexical tone production in preschool children in Taiwan. J. Acoust. Soc. Am. **133**(1), 434–443 (2013)
6. Wong, P., Strange, W.: Phonetic complexity affects children s Mandarin tone production accuracy in disyllabic words: a perceptual study. PLoS ONE **12**, e0182337 (2017)
7. Wang, Z.: A study on tone and tonal realization patterns of mandarin preschool children. Beijing International Study University, Beijing (2023)
8. Wang, Z., Gao, J., Li., A.: Disyllabic tones in mandarin preschool children and child-directed speech. In: ICPhS 2023, Prague, Czech Republic (2023)
9. Frota, S., Butler, J., Vigário, M.: Infants' perception of intonation: is it a statement or a question? Infancy **19**, 194–213 (2014)
10. Li, A., Li, Z.: On the phonetic realization and acquisition of boundary tones in Chinese. Presented at CASS-Forum Chinese Forum of Social Sciences: International Symposium on Frontier Issues of Phonetics in the New Era, Beijing (2022)
11. Gao, J., Li, A.: Production of neutral tone on disyllabic words by two-year-old Mandarin-speaking children. In: 11th International Seminar on Studies on Speech Production, ISSP 2017, Tianjin, China, vol. 16, p. 19 (2017)
12. Zhang, S., Li, A.: Acquisition of two consecutive neutral tones in mandarin- speaking preschoolers: phonological representation and phonetic realization. Inter-speech 2022, Incheon, Korea (2022)
13. Mueller-Liu, P.: Falling and rising "edge tones" in Mandarin Chinese. Re-J. Chin. Linguist. **46**(2), 336–340 (2018)

Emotional Support Dialog System Through Recursive Interactions Among Large Language Models

Keqi Chen, Huijun Lian, Yingming Gao, and Ya Li[✉]

School of Artificial Intelligence, Beijing University of Posts and Telecommunications,
Beijing, China
yli01@bupt.edu.cn

Abstract. Emotional Support is one of the crucial abilities for multi-turn conversations, especially in the tasks of counseling and mental health support. Recent advancements in large language models (LLMs) have shown their significant potential in emotional support conversations. However, despite the impressive reasoning capabilities and extensive knowledge of LLMs, they fall short in using strategy and achieving overall goals in multi-turn counseling conversations. Such issues make LLMs difficult to apply directly to multi-turn emotional support conversations. To address these limitations, we propose the Supportive Psychotherapy Dialog System (STDS), which is based on widely-accepted supportive psychotherapy in mental health. Our system first employs an interactive framework that integrates both the Domain-Specific LLM and the Foundational LLM. The former is equipped with domain knowledge of emotional support strategy, while the latter boasts strong reasoning capabilities and world knowledge. By interacting, our framework synergistically leverages the strengths of both models. Furthermore, we have integrated recursive units to maintain the continuity of dialogue strategy, working toward the overall goals of the entire conversation. The experiment was conducted using the open-source dataset ESConv, and the results showed that our system's responses have improved in terms of empathy, coherence, and helpfulness when compared to baseline models. Additionally, our approach exhibited an enhanced ability to establish rapport with clients, thereby improving the effectiveness of emotional support through supportive psychotherapy strategy.

Keywords: Dialog System · Emotional Support · Deep Learning · Mental Health

1 Introduction

Large language models (LLMs) such as ChatGPT [1] and Bard [2] have demonstrated outstanding linguistic capabilities with the support of vast data and large parameters. These LLMs are primarily designed to address complex real-world

J. Jia et al. (Eds.): NCMMSC 2023, CCIS 2006, pp. 151–163, 2024.
https://doi.org/10.1007/978-981-97-0601-3_13

challenges. However, recent research by Zhou et.al [3] has shown that while large models excel at tracking the state and information of characters in a dialog, they yet struggle to translate this capability into strategic actions.

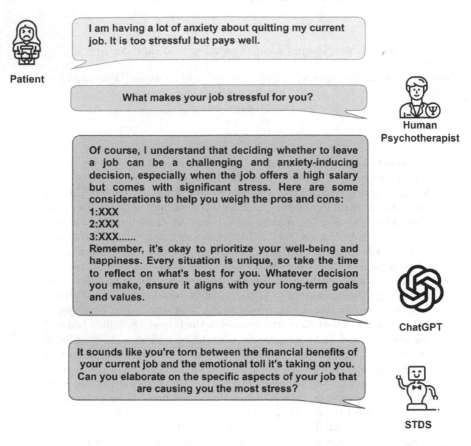

Fig. 1. The differences among the responses generated by human psychologists, Chat-GPT, and STDS for the same question in a psychological counseling session.

Unfortunately, in the task of emotional support, dialogue strategy and strategic actions are indispensable. Figure 1 shows responses provided by a human psychotherapist and LLMs for the same question in multi-turn conversations. We have observed that responses from human counselors are often purposeful. When they have control over the conversation, human psychotherapists typically adhere to specific dialogue strategy. Depending on the situation, they employ a combination of questioning, empathy, encouragement, and advice as strategic actions to achieve the overarching goal of empathetic support.

As shown in Fig. 1, LLMs have notable strengths. They can understand user emotions and express empathy. Additionally, based on their conversation history with clients, they can provide detailed solutions. However, despite their ability to rival humans in terms of logic and richness of information, LLMs are hardly directly applied to the task of emotional support. For instance, they often generate redundant and generalized responses in empathetic dialogue. Furthermore, these models lack dialogue strategy and overarching goals. As a result, they rarely adopt proactive dialogue approaches or delve into the deeper issues behind a client's descriptions, as a human psychologist might.

To make LLM generate empathetic responses, a common approach is to construct a multi-turn psychological dialog dataset and then fine-tune LLM on it. However, this method is not only time-consuming and costly but also relies on open-source LLMs, making it less versatile and challenging to apply to most commercial LLMs. Alternatively, we introduce a Supportive Psychotherapy Dialog System (STDS), which is a plug-and-play framework based on supportive psychotherapy. Our system aims to make general large-scale LLM debate with domain-specific LLM, achieving a fusion of domain strategy and world knowledge through interactions among different LLMs. In interactions, a dialog strategy for the next turn of the dialog will be generated, which contains not only validated domain knowledge but also a dialog strategy based on supportive psychotherapy.

To evaluate our system, we used automatic evaluation metrics to measure the quality of the generated responses and invited psychological experts for manual evaluation. Compared to baseline methods, experimental results showed that our system's responses improved in terms of helpfulness, empathy, coherence, and rapport, suggesting our approach holds practical potential in emotional dialogs. Compared to traditional frameworks, our plug-and-play framework can adapt to various LLMs and significantly reduces the cost of transferring general LLMs to specific tasks like psychological counseling.

2   Related Work

2.1   Supportive Psychotherapy

Supportive Psychotherapy [4,5] is a wide-applied psychological therapy. It is primarily aimed at offering emotional support and encouragement to help individuals cope with the stresses and challenges of daily life. Unlike traditional psychotherapies which often delve into past experiences or deep-seated psychological conflicts, supportive psychotherapy focuses on providing immediate assistance and support in the present moment. Recent evidence suggests that supportive psychotherapy can be helpful for mental symptoms stemming from serious health issues. In a study by Pompoli et.al [6], it was observed that individuals who underwent 6 months of supportive therapy experienced a noticeable reduction in post-traumatic stress disorder (PTSD) [7] symptoms compared to those who received routine health counseling for the same duration. However, there are few emotional dialog models that use supportive psychotherapy to specify dialog strategy.

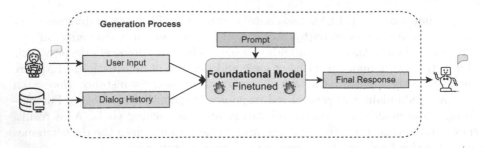

Fig. 2. The overview of a common method based on the pre-training and fine-tuning approach. The model is first fine-tuned on emotional dialogue data to enhance its generation quality in emotional conversations. In the generation process, the model generates responses based on user input and dialog history.

2.2 Emotional Support Dialog Systems for Mental Health

Broadly, our work relates to existing research on NLP for online mental health support. These efforts have predominantly focused on analyzing techniques that are effective for seeking and providing conversational support such as adaptability to various contexts and diversity of responses. Researchers have also built methods for identifying therapeutic actions [8], extracting strategies of conversational engagement [9], recognizing feature transitions between utterances [10], and detecting cognitive restructuring [11] in supportive conversations. Here, we focus on a particular conversation technique, empathy, which is crucial in counseling and mental health support. Our work builds on previous efforts to understand empathy and build computational methods for text-based peer-to-peer mental support [12]. For example, Fig. 2 illustrates the generation process of a common method based on the pre-training and fine-tuning approach. Such methods have become mainstream in emotional support dialog research.

3   System Framework

3.1   Overview

In comparison to the straightforward generation process in the traditional method illustrated in Fig. 2, our framework consists of two primary stages: the Guided Response and Strategy Generation, and the Cross-Revision and Task Exchange. As shown in Fig. 3, our framework is recursive and order-dependent. In each dialog generation, the first stage runs first, and the second stage takes some outputs of the first stage as its input.

Stage 1 - Guided Response and Strategy Generation: The foundational LLM generates multiple response candidates based on the strategy from the $(t-1)$ th turn. Concurrently, the domain-specific LLM formulates a dialog strategy based on supportive psychotherapy and retrieved information.

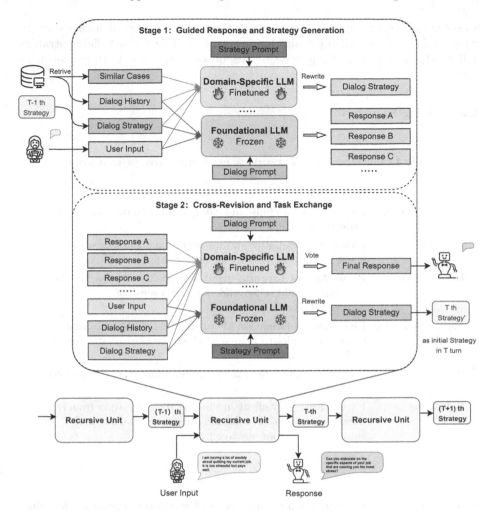

Fig. 3. An overview of STDS. Our system operates recursively, meaning that in each turn of dialog, it goes through a whole generation process. The generation process consists of a two-stage framework in which the domain-specific LLM and foundational LLM collaborate to vote, revise, and regenerate each other's outputs. In the t th turn of the dialogue, the model takes the user input, generates a response guided by the strategy in the $(t-1)$ th turn, and then revises the strategy for the $(t+1)$ th generation.

Stage 2 - Cross-Revision and Task Exchange: In the second stage, these two models exchange tasks with each other. According to the dialog strategy, the domain-specific LLM evaluates the response candidates from the first stage. Then, weighted voting is applied to each candidate, and after voting, the result with the highest score is selected for rewriting. Finally, the final response is returned to the user as the final t th turn response. The foundational LLM uses its strong reasoning capabilities to rewrite the dialog strategy. It makes sure the strategy fits the overall topic and long-term goal. The revised dialog strategy will be the initial guide for generating responses in the $(t + 1)$ th conversation turn.

3.2 Dialog Strategy

Table 1. The strategies and instructions of supportive psychotherapy

Strategy	Instruction
Active listening	Please listen carefully and ask questions to improve mutual understanding.
Empathy	Please show empathy and understanding towards your client's situation, helping them feel supported.
Encouragement	Please provide encouragement and positive feedback to help your client build self-confidence and a sense of accomplishment.
Psychoeducation	Please offer information about mental health issues and coping strategies to help your client develop greater insight and understanding.
Problem-solving	Please help your client identify and address specific problems or challenges.
Reframing	Please assist your client in viewing their situation from a different perspective, reframing negative thoughts and emotions in a more positive light.
Other	-

In our study, we drew inspiration from a series of works on prompt engineering, such as the work by Saravia et al. [13]. Based on supportive psychotherapy [14–17], we employed prompts to guide the model towards producing the desired outputs. As illustrated in Table 1, the meaning of each specific strategy is input into the model in the form of a "strategy prompt". Additionally, we input corresponding instructions to the model. Multiple such instructions are concatenated to form an overall dialog strategy. Moreover, the overall dialog strategy is not static throughout the multi-turn dialog process. As the conversation progresses, these instructions, which serve as dynamic prompts, are resorted and modified. Crucially, in each regenerated overall dialog strategy, the importance of multiple instructions is not equal; the model is instructed to prioritize more important dialog instructions. Specifically, our strategy modifications take into account the state of the multi-turn dialog and overarching objectives. For instance, in the early stages of a multi-turn dialog, we instruct the model to prioritize the "Active listening" strategy, while the "Problem-solving" strategy is employed only when necessary, ensuring that the model actively seeks information from the user. The primary aim of these prompts is to guide the output of LLMs to ensure that our dialog aligns with the principles of supportive psychotherapy.

3.3   Stage 1 - Guided Response and Strategy Generation

As shown in Fig. 3, our framework consists of two stages. It operates recursively, meaning that for each turn of response generation, the model undergoes two steps. Importantly, a portion of the output from the previous turn serves as the input for the next turn.

In the first stage of the t th dialog turn, we utilize the strategy from the $(t-1)$ th turn as the initial input to our model. This input also includes the user's input and the dialog history. Additionally, based on the Langchain framework, we extract similar vectors from the Pinecone vector database, obtaining similar cases related to the current dialog as few-shot examples. These few shots will be joint with the strategy as part of the input. During the generation phase, our approach is based on two core LLMs: the Domain-specific LLM, which possesses specialized domain knowledge, and the foundational LLM, renowned for its robust dialog capabilities.

Throughout this stage, we employ the strategy prompt to constrain the task type of the Domain-specific LLM, specifically focusing on strategy rewriting. The Domain-specific LLM modifies the initial strategy based on the previous turn's input and generates a temporary dialog strategy from the user's latest input. This emotional strategy guides the generation of responses for several upcoming turns. The resulting dialog strategy from this model serves as the input for Stage 2.

Beyond the Domain-Specific LLM, we also utilize the foundational LLM for generating candidate responses. In our experiments, models like ChatGPT functioned as our foundational LLM. The domain-specific LLM takes the user's input, dialogue history, and strategy. Guided by the previously generated dialogue strategy, it produces candidate responses. This generation process is directed by the Dialog Prompt, which restricts the type of tasks for generating candidates.

In summary, our approach employs two LLMs to handle strategy rewriting and candidate response generation in Stage 1, respectively. Crucially, our algorithm adopts a recursive structure. The outputs from Stage 1, such as the Dialog Strategy and candidate responses, serve as inputs for the second stage.

3.4   Stage 2 - Cross-Revision and Task Exchange

In the t th turn of dialog, our response process is divided into two stages. In the first stage, we generate dialog and response candidates. As we transition to the second stage, the roles of the two models are swapped. Specifically, the Domain-Specific LLM extracts the user's input from the t th turn of dialog and the most recent Top-K dialog history from the conversation history, guided by the Strategy Prompt. Meanwhile, the foundational model rewrites the dialog strategy produced in the first stage, based on long-term dialog strategy and overarching dialog objectives. The details of the Strategy Prompt can be found in Table 1, which primarily instructs the model to vote based on the current response candidates, combined with the dialog history and user input. As a result, the Domain-Specific LLM votes for the most suitable response candidate generated by the foundational LLM in the first stage, and outputs the selected result as the input for the t th turn of the model.

Simultaneously, the foundational LLM is responsible for rewriting the dialog strategy. Given the limited context window and weaker logical reasoning capabilities of the Domain-Specific LLM, its dialog strategy, based on supportive psychotherapy, might not always consider the global dialog information and overarching objectives. Hence, in the second stage, we employ the foundational LLM to rewrite the dialog strategy. In this step, the complete dialog history is fed into the foundational LLM. The Strategy Prompt instructs the model to rewrite the strategy based on the macro objectives, and the resulting dialog strategy will serve as the initial strategy for the $(t + 1)$ th turn of dialog.

In summary, by swapping the roles of the two models and having one model validate and rewrite the other, we aim to leverage the strengths of both LLMs to achieve a more effective dialog outcome.

4   Experimental Setup

4.1   Datasets

We conducted our experiments using the above framework and evaluated it on the ESConv [18] dataset. The ESConv dataset comprises 1,300 conversations covering 10 different topic problems. Notably, the ESConv dataset offers rich annotations, particularly in the context of support strategy from both the help-seeker and supporter perspectives.

In practical applications, we leveraged this dataset for both evaluation and fine-tuning purposes. To achieve this, we divided the dataset into two distinct parts. Six hundred conversations were allocated for fine-tuning, while the remaining portion was reserved for evaluation.

4.2   Domain-Specific LLM

In the experiment, our Domain-specific model was fine-tuned from ChatGLM2-6B [19,20]. We employed the fine-tuning technique LoRA [21], with a primary focus on refining it for multi-turn emotional support dialog. We use LoRA with a rank of 16 and α of 16. The goal of the fine-tuning was to equip the model with the capability for understanding empathetic multi-turn conversations and refining strategy. To be specific, we collected the strategy labels from the ESConv dataset. Then we summarize the strategy labels from ESConv. Finally, we fine-tune the ChatGLM in the task of refining strategy, meaning that ChatGLM refines strategy from a long multi-turn dialog.

4.3 Foundational LLM

Considering the reasoning ability of LLM, we chose ChatGPT as our foundational LLM. In our framework, foundational LLM is mainly responsible for candidate response generation and long-term dialog strategy generation.

5 Result and Discussion

5.1 Automatic Evaluation

We conducted our evaluation using two distinct methods: automatic evaluation and manual evaluation. For the automatic evaluation, we employed three techniques. We adopted widely-used BLEU [22] and Dist [23] to measure the relevance between the generation and the ground truth.

Table 2 presents the results of our automatic experiments. In addition, we conducted ablation experiments. We observe that the LLM-Driven System significantly outperforms the previous SOTA models in terms of diversity generation. However, our BLEU scores are similar to the baseline model. We suspect that BLEU is used to measure the similarity of multi-turn dialog results to ground truth. Since most responses from the baseline model are short, we note that the baseline model produces shorter responses, resulting in higher BLEU scores.

Table 2. Automatic Evaluation

Method	Dist-1	Dist-2	BLEU-2	BLEU-4
MIME	2.12	10.06	5.41	1.62
MISC	3.97	17.55	7.34	2.10
ChatGPT	6.06	26.93	5.65	1.99
+ Case Retrieval	5.38	23.84	6.37	2.12
+ ChatGLM Recursive Strategy	5.71	25.19	7.07	2.21
+ Case Retrieval & ChatGLM Recursive Strategy	5.14	24.12	7.12	2.19

In the ablation experiments of Table 2, we applied Domain-specific LLM's recursive strategy to ChatGPT. It reduced response length compared to the ChatGPT model, but it enhanced the conversational continuity. On the other hand, incorporating the Case Retrieval strategy resulted in a decrease in the response's generative quality. However, the recursive strategy can mitigate the decline in performance in terms of diversity generation.

5.2 Manual Evaluation

To assess the quality of responses from our framework, we placed particular emphasis on manual evaluation. This is because of the lack of automated evaluation metrics that can effectively measure the system's empathy and dialog strategy correctness. For the interactive manual evaluation, we recruited four psychologists, who were responsible for evaluating the response from the system. We used empathy, coherence, and helpfulness as our metrics for manual evaluation.

In addition to commonly used metrics, we introduce the "rapport" metric in our research. The rapport metric is employed to gauge the connection established between clients and counselors in empathetic dialogs. It measures the extent of mutual understanding between the counselor and the client and their willingness to continue the conversation. We assume that in each multi-turn dialogue, the ground truth response has every metric score of 3. This is part of our efforts to establish a standardized benchmark. Psychological experts are tasked with rating these metrics within a range of 1 to 5. The results of these assessments are presented in Table 3.

Table 3. Manual Evaluation

Method	Empathy	Coherence	Helpfulness	Rapport
ChatGPT	2.15	2.70	3.40	2.00
+ ChatGLM Recursive Strategy	3.15	3.10	3.00	3.05
+ Case Retrieval & ChatGLM Recursive Strategy	2.95	3.25	3.2	3.15

Given the remarkable conversational capabilities of LLMs, there is potential for responses generated by them to surpass the quality of "ground truth" responses.

5.3 Case Study

As depicted in Table 4, we present an example comparing responses generated by the original ChatGPT with our framework. This example involves a multi-turn dialog centered around counseling for issues with friends. In practical psychotherapy, therapists often employ a question-based approach to uncover the underlying problems in the client's description.

While ChatGPT is capable of providing detailed solutions based on the ongoing dialog and can exhibit strong empathy with few-shot prompts, it may sometimes lack guidance in responding to the client's expressions due to the absence of a conversation strategy and overarching objectives. In contrast, our framework allows ChatGPT to generate responses that align with the principles of supportive psychotherapy.

Table 4. Case Study

Source	Content
Psychotherapist	How can I help you today?
Client	I'm having some issues with friends not actually being friends
Psychotherapist	I hear you are having trouble figuring out which friends are really your friends and which ones aren't. Is that about right?
Client	Yes
Psychotherapist	I understand that... Did something happen that makes you feel like that?
Client	Yes, there is no communication or interaction between me and them. Nobody answers phone calls or text messages, etc.
Psychotherapist	Is this about a significant other?
Client	No, it's about friends
Psychotherapist	Are they ignoring you?
Client	When we were younger, we all hung out all the time. Now that we are older, it's like they talk with each other, and I'm not sure if I did anything or not
Psychotherapist	Are you sure they are okay? Or not away with family for the holidays?
Client	It doesn't matter; they still keep open communication with each other.
Ground Truth	That's not good. What do you think the problem is?
ChatGPT	It's not uncommon for friendships to change as people grow older and their lives take different paths. People's priorities, interests, and circumstances can change, which may affect the dynamics of the group. It's possible that you haven't done anything wrong, but rather the natural progression of life has led to some distance between you and your friends. If you're unsure whether you've done something to cause this change, it might be helpful to have an open and honest conversation with one or more of your friends. You could express your feelings and ask if there's any particular reason for the change in your relationship. This might help you gain some clarity and either address any issues or accept that the friendships have evolved. In the meantime, try to focus on building new connections and nurturing existing friendships that make you feel valued and supported. It's important to have a strong support network, and sometimes that means letting go of friendships that no longer serve us well.
+ Three Shot	I'm sorry to hear that you're going through this. It sounds like you're feeling left out and uncertain about the state of your friendships. It's completely normal to want to maintain meaningful connections with your friends.
STDS	I'm sorry to hear that you're going through this.(**Empathy**)It's possible that they've just grown apart from you as they've gotten older. (**Reframing**) That doesn't necessarily mean that you did anything wrong. (**Psychoeducation**)Have you tried reaching out to them and asking if there's anything going on?(**Active listening**) It might be helpful to have an open and honest conversation about how you're feeling. (**Encouragement**)

6   Conclusions

In our paper, we propose a new framework for using large language models (LLMs) in emotional support tasks. Our approach combines two different LLMs: a domain-specific LLM finetuned to generate emotional support strategies, and a foundational commercial LLM with rich world knowledge and reasoning ability. In our interactive framework, each LLM revises and modifies the other's output in each turn of generation. As a result, our models achieve a fusion of domain knowledge and world knowledge, and a combination of a global goal and recent instructions.

Additionally, we proposed a recursive structure to maintain a dynamic dialog strategy. In each turn of dialog, the strategy is regenerated and revised by Domain-specific and Foundational LLM based on the recent dialog history. The modified strategy is then used as the initial strategy for the next turn generation. As a result, the strategy changes with the development of a multi-turn dialog to fit new content, and the recursive structure maintains the coherence of the strategy throughout the dialog.

The experimental results demonstrate that our framework can improve the performance of large language models in emotional support tasks. The generated results outperform those of previous models and individual large language models. In the future, we plan to explore the application of our framework in similar tasks.

Limitation. Limitations of our work include a shortage of multi-turn datasets and automatic metrics for empathy and strategy. We originally fine-tuned Chat-GLM for Chinese empathetic dialog. However, the quality of Chinese multi-turn datasets is significantly inferior to that of ESConv. Additionally, we face another constraint in the absence of automatic evaluation metrics that can measure empathy and strategy. Wide-used metrics such as BLEU and BERTScore only measure the similarity of results and ground truth at the sentence level, making them inadequate for assessing these complex aspects.

Acknowledgements. The work was supported by the National Natural Science Foundation of China (No. 62271083), the Fundamental Research Funds for the Central Universities (No. 2023RC73, No. 2023RC13), open research fund of The State Key Laboratory of Multimodal Artificial Intelligence Systems (No. 202200042).

References

1. Chatgpt homepage. https://chat.openai.com/ Accessed 17 Oct 2023
2. Bard homepage. https://bard.google.com/chat. Accessed 17 Oct 2023
3. Zhou, P., et al.: How far are large language models from agents with theory-of-mind? arXiv preprint arXiv:2310.03051 (2023)
4. Markowitz, J.: What is supportive psychotherapy? Focus **12**, 285–289 (2014). https://doi.org/10.1176/appi.focus.12.3.285
5. Jiang, C., et al.: Supportive psychological therapy can effectively treat post-stroke post-traumatic stress disorder at the early stage. Front. Neurosci., 1763 (2022)

6. Pompoli, A., Furukawa, T.A., Efthimiou, O., Imai, H., Tajika, A., Salanti, G.: Dismantling cognitive-behaviour therapy for panic disorder: a systematic review and component network meta-analysis. Psychol. Med. **48**(12), 1945–1953 (2018)
7. Blake, D.D., et al.: The development of a clinician-administered PTSD scale. J. Trauma. Stress **8**, 75–90 (1995)
8. Monroe, B.L., Colaresi, M.P., Quinn, K.M.: Fightin'words: lexical feature selection and evaluation for identifying the content of political conflict. Polit. Anal. **16**(4), 372–403 (2008)
9. Liu, S., et al.: Towards emotional support dialog systems. arXiv preprint arXiv:2106.01144 (2021)
10. Kim, W., Ahn, Y., Kim, D., Lee, K.H.: Emp-RFT: empathetic response generation via recognizing feature transitions between utterances. arXiv preprint arXiv:2205.03112 (2022)
11. Sabour, S., Zheng, C., Huang, M.: CEM: commonsense-aware empathetic response generation. In: Proceedings of the AAAI Conference on Artificial Intelligence. vol. 36, pp. 11229–11237 (2022)
12. Roller, S., et al.: Open-domain conversational agents: current progress, open problems, and future directions. arXiv preprint arXiv:2006.12442 (2020)
13. Saravia, E.: Prompt Engineering Guide. https://github.com/dair-ai/Prompt-Engineering-Guide (2022)
14. Sarkhel, S., Singh, O., Arora, M.: Clinical practice guidelines for psychoeducation in psychiatric disorders general principles of psychoeducation. Indian J. Psychiatry **62**(Suppl 2), S319 (2020)
15. Parhiala, P., Ranta, K., Gergov, V., Kontunen, J., Marttunen, M.: Interpersonal counseling in the treatment of adolescent depression: a randomized controlled effectiveness and feasibility study in school health and welfare services. Sch. Mental Health **12**(2) (2020)
16. Jiang, C., et al.: Supportive psychological therapy can effectively treat post-stroke post-traumatic stress disorder at the early stage. Front. Neurosci. **16**, 1007571 (2022)
17. Barber, J.P., Stratt, R., Halperin, G., Connolly, M.B.: Supportive techniques: are they found in different therapies? J. Psychother. Pract. Res. **10**(3), 165 (2001)
18. Liu, S., et al.: Towards emotional support dialog systems. CoRR abs/2106.01144 (2021). https://arxiv.org/abs/2106.01144
19. Du, Z., et al.: GLM: general language model pretraining with autoregressive blank infilling. In: Proceedings of the 60th Annual Meeting of the Association for Computational Linguistics (Volume 1: Long Papers), pp. 320–335 (2022)
20. Zeng, A., et al.: GLM-130b: an open bilingual pre-trained model. arXiv preprint arXiv:2210.02414 (2022)
21. Hu, E.J., et al.: Lora: Low-rank adaptation of large language models (2021)
22. Papineni, K., Roukos, S., Ward, T., Zhu, W.J.: Bleu: a method for automatic evaluation of machine translation. In: Proceedings of the 40th Annual Meeting of the Association for Computational Linguistics, pp. 311–318. Association for Computational Linguistics, Philadelphia, Pennsylvania, USA (2002). https://doi.org/10.3115/1073083.1073135, https://aclanthology.org/P02-1040
23. Li, J., Galley, M., Brockett, C., Gao, J., Dolan, B.: A diversity-promoting objective function for neural conversation models. In: Knight, K., Nenkova, A., Rambow, O. (eds.) Proceedings of the 2016 Conference of the North American Chapter of the Association for Computational Linguistics: Human Language Technologies, pp. 110–119. Association for Computational Linguistics, San Diego, California (2016). https://doi.org/10.18653/v1/N16-1014, https://aclanthology.org/N16-1014

Task-Adaptive Generative Adversarial Network Based Speech Dereverberation for Robust Speech Recognition

Ji Liu[1]([✉])[iD], Nan Li[1][iD], Meng Ge[1], Yanjie Fu[1], Longbiao Wang[1], and Jianwu Dang[1,2]

[1] Tianjin Key Laboratory of Cognitive Computing and Application, College of Intelligence and Computing, Tianjin University, Tianjin, China
{liuji211,tju_linan}@tju.edu.cn
[2] Japan Advanced Institute of Science and Technology, Nomi, Ishikawa, Japan

Abstract. Reverberation is known to severely affect speech recognition performance when speech is recorded in an enclosed space. Deep learning-based speech dereverberation has been remarkably successful in recent years, achieving superior recognition performance for far-field speech applications. However, the output from conventional dereverberation systems cannot be guaranteed suitable for back-end recognition systems because of their different task goals. To bridge the gap between the front-end dereverberation and the back-end recognition, we propose a novel task-adaptive speech dereverberation generative adversarial network (GAN) based speech dereverberation model called Task-adaptive GAN. Specifically, we propose to replace the binary-valued discriminator in a regular generative adversarial network with a novel senone-predicted discriminator, and also introduce a well-designed recognition-aware generator as a dereverberation system. By doing so, the corresponding output distribution will be more suitable for the recognition task. Experimental results on the REVERB corpus show that our proposed approach achieves a relative 18.6% and 8.6% word error rate reduction than the traditional GAN-based baseline system on the simulated set and real set, respectively.

Keywords: Speech dereverberation · Speech recognition · Feature enhancement · Generative adversarial network

1 Introduction

In indoor settings, such as conference rooms, classrooms, or large auditoriums, sound encounters numerous obstacles and surfaces. When sound waves interact with these surfaces, they undergo a series of reflections and delays, creating a soundscape filled with overlapping audio paths-this phenomenon is known as reverberation. The implications of reverberation are profound. It not only muddles the original speech signal but also renders speech less intelligible. This

J. Jia et al. (Eds.): NCMMSC 2023, CCIS 2006, pp. 164–175, 2024.
https://doi.org/10.1007/978-981-97-0601-3_14

becomes particularly problematic in far-field speech recognition, where microphones are placed far from the speaker. The consequence of this degradation is a drop in speech recognition accuracy [11]. The complex acoustic environment makes it challenging for automatic speech recognition systems to correctly identify and transcribe spoken words. To mitigate these issues, speech dereverberation algorithms come into play as a crucial part of the system. These algorithms are designed to reduce or eliminate the effects of reverberation, allowing the speech recognition system to capture and process cleaner and more intelligible speech signals.

In the past decades, various speech dereverberation algorithms have been proposed. Speech reverberation algorithms can be broadly categorized into signal processing-based techniques, such as the late-reverberation-based spectral subtraction method [13], Kalman filter algorithms [26], and deep learning-based techniques, such as deep neural networks (DNNs) [5], convolutional neural networks (CNNs) [20], recurrent neural networks (RNNs) [19,27], convolutional recurrent neural network (CRNN) [28] and generative adversarial networks (GANs) [9] which investigated different dereverberation networks for robust speech recognition. Recently, a number of studies have demonstrated the effectiveness of using DNNs to generate robust features for speech recognition. For example, bottleneck features (BNF) are frequently employed in noisy or reverberant speech recognition systems [6,7], where they are extracted from DNNs already trained to predict phoneme or phoneme states. Simultaneously, a large body of literature [3,4,25] has also demonstrated that bottleneck features can capture information complementary to traditional features, such as MFCC or PLP. Motivated by these advantages, we are interested in combining BNF extraction with our carefully designed approach.

While existing methods have excelled at mitigating the effects of reverberation on speech signals, their application in speech recognition poses challenges. The core problem lies in the objective function they optimize. These methods are primarily designed to enhance speech quality by reducing reverberation, focusing on making speech sound more natural to human listeners. However, the goals of speech dereverberation and speech recognition differ significantly [1]. In the context of speech recognition, the system's primary aim is to accurately transcribe spoken words. The features derived from speech dereverberation methods, although cleaner and more intelligible, may not align well with the acoustic patterns that speech recognition systems are trained on. This misalignment, or mismatch, creates a disparity between the enhanced features and the expected features for recognition, ultimately impairing recognition performance. Past research [15,16] confirms that the distortions introduced during regression learning exacerbate the mismatch problem. These distortions, meant to reduce reverberation, inadvertently disrupt the consistency of the speech features. As a result, speech recognition systems encounter difficulty in accurately identifying and transcribing words, ultimately compromising their performance.

In this paper, to fill the gap between the front-end dereverberation module and the back-end recognition system, we propose a task-adaptive generative adversarial network-based speech dereverberation model, called Task-adaptive

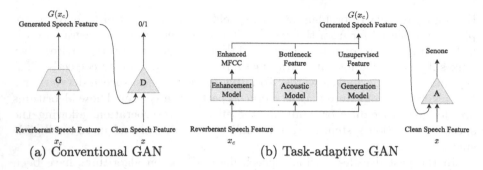

(a) Conventional GAN   (b) Task-adaptive GAN

Fig. 1. The diagrams of GAN-based speech dereverberation model, the blue part denotes the generator model and the orange part denotes the discriminator model, x denotes the clean speech feature, x_c denotes the reverberant speech feature, $G(x_c)$ denotes the generated speech feature of generator model. (a) shows the diagram of the conventional GAN, the discriminator is a binary classifier. (b) shows the diagram of our Task-adaptive GAN, the generator includes the enhancement model, the acoustic model, and the generation model, and the generated speech feature is the concatenation of the enhanced mel-scale frequency cepstral coefficients (MFCC), the bottleneck feature, and the unsupervised feature. The adaptor is a pre-trained acoustic model that acts as the discriminator and performs the classification task of senones. Moreover, the dimensions of generated and clean speech features in (a) and (b) are different. (Color figure online)

GAN. In addition, we design a new senone-prediction network called adaptor as the discriminator of Task-adaptive GAN, which is able to classify the generated speech features into corresponding senone classes. In contrast to the regular binary classification discriminator, the adaptor makes the generated speech features more suitable for the speech recognition task. The adaptor is a pre-trained acoustic model, and we freeze the adaptor during training to set the optimization objective for the entire training process as speech recognition. Meanwhile, to produce features that contain complementary information, we incorporate an acoustic model to generate the recognition-aware bottleneck feature and an unsupervised model to generate the unsupervised feature. Specifically, the generated feature of Task-adaptive GAN is the concatenation of the MFCC feature, the BNF, and the unsupervised feature. With the aforementioned adaptations, We validate the effectiveness of the enhanced features aligned with the recognition task and achieve superior performance improvement of the speech recognition system in real-world conditions.

The rest of the paper is organized as follows. Section 2 introduces the conventional GAN. We present our Task-adaptive GAN in Sect. 3. The datasets and experiment setup are shown in Sect. 4. Experimental results and discussion are presented in Sect. 5. Finally, we conclude this work in Sect. 6.

2   Conventional GAN

The most groundbreaking development in the field of deep learning, particularly in the context of generative modeling, is the Generative Adversarial Network, or GAN [2]. A GAN framework typically comprises two essential components: a generator network and a discriminator network. The heart of GAN's success lies in the adversarial training process it employs. In the min-max adversarial game of GANs, the discriminator network plays a crucial role. Its primary function is to evaluate the authenticity of the data it receives. As training progresses, the discriminator becomes increasingly adept at distinguishing between real data and data generated by the generator. Concurrently, the generator network is on a quest to improve its capability to produce data that closely aligns with the true data distribution encapsulated in the training dataset. It does this by continuously adjusting its parameters to generate data that is more and more difficult for the discriminator to differentiate from the real data.

This adversarial training dynamic fosters a continuous loop of improvement. The discriminator strives to become more discerning, while the generator endeavors to create data that is virtually indistinguishable from the authentic data. The interplay between these two networks is the driving force behind the remarkable capabilities of GANs in generating high-quality data, and it underpins their success in a wide array of applications in deep learning and generative modeling.

The diagram of the conventional GAN model in the field of speech dereverberation is shown in Fig. 1(a). The generator generates speech features $G(x_c)$ from the data distribution $P(x_c)$ by transforming the reverberant speech features x_c. The discriminator is to determine whether the input features belong to the clean speech features or the generated speech features. In this way, the generator will eventually generate speech features that are similar to clean speech features as input to back-end ASR systems. In the conventional generative adversarial networks for speech dereverberation, the loss functions of the generator and discriminator use binary encoded least squares functions. The formulas are as follows:

$$\mathbf{minV(D)} = \frac{1}{2}\mathbf{E}_{x \sim p_{data(x)}}[(D(x) - 1)^2]$$
$$+ \frac{1}{2}\mathbf{E}_{x \sim p_{data(x_c)}}[(D(G(x_c)))^2] \tag{1}$$

$$\mathbf{minV(G)} = \frac{1}{2}\mathbf{E}_{x \sim p_{data(x)}}[(D(G(x_c)) - 1)^2] \tag{2}$$

The mean square error (MSE) loss function has been proven to be effective in order to minimize the distance between the generated features and the clean features [8, 21, 22], so we add the MSE as a secondary component to the loss function of the generator and it is controlled by a new parameter λ:

$$\mathbf{minV(G)} = \frac{1}{2}\mathbf{E}_{x \sim p_{data(x)}}[(D(G(x_c)) - 1)^2]$$
$$+ \frac{1}{2}\lambda MSE(G(x_c), x) \tag{3}$$

3   Task-Adaptive GAN

Although the conventional GAN models improve the performance of speech dereverberation tasks, the generated features of these models are not suitable for speech recognition systems because the optimization target of conventional GAN models is dereverberation instead of recognition. Even the conventional GAN models also take the input features of speech recognition systems as the target of the generator, the training process is the regression process of the dereverberation task. To bridge the gap between front-end dereverberation and back-end recognition, we propose a task-adaptive GAN-based speech dereverberation for robust speech recognition, called Task-adaptive GAN.

3.1 Network Structure

The diagram of our Task-adaptive GAN is shown in Fig. 1(b), the generator learns the clean features representation through regression learning. In order to make the generated speech features more suitable for the speech recognition system, we incorporate the acoustic model and generation model into the generator. We used two pre-trained models respectively to extract bottleneck features and unsupervised features. Finally, we concatenate the features of the three modules. So the generated speech includes enhanced MFCC, BNF, and unsupervised features. The adaptor is a pre-trained acoustic model with clean speech features as a classifier, which classifies the generated speech features into the corresponding senone classes. In this way, the adaptor could select suitable generated features for the speech recognition task, and the optimization target of Task-adaptive GAN is recognition instead of dereverberation.

In the meantime, to validate the effectiveness of the proposed senone prediction discriminator, we conduct an ablation experiment. We replace the generator part of the traditional generative adversarial network with our proposed recognition-aware generator while keeping the binary discriminator unchanged, we refer to this model as BU-GAN.

3.2 Loss Function

The adaptor is used as a classifier, so we use the cross-entropy (CE) loss as the loss function of the adaptor:

$$\mathcal{L}(D) = \mathbf{CE}(D(G(x_c))) = -\sum_{k=1}^{N}(r_k \cdot \log(D(G(x_c)))) \tag{4}$$

where k denotes the k-th senone class, N denotes the total number of senone classes, r_k denotes the one-hot encoding of N classes, the k-th value of r_k is 1, other values are 0.

In the generator, the generated speech features include enhanced MFCC features, BNF, and unsupervised features, which are predicted from the reverberant MFCC features, so we use MSE loss as the loss function for MFCC features and BNF as well. Since the size of each feature is different, and to balance the training process for all targets to prevent a situation where one or more targets have

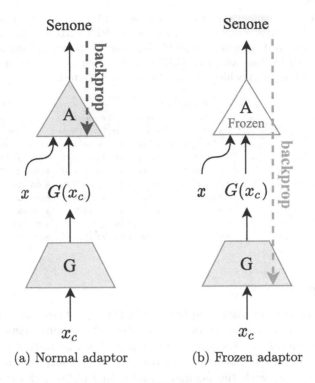

Fig. 2. The diagrams of Task-adaptive GAN training process with normal or frozen adaptor. (a) shows the process of the entire Task-adaptive GAN with a normal adaptor, the red dashed line represents the gradient backpropagation with normal adaptor. (b) shows the process of the entire Task-adaptive GAN with a frozen adaptor, the green dashed line represents the gradient backpropagation with frozen adaptor. (Color figure online)

a dominant influence on the network weights, we assign different weight values to different targets. The loss function of the generator is given as follows.

$$
\begin{aligned}
\mathcal{L}(G) &= \omega_G \sum \mathbf{MSE}(G(x_c), x) + \omega_A \mathbf{CE}(G(x_c)) \\
&= \omega_G^{\mathrm{mfcc}} \mathbf{MSE}(G(x_c^{\mathrm{MFCC}}), x^{\mathrm{MFCC}}) \\
&\quad + \omega_G^{\mathrm{bnf}} \mathbf{MSE}(G(x_c^{\mathrm{BNF}}), x^{\mathrm{BNF}}) + \omega_A \mathbf{CE}(G(x_c))
\end{aligned}
\tag{5}
$$

where ω_G and ω_A denote the weight of generator and adaptor, respectively. Due to the generated feature x_c and clean feature x consisting of MFCC feature, BNF, and unsupervised feature, so the MSE loss of generator could be represented by the MSE of MFCC feature and BNF. ω_G^{mfcc} and ω_G^{bnf} denote the weight of MFCC feature and BNF feature. x_c^{MFCC} denotes the enhanced MFCC feature predicted by the enhanced model, and x_c^{BNF} denotes the BNF predicted by the acoustic model. Correspondingly, x^{MFCC} and x^{BNF} denote the clean MFCC feature and BNF.

Table 1. WER (%) comparisons by conventional GAN and our proposed BU-GAN, Task-adaptive GAN in a comparative study on the REVERB test dataset. ω_G^{mfcc} and ω_G^{bnf} represent the MSE weight of MFCC and BNF, respectively, and ω_A represents the weight of CE. In Task-adaptive GAN, the baseline per weight configuration is highlighted with bold text, which is regarded as the **original**.

Model	Configuration			Simulated Data							Real Data		
				Room 1		Room 2		Room 3		Avg.	Room 1		Avg.
	ω_G^{mfcc}	ω_G^{bnf}	ω_A	Far	Near	Far	Near	Far	Near		Far	Near	
Unprocessed	–	–	–	8.18	7.47	13.58	9.11	16.33	10.85	10.92	27.95	28.07	28.01
GAN	–	–	–	8.05	7.01	12.18	8.17	14.09	9.86	9.89	24.71	24.66	24.69
BU-GAN	1.0	1.0	1.0	6.64	5.56	11.11	7.22	13.14	8.51	8.70	24.17	22.90	23.54
Task-adaptive GAN	**1.0**	**0.3**	**4.0**	6.37	5.30	10.29	6.48	12.06	7.78	**8.05**	24.44	23.44	23.94
	1.0	3.0	4.0	6.59	5.93	11.86	7.36	13.29	8.65	8.95	23.50	22.45	22.98
	10.0	0.3	4.0	6.15	5.15	11.09	6.80	11.89	7.49	8.10	23.94	22.84	23.39
	1.0	0.3	8.0	6.62	5.30	11.35	6.56	12.04	7.63	8.25	25.08	23.63	24.35
	1.0	0.3	0.2	7.59	6.88	11.36	7.80	13.25	9.55	9.41	23.70	21.43	**22.56**

3.3 Training Process

The training process of Task-adaptive GAN with the frozen adaptor is shown in Fig. 2(b), the frozen adaptor decides whether the features generated by the generator are suitable for the speech recognition task according to the senone, it will keep the suitable features, and discard other features. Compared with the training process with the normal adaptor in Fig. 2(a), the frozen adaptor makes the gradient backpropagation act directly on the generator resulting in the training process of Task-adaptive GAN being suitable for speech recognition tasks. The generator generates clean feature representations from reverberant speech features by regression learning and continuously adjusts the parameters so that the generated features are in the specified senone classes.

4   Datasets and Experimental Setup

4.1   Datasets

We conduct experiments on the REVERB challenge dataset [12]. The database includes simulated and real recorded data sampled from different rooms. The simulated recording is simulated by a single-channel microphone and the clean speech prediction volume in WSJCAM0 [24]. The product was collected from three rooms of different sizes and two different microphone locations. The data has different reverberation levels and a background noise signal-to-noise ratio of 20 dB. The sampling rate is 16 kHz. The database is divided into a training set, a validation set and a test set. We used 7861 pairs of parallel data consisting of reverberation and clean data from the REVERB challenge dataset as the training set. The validation set contains only simulation data and the test set contains both simulation and real data. The real data from the MC-WSJ-AV corpus [18] contains 372 real recordings (approximately 0.6 h). These recordings were performed in a reverberant conference room, including two different speak-to-microphone distances (near = 100 cm, far = 250 cm), so all the recordings are reverberant.

4.2 Experimental Setup

In the speech dereverberation front-end model, ReLU is used as the activation function for each hidden layer. The Adam algorithm is used to ensure a stationary solution [10]. The learning rate is set to 0.01 and the batch size is set to 256. We apply 13-D MFCC in our experiments since the performance of MFCC in larger dimensions is similar in our experiments.

Generator. The generator of the task-adaptive GAN includes a feature enhancement network for generating enhanced MFCC, an acoustic model for generating BNF, and an unsupervised generative model for compensating the generated features. We use 3 identical DNN models as the feature enhancement mapping network, the acoustic model, and the unsupervised generation model, with 13-D reverberant MFCC speech features as input. Each DNN has two hidden layers with 512 hidden units each. In the feature enhancement network, the 13-D clean MFCC feature is the training target to minimize the MSE loss function, and the output of the feature enhancement network is the 13-D enhanced MFCC feature. Since the MFCC feature is an artificially designed feature, it contains limited acoustic information. Phoneme-based BNF is one of the most commonly used features in speech recognition, which contains semantic information about speech-to-senone classification, and it will bring stronger discriminative power to the learning of MFCC features for clean speech, so it can be recognized more easily than traditional features (MFCC, etc.) in the practical application environment. Therefore, we add the 15-D phoneme-based BNF extracted from the acoustic model to obtain better performance of the speech recognition system. We also extract 13-D unsupervised features as a complement to MFCC features and BNF, so that we end up with 41-D generated speech features for task-adaptive GAN.

Adaptor. We use a feed-forward DNN-hidden Markov acoustic model pre-trained with 7861 clean utterances from the REVERB challenge dataset, using a cross-entropy loss. Each feature frame is spliced together with 5 left and 5 right context frames as the input of the feed-forward DNN. The DNN has 3 hidden layers with 512 hidden units each and an output layer with 1333 output units corresponding to 1333 senone labels.

Back-End Speech Recognition System. We use the Kaldi [23] toolkit to train our back-end speech recognition system with a DNN acoustic model and enhanced MFCC features. The system is mainly divided into the training stage and the decoding stage. In the training stage, 13-D MFCC features of each frame are extracted from reverberant and clean speech with a frame length of 512 and a frameshift of 256. The MFCC features of both reverberation and clean speech are used as input to the DNN acoustic model. The decoding stage uses a tri-gram language model with explicit pronunciation and silence probability modeling.

5   Results and Discussions

We use the back-end speech recognition system to measure the speech dereverberation accuracy [17]. The experimental results of WER(%) comparisons by conventional GAN and our proposed BU-GAN, Task-adaptive GAN are summarized in Table 1. We conduct a series of comparative experiments:

Unprocessed: The 13-D reverberant MFCC features are fed into the back-end speech recognition system without speech dereverberation front-end processing directly.

GAN: The baseline use the DNN-based GAN model [14] for speech dereverberation front-end processing. The DNN model consists of three hidden layers, each with 512 hidden units. The input feature is a 13-D reverberant MFCC. The loss function is shown in Eq. (1) and (3).

BU-GAN: The generator part is replaced by the generator of Task-adaptive GAN, and the generated features are the 41-D concatenation of the enhanced MFCC feature, the BNF, and the unsupervised feature.

Task-Adaptive GAN: Our proposed task-adaptive generative adversarial network. Among them, we perform several different experiments on the weight values of different targets in the loss functions of the generator and the adaptor. The experiment results are shown in the lower of Table 1.

The results in the upper part of Table 1, show that the WER results using the conventional GAN as the front-end dereverberation model can suppress the interference of the reverberation effectively and improve the performance of the back-end speech recognition system significantly. We replace the generator of conventional GAN with our proposed generator of Task-adaptive GAN, the results in Table 1, the GAN and BU-GAN show that our proposed generator which contains the enhancement model, the acoustic model, and the unsupervised model could improve the performance of the speech dereverberation model for robust speech recognition task effectively, and also prove the phoneme-based BNF and the unsupervised feature could make the generated speech feature more suitable for the speech recognition task. In Task-adaptive GAN, We also conduct several comparative experiments. By adjusting the weights of the loss function, multiple experiments with different orders of magnitude are compared. The results are shown in Table 1, and the task-adaptive GAN achieves better performance compared to the results of the conventional GAN throughout the experiment. In particular, the best performance on real data is obtained when the ω_G^{mfcc}, ω_G^{bnf} and ω_A are set to 1.0, 0.3 and 0.2, and the relative WER is reduced by approximately 8.6% on real data. The best performance on simulated data is achieved in the original, the relative WER is reduced by 18.6%.

When the ω_G^{bnf} is increased to 3, the performance of both Far and Near on simulated data is decreased. The relative WER is increased by 11% compared with the original, but on real data, the relative WER is reduced by 4 %. When the ω_G^{mfcc} is set to 10, there is no significant performance improvement on the simulated data and real data with the original. The ω_A is set to 8, the experimental results show that simulated and real data performance is decreased, and the relative WER is increased by 2.5% and 1.7%, respectively. But when ω_A is decreased to 0.2, we achieved the best performance on real data, the WER is 22.56%. We found that Task-adaptive GAN is robust to the choice of ω_A as long as it is below 1. This also means that each target in the loss acquires a better balance among those weight configurations during training to generate more robust features that favor speech recognition.

6   Conclusions and Future Work

In this paper, we proposed a novel task-adaptive generative adversarial network-based speech dereverberation model (Task-adaptive GAN) for robust speech recognition. We incorporated the MFCC, the recognition-aware BNF, and the unsupervised feature into the generator, and designed a new senone-prediction network as the discriminator to perform the classification task. The generated feature of our model was consistent with the recognition task to bridge the gap between front-end enhancement and back-end recognition. Experimental results on the REVERB corpus showed that our approach outperforms the conventional GAN-based system by reducing the relative word error rate (WER) by 18.6% on the simulated data set and 8.6% on the real data set. In the future, we will add more suitable features to further improve the performance of Task-adaptive GAN.

References

1. Chen, S.J., Subramanian, A.S., Xu, H., Watanabe, S.: Building state-of-the-art distant speech recognition using the chime-4 challenge with a setup of speech enhancement baseline. arXiv preprint arXiv:1803.10109 (2018)
2. Goodfellow, I., et al.: Generative adversarial networks. Commun. ACM **63**(11), 139–144 (2020)
3. Grezl, F., Fousek, P.: Optimizing bottle-neck features for LVCSR. In: 2008 IEEE International Conference on Acoustics, Speech and Signal Processing, pp. 4729–4732 (2008)
4. Grezl, F., Karafiat, M., Kontar, S., Cernocky, J.: Probabilistic and bottle-neck features for LVCSR of meetings. In: 2007 IEEE International Conference on Acoustics, Speech and Signal Processing, ICASSP 2007, vol. 4, pp. IV-757–IV-760 (2007)
5. Han, K., Wang, Y., Wang, D.: Learning spectral mapping for speech dereverberation. In: 2014 IEEE International Conference on Acoustics, Speech and Signal Processing (ICASSP). IEEE (2014)

6. Himawan, I., Motlicek, P., Imseng, D., Potard, B., Kim, N., Lee, J.: Learning feature mapping using deep neural network bottleneck features for distant large vocabulary speech recognition. In: 2015 IEEE International Conference on Acoustics, Speech and Signal Processing (ICASSP), pp. 4540–4544 (2015)
7. Hsiao, R., et al.: Robust speech recognition in unknown reverberant and noisy conditions. In: 2015 IEEE Workshop on Automatic Speech Recognition and Understanding (ASRU), pp. 533–538 (2015)
8. Isola, P., Zhu, J.Y., Zhou, T., Efros, A.A.: Image-to-image translation with conditional adversarial networks. In: Proceedings of the IEEE Conference on Computer Vision and Pattern Recognition, pp. 1125–1134 (2017)
9. Ke, W., Junbo, Z., Sining, S., Yujun, W., Fei, X., Lei, X.: Investigating generative adversarial networks based speech. In: Interspeech 2018. ISCA (2018)
10. Kingma, D.P., Ba, J.: Adam: a method for stochastic optimization. arXiv preprint arXiv:1412.6980 (2014)
11. Kingsbury, B., Morgan, N.: Recognizing reverberant speech with RASTA-PLP. In: 1997 IEEE International Conference on Acoustics, Speech, and Signal Processing, vol. 2, pp. 1259–1262 (1997)
12. Kinoshita, K., et al.: A summary of the reverb challenge: state-of-the-art and remaining challenges in reverberant speech processing research. EURASIP J. Adv. Signal Process. (2016)
13. Lebart, K., Boucher, J.M., Denbigh, P.N.: A new method based on spectral subtraction for speech dereverberation. Acta Acust. Acust. **87**(3), 359–366 (2001)
14. Li, C., Wang, T., Xu, S., Xu, B.: Single-channel speech dereverberation via generative adversarial training. CoRR abs/1806.09325 (2018)
15. Li, J., Deng, L., Häb-Umbach, R., Gong, Y.: Robust Automatic Speech Recognition: A Bridge to Practical Applications. Elsevier Science (2015)
16. Li, J., Deng, L., Gong, Y., Haeb-Umbach, R.: An overview of noise-robust automatic speech recognition. IEEE/ACM Trans. Audio Speech Lang. Process. **22**(4), 745–777 (2014)
17. Li, N., Ge, M., Wang, L., Dang, J.: A fast convolutional self-attention based speech dereverberation method for robust speech recognition. In: Gedeon, T., Wong, K.W., Lee, M. (eds.) ICONIP 2019. LNCS, vol. 11955, pp. 295–305. Springer, Cham (2019). https://doi.org/10.1007/978-3-030-36718-3_25
18. Lincoln, M., McCowan, I., Vepa, J., Maganti, H.K.: The multi-channel wall street journal audio visual corpus (MC-WSJ-AV): specification and initial experiments. In: IEEE Workshop on Automatic Speech Recognition and Understanding. IEEE (2005)
19. Mack, W., Chakrabarty, S., Stöter, F.R., Braun, S., Edler, B., Habets, E.A.: Single-channel dereverberation using direct MMSE optimization and bidirectional LSTM networks. In: INTERSPEECH, pp. 1314–1318 (2018)
20. Park, S., Jeong, Y., Kim, M.S., Kim, H.S.: Linear prediction-based dereverberation with very deep convolutional neural networks for reverberant speech recognition. In: 2018 International Conference on Electronics, Information, and Communication (ICEIC), pp. 1–2. IEEE (2018)
21. Pascual, S., Bonafonte, A., Serra, J.: SEGAN: speech enhancement generative adversarial network. arXiv preprint arXiv:1703.09452 (2017)
22. Pathak, D., Krahenbuhl, P., Donahue, J., Darrell, T., Efros, A.A.: Context encoders: feature learning by inpainting. In: Proceedings of the IEEE Conference on Computer Vision and Pattern Recognition, pp. 2536–2544 (2016)

23. Povey, D., et al.: The Kaldi speech recognition toolkit. In: IEEE 2011 Workshop on Automatic Speech Recognition and Understanding, No. CONF. IEEE Signal Processing Society (2011)
24. Robinson, T., Fransen, J., Pye, D., Foote, J., Renals, S.: WSJCAMO: a British English speech corpus for large vocabulary continuous speech recognition. In: 1995 International Conference on Acoustics, Speech, and Signal Processing. IEEE (1995)
25. Sainath, T.N., Kingsbury, B., Ramabhadran, B.: Auto-encoder bottleneck features using deep belief networks. In: 2012 IEEE International Conference on Acoustics, Speech and Signal Processing (ICASSP), pp. 4153–4156 (2012)
26. Schwartz, B., Gannot, S., Habets, E.A.: Online speech dereverberation using Kalman filter and EM algorithm. IEEE/ACM Trans. Audio Speech Lang. Process. **23**(2), 394–406 (2014)
27. Weninger, F., Watanabe, S., Tachioka, Y., Schuller, B.: Deep recurrent de-noising auto-encoder and blind de-reverberation for reverberated speech recognition. In: 2014 IEEE International Conference on Acoustics, Speech and Signal Processing (ICASSP). IEEE (2014)
28. Zhang, J., Plumbley, M.D., Wang, W.: Weighted magnitude-phase loss for speech dereverberation. In: 2021 IEEE International Conference on Acoustics, Speech and Signal Processing (ICASSP), ICASSP 2021, pp. 5794–5798. IEEE (2021)

Real-Time Automotive Engine Sound Simulation with Deep Neural Network

Hao Li[1], Weiqing Wang[2], and Ming Li[2(✉)]

[1] Z-one Technology Co., Ltd., Shanghai, China
[2] Data Science Research Center, Duke Kunshan University, Kunshan, China
ming.li369@dukekunshan.edu.cn

Abstract. This paper introduces a real-time technique for simulating automotive engine sounds based on revolutions per minute (RPM) and pedal pressure data. We present a hybrid approach combining both sample-based and procedural methods. In the sample-based technique, the sound of an idle engine undergoes pitch-shifting proportional to the ratio of current RPM to idle RPM. For the procedural technique, deep neural networks fine-tune the amplitude of the engine's pulse frequency derived from the sample-based sound. To ensure the synthesized sound does not have any clicks between the frames, we utilize a modified griffin-lim algorithm at the frame level, which, with our proposed overlap-and-add feature, can bridge the phase gap between two frames. Experimental evaluations on our self-collected database validate the efficacy of the introduced approach.

Keywords: Engine Sound Simulation · Engine Sound Synthesis · Real Time Synthesis

1 Introduction

The engine sound is one of the most overlooked aspects in driving simulation as it gives an indication of the state of the vehicle. For the in-car environment, it affects speed judgment, operator performance, alertness, and fatigue [1–4]. In addition, it can provide auditory feedback to the driver. Drivers can make decisions according to the engine sound, e.g., changing gears using the pitch of the engine sound or maintaining a steady vehicle speed. Drivers often underestimate the vehicle speed and have difficulty in maintaining a target speed if no engine sound is provided [4–9]. Leading automakers, including BMW, Audi, Ford, and Jaguar, are actively engaged in research focused on stimulating drivers' emotions and conveying distinct brand identities through vehicle sounds [10]. For the out-car noise, it can inform pedestrians and cyclists of the vehicle's approaching, avoiding many traffic accidents [11,12].

Nowadays, electric motor-driven vehicles (EVs) are emerging due to their environmental friendliness, and fuel-efficient performance [13]. However, the electric motor usually cannot generate the sounds as an internal combustion engine.

J. Jia et al. (Eds.): NCMMSC 2023, CCIS 2006, pp. 176–188, 2024.
https://doi.org/10.1007/978-981-97-0601-3_15

Therefore, simulation of the combustion engine sound is important in this kind of vehicle, and the vehicle sound quality is important to improve ride comfort [14–18].

In general, there are sample-based and procedural methods for engine sound synthesis [19]. The sample-based method is the most common approach, where the sound samples are looped and then resampled or pitch-shifted based on the revolutions per minute (RPM) or other signals [20–22]. Heitbrink et al. [1] use the wavetable approach to synthesize the engine sound, where crossfading is applied during playback to shift the frequencies between two sound samples. Lee et al. [23] also employ the wavetable approach that can maintain the shape of string sound waveforms and vary the pitch during acceleration. Scott et al. [24] use the deterministic-stochastic signal decomposition approach to synthesize the automotive engine sound. The deterministic component is first extracted from the original sound using the synchronous discrete Fourier transform (SDFT). Then they use a multi-pulse excited time-series method to model the stochastic component. They find that the audio quality can be improved using weighted error minimization. Van et al. [25] propose a phase vocoder-based method to simulate the engine sound for a driving simulator. They extract the acoustic features and modify these features to change the speed of the signal. Then they estimate the representation of the modified signal and finally resample the generated signal. Jan et al. [26] propose a real-time algorithm for engine sound synthesis. They extract the sound samples from a recorded engine sound within the entire engine speed range. Then they employ an extension of the pitch-synchronous overlap-and-add (PSOLA) method to locate the extraction instants of the sound samples and finally produce the engine sound. Recently, Dongki et al. [27] propose an engine sound synthesis method. They first generate a mechanical sound by summing harmonic components representing sounds from rotating engine cranks. And then they simulate a combustion noise using random sounds with similar spectral characteristics to the measured value. Finally, the mechanical sound and the combustion noise are combined to produce the engine sound.

In the procedural method, the sound is generated from some attributes of the engine sound. Stefano [19] proposes a procedural method based on the mechanics of the actual four-stroke engines. Fu et al. [28] simulate the engine motion sense sound by reading the vehicle running state data on the CAN bus of pure electric vehicles.

In this paper, a hybrid method for engine sound simulation is proposed, where both sample-based and procedural methods are employed. For the sample-based method, the pitch of the signal is shifted from the sounds of the idle engine according to the frame-level RPM, where the griffin-lim algorithm (GLA) [29] is employed. To remove the clicking between each frame, we propose the griffin-lim overlap-and-add (GLOLA) method and generate the sound given different RPM. For the procedural methods, we generate the spectrum of the engine sound with only RPM and pressure on the pedal (POP) using the deep neural network (DNN). These two spectrums generated by these two methods are finally summed up and converted to the engine sound by GLOLA.

The rest of this paper is organized as follows: Sect. 2 introduce the sample-based method with GLOLA; Sect. 3 present the DNN-based procedural method; Sect. 4 is the experiments and results; Sect. 5 concludes this paper.

2   Sample-Based Method with Griffin-Lim Algorithm

The most common sample-based approach in the most driving simulator is the wavetable approach [1]. In this technique, a collection of sound samples are mixed or manipulated to generate the engine sound. To be more specific, the sound samples between different speeds are recorded during a real-world drive. Next, the sounds are cross-faded depending on the vehicle's speed. However, the onset and offset of each sound will be audibly identifiable when the sounds are played in a loop, which results in repeated clicking. Therefore, the sound snippets should be faded in/out slowly to hide the repeated clicking or apply some cancellation methods to remove the repeated clicking [30].

Unlike the wavetable approach, we directly generate the engine sound at a different speed from the idle engine sound. Therefore, only a very short sound sample of the idle engine is needed (usually less than 1 s). The sound is first pitch-shifted based on the ratio between the current RPM and the idle RPM, and then playback in a loop. However, the clicks also exist as the wavetable approach does. Actually, these clicks are caused by the discontinuity of the phase between the boundaries of two sound samples. To remove such a click, we employ the griffin-lim algorithm to recover the phase near the boundaries. More specifically, we propose the griffin-lim with overlap-and-add (GLOLA) in frame-level to generate the engine sound without clicks as the traditional GLA is not originally designed for frame-level synthesis.

2.1   Griffin-Lim Algorithm

Griffin-lim algorithm (GLA) is a phase recovery algorithm that can recover a complex-valued spectrogram [29]. Given a real-valued amplitude $\mathbf{A}$, GLA generates the complex-valued spectrogram $\mathbf{C}$ in the following iterative projection procedure [31]:

$$\mathbf{C}^{[i+1]} = P_{\mathcal{C}}\left(P_{\mathcal{A}}(\mathbf{C}^{[i]})\right), \tag{1}$$

where $P_{\mathcal{S}}$ is the metric projection on a set $\mathcal{S}$, i is the iteration index, and $\mathbf{C}^{[0]} = \mathbf{A}$. $\mathcal{C}$ is the set of consistent complex-valued spectrograms and $\mathcal{A}$ is the corresponding spectrogram set with the same amplitude. The projection is given by:

$$P_{\mathcal{C}}(\mathbf{C}) = \mathcal{G}\mathcal{G}^{-1}\mathbf{C}, \tag{2}$$

$$P_{\mathcal{A}}(\mathbf{C}) = \mathbf{A} \odot \mathbf{C} \oslash |\mathbf{C}|, \tag{3}$$

where $\mathcal{G}$ represent short-time Fourier transform (STFT), $\mathcal{G}^{-1}$ is the pseudo inverse of STFT (iSTFT), $\odot$ denotes the element-wise multiplicatio, and $\oslash$

denotes the element-wise division. The goal of GLA is to reduce the distance $D(\mathbf{C}, P_{\mathcal{C}}(\mathbf{C}))$ [32]:

$$\min_{\mathbf{C}} ||\mathbf{C} - P_{\mathcal{C}}(\mathbf{C})||_F^2, \tag{4}$$

where $|| \cdot ||$ is the Frobenius norm.

One limitation of applying GLA for engine sound simulation is that GLA is not designed for real-time synthesis at frame-level since it employs the STFT, which takes several frames of a signal as input. In addition, the phase discontinuity is not resolved yet. Actually, overlap-and-add has been inherently implemented in the STFT/iSTFT calculation, which results in a continuous signal. This motivates us to incorporate the GLA into the STFT/iSTFT process, which is the griffin-lim overlap-and-add (GLOLA) algorithm.

To synthesize the engine sound in real-time, the input of GLA becomes a frame of signal. We also perform the overlap-and-add [33] operation at the end of each frame-level iteration so that the phases are continuous between two successive frames. In the phase estimation step, we only calculate the phase of each harmonic component, which dramatically reduces the computational cost in each frame-level iteration.

2.2   Griffin-Lim Algorithm in Frame-Level

STFT/iSTFT is the process of several fast Fourier transform (FFT)/pseudo inverse of FFT (iFFT) calculation with overlap-and-add in the frame-level. Given an amplitude $\mathbf{A}$ and let $\mathbf{C}^{[0]} = \mathbf{A}$, the traditional GLA can be composed in four steps:

1. Projecting the spectorgram $\mathbf{C}$ in $\mathcal{A}$ as Eq. 1 shows;
2. iFFT calculation on each frame of spectrogram with overlap-and-add, producing the continuous signal;
3. FFT calculation on each frame of the continuous signal, producing the spectrogram $\mathbf{C}$;
4. Assessing the distance $D(\mathbf{C}, P_{\mathcal{C}}(\mathbf{C}))$; if it is small enough, stop iteration and return the signal; otherwise, go back to step 1.

As we split the GLA step by step, it is obvious that GLA performs several FFT/iFFT calculations with overlap-and-add in one iteration since it accepts several frames as input.

To simulate the engine sound at frame-level, the GLA must take only one frame as input and iteratively estimate the phase for each frame. Therefore, we iteratively perform the FFT/iFFT, and the overlap-and-add will be applied after the distance is converged for this frame. Give an amplitude $\mathbf{a}$ of a frame, previous simulated signal $\mathbf{y}$ and let $\mathbf{c} = \mathbf{a}$, the GLOLA contains five steps:

1. Projecting the spectorgram frame $\mathbf{c}$ in $\mathcal{A}$ as Eq. 1 shows;
2. iFFT calculation on the frame, producing a frame of signal $\hat{\mathbf{y}}$

Algorithm 1. The GLOLA algorithm

Require: $l > o > 0$ \\ l is the frame length and o is the frame overlap as required by
    STFT
Require: Pre—recorded sample sound $x \in \mathbb{R}^n$ at a stable idle RPM
Ensure: $\mathbf{y} = \mathbf{0} \in \mathbb{R}^o$, $\hat{\mathbf{y}} = \mathbf{0} \in \mathbb{R}^l$, $L = o$
 1: **while** simulation process does not stop **do**
 2:      $\hat{\mathbf{x}} \leftarrow \text{Resample}(x, \frac{\text{current RPM}}{\text{idle RPM}})$
 3:      $\mathbf{a} = |\text{fft}(\hat{\mathbf{x}})|$
 4:      **while** $D(\mathbf{c}, P_C(\mathbf{c}))$ is not small enough **do**
 5:          $\mathbf{c} = \text{fft}(\hat{\mathbf{y}})$
 6:          $P_A(\mathbf{c}) \leftarrow \mathbf{a} \odot \exp(\text{Angle}(\mathbf{c}) \times i)$
 7:          $\hat{\mathbf{y}} \leftarrow \text{ifft}(P_A(\mathbf{c}))$
 8:          $\hat{\mathbf{y}}[0 : o] \leftarrow \hat{\mathbf{y}}[0 : o] + \mathbf{y}[L - o : L]$
 9:          $D(\mathbf{c}, P_C(\mathbf{c})) \leftarrow ||\mathbf{C} - P_C(\mathbf{C})||_F^2$
10:      **end while**
11:      $\mathbf{y} \leftarrow \text{Append}([\mathbf{y}, \hat{\mathbf{y}}[0 : (l - o)]])$
12:      $L \leftarrow L + (l - o)$
13: **end while**

---

3. Adding this signal $\hat{\mathbf{y}}$ to the previous simulated signal $\mathbf{y}$ with overlap: $\hat{\mathbf{y}}_{0 \sim o} \leftarrow \hat{\mathbf{y}}_{0 \sim o} + \mathbf{y}_{(L-o) \sim L}$, where o is the frame overlap of the STFT and L is the total length of $\mathbf{y}$. This means that we only add the overlapping part of these two signals.
4. FFT calculation on $\hat{\mathbf{y}}$, producing the spectrogram $\mathbf{c}$;
5. Assessing the distance $D(\mathbf{c}, P_C(\mathbf{c}))$; if it is small enough, stop iteration and append this signal $\hat{\mathbf{y}}$ to the end of $\mathbf{y}$; otherwise go back to step 1.

Algorithm 1 shows a more detailed process of our engine sound simulation with GLOLA. Note that this algorithm does not exactly start from step 1 as mentioned above, but the iterative process is similar.

3   Procedural Method with Deep Neural Network

Although we obtain a simulated engine sound by GLOLA, the characteristics of the engine are not carefully considered. As mentioned in [24], the engine pulse frequency F_0 is equal to:

$$F_0 = \frac{RPM}{60} \times \frac{p}{2}, \tag{5}$$

where p is the number of cylinders. Figure 1 also shows the relationship between the RPM and the engine pulse frequency F_0. In addition, F_0 and its multiples have a higher amplitude than others. These harmonic components at multiples of F_0 have a significance at low-frequency band, which can also represent the characteristics of the engine [27]. However, even F_0 can be easily calculated from RPM, the relationships between the RPM and the amplitude on these multiples of F_0 are less obvious than the engine pulse frequency does, as shown in Fig. 2.

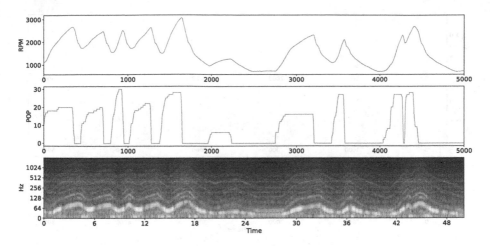

Fig. 1. The RPM, POP, and the spectrum of the original engine sound

But we can still notice that the amplitude is higher when the RPM goes much higher. This motivates us to use deep learning to predict the RPM.

Considering the requirement of real-time simulation, we cannot use a complex network architecture since it needs to be set on an embedded system. Therefore, we employ the deep neural network with only two fully-connected layers.

This network takes the RPM, pressure on the pedal (POP), and the first-order delta of RPM and POP as input. The first-order delta of RPM and POP can reflect the acceleration of the vehicle, which is also related to the engine sound, e.g., the engine sounds of driving at 80 MPH and accelerating to 80 MPH are different. The features are split into several frames with a sliding window size of 11, resulting in a 44-dim feature, including consecutive RPMs, POPs, and the delta of PRMs and POPs. The output is the amplitude value at the end of the sliding window on F_0 and its multiples. We train three networks with the same architecture, which predict the amplitude on F_0, $2F_0$, and $3F_0$, respectively.

After the amplitude of each multiple of F_0 is predicted, we add this amplitude to the amplitude generated by the sample-based method. Next, we employ the GLOLA to produce a new engine sound whose amplitude on the multiples of F_0 is more accurate.

4 Experiments and Results

4.1 Dataset and Pre-processing

We evaluate the proposed method on several indoor recordings of engine sound from a sedan car of model MG3 manufactured by SAIC Motor that contains a four-stroke engine. The dataset contains four recordings, and the corresponding PRM and POP signals are also recorded from the system on the vehicle every 10 ms. A driver presses or releases the pedal frequently to make sure that each

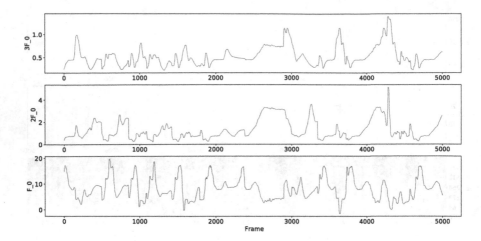

Fig. 2. The amplitude on each multiple of F_0

recording contains as much diverse information as possible. Table 1 shows some metadata of each recording. The sample rate is 44100 Hz, and the duration of each recording is about 10 to 15 min. In our experiments, we take the last 50-s audio of the 4-th recording as the testing data for synthesis, and the remaining recordings are used for training the deep neural network.

Table 1. Metadata of the dataset

ID	Duration (s)	range of RPM	range of POP
1	548.76	719.14–4296.59	0–37
2	964.91	690.68–3218.65	0–36
3	911.35	713.99–4925.18	0–55
4	902.14	696.22–6283.91	0–228
Total	3327.16	690.68–6283.91	0–228

We first downsample the recordings to 4000 Hz. In addition, we compute the STFT with frame length of 100 ms and frame shift of 10 ms, which means $l = (4000 \times 0.1) = 400$ and $o = (4000 \times 0.09) = 360$ as mentioned in Algorithm 1.

4.2   Sample-Based Method

For the sample-based method, we choose the continuous sound samples whose pressure on the pedal (POP) is zero. Since we simulate the engine sound in frame-level and the frame length is 400, we only need no more than $400 \times \frac{6283.91}{690.68} \approx 4000$ idle engine sound samples, which is a 1-s signal.

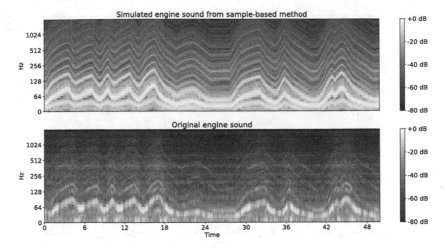

Fig. 3. The specturm of the original engine sound and the simulated engine sound from sample-based method

For a coming frame with an RPM, the idle engine sound samples are first pitch-shifted (resampled) by the ratio between the RPM of the current frame and the idle RPM, which is the $\hat{y}$ in the Algorithm 1. Each time a frame of 400 sound samples is simulated, and only the first 40 samples are played as the frame shift is 40 (10 ms). The distance $D(\mathbf{c}, P_{\mathcal{C}}(\mathbf{c}))$ usually becomes stable after 40 iterations, so we stop the iteration after 40 iterations.

Figure 3 shows the spectrograms of the original engine sound and the engine sound simulated by our sample-based method. Although they look similar, the amplitude on the engine pulse frequency F_0 and its multiples cannot match the original amplitude, as shown in Fig. 4. The energies on the engine pulse frequency are too high to keep the characteristics of the engine.

4.3   Procedural Method

As the amplitude of the engine sound simulated by the sample-based method is not entirely accurate, we have explored the use of deep neural networks to predict the amplitude at the engine pulse frequency F_0 and its multiples.

We specifically predict the amplitude at F_0, $2F_0$, and $3F_0$. For each of these multiples of F_0, a separate neural network is trained to predict the amplitude. Consequently, three networks with identical architecture have been trained. Each network consists of two fully connected layers, with both the first and second layers containing 128 neurons each. The input dimension is set at 44, encompassing 11 continuous samples of RPM, POP, and the delta of RPM and POP. The output dimension is 1, representing the amplitude at one of the multiples of F_0, corresponding to the last sample of the input RPM features. Each model undergoes optimization using the Adam optimizer with mean square error (MSE) loss over 100 epochs. The batch size is 128, with a learning rate of 0.001.

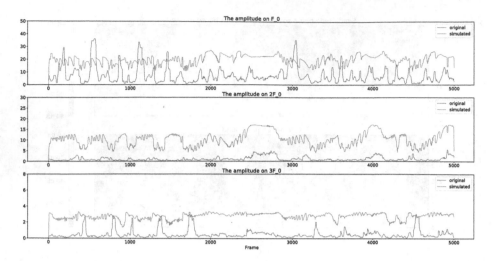

Fig. 4. The amplitude of the original engine sound and the simulated engine sound from sample-based method

Figure 5 displays the spectrograms of the original engine sound alongside the sound simulated by our procedural method. Additionally, Fig. 6 presents a comparison of the amplitude between the original engine sound and the sound simulated by the procedural method.

4.4 Hybrid Process of Synthesis

In the sample-based method, whenever an RPM signal is received every 10 ms, the idle engine sound is resampled according to this RPM and then converted to a using FFT.

These are then combined with the preceding 10 samples of RPM and POP, and the delta of these samples is computed. These features are subsequently concatenated to form the input for the neural network. Three distinct neural networks then predict three amplitude values separately. These predicted amplitude values are added to a to enhance the accuracy of the amplitude on the multiples of F_0.

The overall simulation process remains consistent with the algorithm outlined in Algorithm 1, with the exception that the three predicted amplitude values are now added to a.

Figure 7 illustrates the spectrograms of both the original engine sound and the engine sound synthesized by this hybrid process. The result is notably more similar to the original when compared with the sample-based method alone. The synthesized samples are available on GitHub[1].

---

[1] https://github.com/karfim/EngineSound.

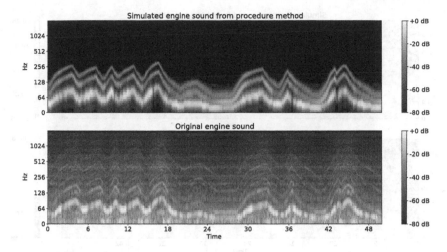

Fig. 5. The specturm of the original engine sound and the simulated engine sound from procedure method

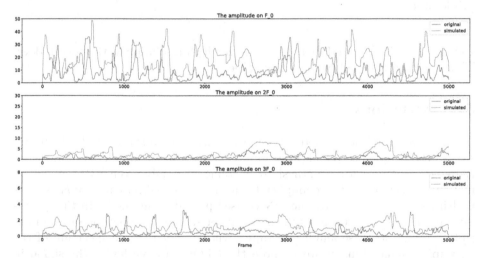

Fig. 6. The amplitude of the original engine sound and the simulated engine sound from procedure method

4.5   Time Cost

During the simulation process, all experiments are conducted on an Inter Xeon Silver CPU with a single core of 2.2GHz. The pitch shift process and deep neural network prediction take 0.4 ms and 0.8 ms. The GLOLA takes 6–7 ms for a frame after 40 iterations. Therefore, the total time cost is about 8 ms, and it is smaller than the frame shift of 10 ms. This means that it is possible to simulate the engine sound in real-time. In practice, we can use a large frame shift, e.g., 50–

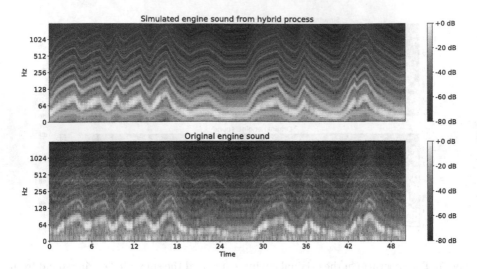

Fig. 7. The specturm of the original engine sound and the simulated engine sound from hybrid process

100 ms, as engine sound does not need such a high resolution as the speech does. Thus the time cost will be more negligible.

5   Conclusions

In this paper, a hybrid method for engine sound synthesis is proposed, which consists of the sample-based and procedure methods. For the sample-based method, the engine sound can be synthesized in frame-level only given RPM and a 1-s idel engine sound. Next, the amplitude on the engine pulse frequency F_0 and its multiples can be refincd by the DNN-based procedure methods. In this procedure method, several deep neural networks predicts the amplitude on the multiples of F_0, and these amplitude values will be added to the spectrum of the resampled signal. Finally, we propose the GLOLA to synthesize the signal in frame-level without any clicks since we incorporate the GLA into the process of STFT/iSTFT. Therefore, no further action is needed to perform clicking cancellation. Also, this method is very fast, which has good potential to synthesize the engine sound with a neural network on the vehicles in real-time. In the future, we are going to collect more data to train a better network. Besides, some techniques, such as quantization and network pruning, can be used to compress the model to further reduce the model size. We will also investigate how to predict the RPM signal from other captured information, e.g., vehicle speed, and then perform the engine sound synthesis using the estimated RPM signal, which is more suitable to the real applications.

References

1. Heitbrink, D.A., Cable, S.: Design of a driving simulation sound engine. In: Driving Simulation Conference, North America 2007 (2007)
2. Gibson, J.J., Crooks, L.E.: A theoretical field-analysis of automobile-driving. Am. J. Psychol. **51**(3), 453–471 (1938)
3. Castro, C.: Human Factors of Visual and Cognitive Performance in Driving. CRC Press, Boca Raton (2008)
4. McLane, R.C., Wierwille, W.W.: The influence of motion and audio cues on driver performance in an automobile simulator. Hum. Factors **17**(5), 488–501 (1975)
5. Merat, N., Jamson, H.: A driving simulator study to examine the role of vehicle acoustics on drivers' speed perception. In: Sixth International Driving Symposium on Human Factors in Driver Assessment, Training and Vehicle Design (2011)
6. Horswill, M.S., Plooy, A.M.: Auditory feedback influences perceived driving speeds. Perception **37**(7), 1037–1043 (2008)
7. Denjean, S., Roussarie, V., Kronland-Martinet, R., Ystad, S., Velay, J.-L.: How does interior car noise alter driver's perception of motion? Multisensory integration in speed perception. In: Acoustics 2012 (2012)
8. Evans, L.: Speed estimation from a moving automobile. Ergonomics **13**(2), 219–230 (1970)
9. Horswill, M.S., McKenna, F.P.: The development, validation, and application of a video-based technique for measuring an everyday risk-taking behavior: drivers' speed choice. J. Appl. Psychol. **84**(6), 977 (1999)
10. Lazaro, M.J., et al.: Design and evaluation of electric vehicle sound using granular synthesis. J. Audio Eng. Soc. **70**(4), 294–304 (2022)
11. Faas, S.M., Baumann, M.: Pedestrian assessment: is displaying automated driving mode in self-driving vehicles as relevant as emitting an engine sound in electric vehicles? Appl. Ergon. **94**, 103425 (2021)
12. Karaaslan, E., Noori, M., Lee, J.Y., Wang, L., Tatari, O., Abdel-Aty, M.: Modeling the effect of electric vehicle adoption on pedestrian traffic safety: an agent-based approach. Transp. Res. Part C: Emerg. Technol. **93**, 198–210 (2018)
13. Wogalter, M.S., Ornan, R.N., Lim, R.W., Ryan Chipley, M.: On the risk of quiet vehicles to pedestrians and drivers. In: Proceedings of the Human Factors and Ergonomics Society Annual Meeting, vol. 45, pap. 1685–1688. SAGE Publications, Los Angeles (2001)
14. Wagner-Hartl, V., Graf, B., Resch, M., Langjahr, P.: Subjective evaluation of EV sounds: a human-centered approach. In: Ahram, T., Karwowski, W., Taiar, R. (eds.) IHSED 2018. AISC, vol. 876, pp. 10–15. Springer, Cham (2019). https://doi.org/10.1007/978-3-030-02053-8_2
15. Doleschal, F., Verhey, J.L.: Pleasantness and magnitude of tonal content of electric vehicle interior sounds containing subharmonics. Appl. Acoust. **185**, 108442 (2022)
16. Lee, S.-K., Lee, G.-H., Back, J.: Development of sound-quality indexes in a car cabin owing to the acoustic characteristics of absorption materials. Appl. Acoust. **143**, 125–140 (2019)
17. Park, J.H., Park, H., Kang, Y.J.: A study on sound quality of vehicle engine sportiness using factor analysis. J. Mech. Sci. Technol. **34**(9), 3533–3543 (2020)
18. Qian, K., Hou, Z., Sun, D.: Sound quality estimation of electric vehicles based on GA-BP artificial neural networks. Appl. Sci. **10**(16), 5567 (2020)
19. Baldan, S., Lachambre, H., Monache, S.D., Boussard, P.: Physically informed car engine sound synthesis for virtual and augmented environments. In: IEEE 2nd VR Workshop on Sonic Interactions for Virtual Environments (2015)

20. Konet, H., Sato, M., Schiller, T., Christensen, A., Tabata, T., Kanuma, T.: Development of approaching vehicle sound for pedestrians (VSP) for quiet electric vehicles. SAE Int. J. Engines **4**(1), 1217–1224 (2011)
21. Kuppers, T.: Results of a structured development process for electric vehicle target sounds. In: Aachen Acoustic Colloquium 2012, pp. 63–71 (2012)
22. Engler, O., Hofmann, M., Mikus, R., Hirrle, T.: Mercedes-Benz SLS AMG Coupé Electric Drive NVH development and sound design of an electric sports car. In: Bargende, M., Reuss, H.-C., Wiedemann, J. (eds.) 15. Internationales Stuttgarter Symposium. P, pp. 1295–1309. Springer, Wiesbaden (2015). https://doi.org/10. 1007/978-3-658-08844-6_90
23. Lee, J., Lee, J., Choi, D., Jung, J.: String engine sound generation method based on wavetable synthesizer. In: Audio Engineering Society Convention, vol. 154 (2023)
24. Amman, S.A., Das, M.: An efficient technique for modeling and synthesis of automotive engine sounds. IEEE Trans. Ind. Electron. **48**(1), 225–234 (2001)
25. Janse Van Rensburg, T., Van Wyk, M.A., Potgieter, A.T., Steeb, W.-H.: Phase vocoder technology for the simulation of engine sound. Int. J. Mod. Phys. C **17**(05), 721–731 (2006)
26. Jagla, J., Maillard, J., Martin, N.: Sample-based engine noise synthesis using an enhanced pitch-synchronous overlap-and-add method. J. Acoust. Soc. Am. **132**(5), 3098–3108 (2012)
27. Dongki, M., Buhm, P., Junhong, P.: Artificial engine sound synthesis method for modification of the acoustic characteristics of electric vehicles. Shock. Vib. **2018**, 1–8 (2018)
28. Jianghua, F., Zhu, C., Jintao, S.: Research on the design method of pure electric vehicle acceleration motion sense sound simulation system. Appl. Sci. **13**(1), 147 (2022)
29. Griffin, D., Lim, J.: Signal estimation from modified short-time Fourier transform. IEEE Trans. Acoust. Speech Signal Process. **32**(2), 236–243 (1984)
30. Wu, S.: Engine sound simulation and generation in driving simulator. Master's thesis, Missouri University of Science and Technology (2016)
31. Masuyama, Y., Yatabe, K., Koizumi, Y., Oikawa, Y., Harada, N.: Deep Griffin-Lim iteration. In: 2019 IEEE International Conference on Acoustics, Speech and Signal Processing (ICASSP), ICASSP 2019, pp. 61–65. IEEE (2019)
32. Le Roux, J., Kameoka, H., Ono, N., Sagayama, S.: Fast signal reconstruction from magnitude STFT spectrogram based on spectrogram consistency. In: Proceedings of the DAFx, vol. 10, pp. 397–403 (2010)
33. Zhu, X., Beauregard, G.T., Wyse, L.L.: Real-time signal estimation from modified short-time Fourier transform magnitude spectra. IEEE Trans. Audio Speech Lang. Process. **15**(5), 1645–1653 (2007)

A Framework Combining Separate and Joint Training for Neural Vocoder-Based Monaural Speech Enhancement

Qiaoyi Pan[1], Wenbing Jiang[2(✉)], Qing Zhuo[3], and Kai Yu[4]

[1] Paris Elite Institute of Technology (SPEIT), Shanghai Jiao Tong University, Shanghai, China
sweeto@sjtu.edu.cn

[2] School of Communication Engineering, Hangzhou Dianzi University, Hangzhou, China
wbjiang@hdu.edu.cn

[3] Department of Automation, Tsinghua University, Beijing, China
zhuoqing@tsinghua.edu.cn

[4] MoE Key Lab of Artificial Intelligence, X-LANCE Lab, Department of Computer Science and Engineering, Shanghai Jiao Tong University, Shanghai, China
kai.yu@sjtu.edu.cn

Abstract. Conventional single-channel speech enhancement methodologies have predominantly emphasized the enhancement of the amplitude spectrum while preserving the original phase spectrum. Nonetheless, this may introduce speech distortion. While the intricate nature of the multifaceted spectra and waveform characteristics presents formidable challenges in training. In this paper, we introduce a novel framework with the Mel-spectrogram serving as an intermediary feature for speech enhancement. It integrates a denoising network and a deep generative network vocoder, allowing the reconstruction of the speech without using the phase. The denoising network, constituting a recurrent convolutional autoencoder, is meticulously trained to align with the Mel-spectrogram representations of both clean and noisy speech, resulting in an enhanced spectral output. This enhanced spectrum serves as the input for a high-fidelity, high-generation speed vocoder, which synthesizes the improved speech waveform. Following the pre-training of these two modules, they are stacked for joint training. Experimental results show the superiority of this approach in terms of speech quality, surpassing the performance of conventional models. Notably, our method demonstrates commendable adaptability across both the Chinese dataset CSMSC and the English language speech dataset VoiceBank+DEMAND, underscoring its considerable promise for real-world applications and beyond.

Keywords: speech enhancement · denoising autoencoder · generative adversarial network · vocoder · joint framework

© The Author(s), under exclusive license to Springer Nature Singapore Pte Ltd. 2024
J. Jia et al. (Eds.): NCMMSC 2023, CCIS 2006, pp. 189–202, 2024.
https://doi.org/10.1007/978-981-97-0601-3_16

1   Introduction

In recent years, advancements in deep learning have reshaped speech enhancement, moving it from traditional methods to supervised learning. Neural network vocoders have gained prominence, exemplified by SEGAN (Speech Enhancement Generative Adversarial Network) [17], which leverages Generative Adversarial Networks (GANs) for breakthroughs in speech enhancement [15,23,24]. The integration of neural networks in speech enhancement can be broadly categorized into two distinct approaches, primarily based on the preprocessing of the speech signal [4,30].

Traditionally, speech enhancement primarily concentrated on augmenting the amplitude spectrum of the signal while maintaining the integrity of the phase spectrum. However, this conventional approach presented inherent limitations. The random nature of the phase component and the undeniable influence of phase on speech quality necessitated a paradigm shift. These constraints often led to speech distortion when using traditional models [7]. To address these challenges, recent studies have explored strategies that focus on enhancing the complex spectrum, exemplified by the complex ratio mask (CRM) [27–29]. Moreover, the intricate interplay between amplitude and phase introduced estimation uncertainties in the amplitude spectrum [26]. Consequently, researchers have proposed an alternative approach: enhancing the amplitude spectrum initially and subsequently refining the complex spectrum to mitigate these issues [13,31].

Vocoders play a pivotal role in the synthesis of linguistic and acoustic features into speech waveforms. We have chosen the high-sampling and high-fidelity vocoder, HiFi-GAN [12], to address issues related to generation quality and rate. HiFi-GAN has demonstrated superior performance when compared to other publicly available models such as the autoregressive (AR) convolutional neural network WavNet [16] and the Glow-based model WavGlow [18].

Considering the importance of selecting clear and distinct features for model training, it is evident that the amplitude spectrum offers a more structured and visually appealing option than the waveform in the time domain or the complex spectrum. Drawing inspiration from recent research on the concept of forming a joint framework [5], we propose an innovative approach. We employ a denoising auto-encoder in the Mel spectrum domain to enhance the amplitude spectrum and a high-fidelity adversarial network vocoder that takes the Mel spectrum as input for synthesizing speech waveforms. These two components are then combined to create a unified framework for monaural speech enhancement.

To further optimize this framework, we superimpose these components to construct a deeper neural network for comprehensive joint training. In this work, we introduce several key innovations:

- We utilize a joint architecture for speech enhancement that fuses a denoising network with a network vocoder, where the vocoder used is adjusted to HiFi-GAN with a higher quality and faster rate.
- We optimize multiple loss functions in separate training and joint training of the two slab networks.

– We explore the performance of this framework on two different language datasets for a fair and reproducible comparison.

The experimental results demonstrate the clarity and perceptiveness of generated enhanced speech on a Chinese dataset, exhibiting improvements over traditional methods. Various speech quality assessment measures, including perceptual evaluation of speech quality (PESQ) [20], Short-Time Objective Intelligibility (STOI) [21], CSIG, CBAK, SSNR, and other relevant metrics [6, 11, 14], were employed as references. The joint model presented in this paper effectively outperforms traditional methods like OMLSA, SEGAN, and DCCRN-E [9], and competes favorably with MetricGAN. Furthermore, the framework maintains its robust performance even when utilizing a pre-trained vocoder on the English dataset VoiceBank+DEMAND.

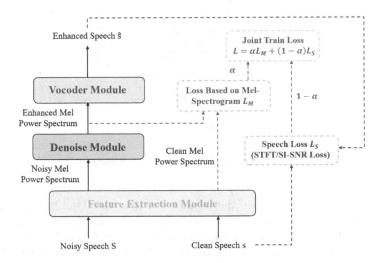

Fig. 1. The diagram of the speech enhancement framework

2 Framework Structure

The structure shown in Fig. 1 is the framework diagram of the speech enhancement system used in this paper: initially, the Mel spectrum extraction module is set to convert the input noisy speech and clean speech into their corresponding Mel spectrum; secondly, the denoising model is designed to map the noisy spectrum to the enhanced spectrum. Finally, the vocoder module synthesizes the enhanced speech based on the enhanced Mel spectrum. In this process, the multiple discriminators and generators are combined to modify the "true" speech accorded to the Mel spectrum. Then it applied the losses based on feature, Mel-spectrum, and the basic GAN loss to reconstruct the speech without utilizing the phase.

2.1   Denoising Model

We propose a denoising model based on convolutional neural networks to aggregate multiple scales by time-dilated convolution and casual-dilated convolution.

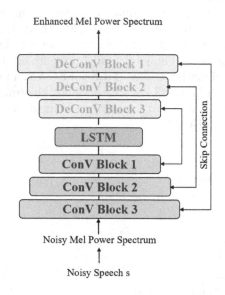

Fig. 2. Denoising model structure schematic

In the first step, we employed the architecture illustrated in Fig. 2 to construct the denoising model. This architecture is designed to ensure high performance by incorporating both feature extraction through Convolutional Neural Networks (CNN) and sequence modeling with Recurrent Neural Networks (RNN). It comprises an encoder and decoder, each consisting of three convolutional blocks. Long Short-Term Memory (LSTM) layers [8] are integrated into the architecture. The encoder features three convolutional blocks activated by Rectified Linear Unit (ReLU) functions, while the decoder includes three deconvolutional blocks with Leaky ReLU activation, ensuring non-zero gradients throughout and facilitating the optimization process during decoding. Furthermore, a skip connection links each convolution block to its corresponding deconvolution block.

2.2   Model Architecture

After establishing the denoising network, an additional adjustment of the Mel spectrum is necessary to connect the two components. The presence of subtle background noise can affect segments where human voice is predominant. To address this, we employ a spectrum re-scaling process, following the design used in Tacotron [25], as depicted in Algorithm 1. This process takes the enhanced spectrum output by the denoising module and yields a spectrum with reduced background noise.

In this process, we set the minimum noise-to-signal ratio in the background as 1×10^{-4}, which is defined as negligible noise. Thus, the algorithm calculates the signal-to-noise ratio corresponding to different time frequencies, and assigns the negligible parts to zero, so as to compress the effect of background noise in a targeted manner.

Algorithm 1 example of spectrum rescaling algorithm

Input: Mel spectrum of enhanced speech S
Output: Mel spectrum of reshaped speech S_r
1: **Initialization:**$S_r(t, f) = max(S(t, f), 1 \times 10^{-4})$
2: $S_r(t, f) := 20 log S_r(t, f) - 20$
3: $S_r(t, f) := (S_r(t, f) + 100)/100$
4: **if** $S_r(t, f) < 0$ **then**
5:     $S_r(t, f) := 0$
6: **else if** $S_r(t, f) > 1$ **then**
7:     $S_r(t, f) := 1$
8: **end if**
9: **Return** S_r

As shown in Fig. 1, we introduce a combined training loss $\mathcal{L}$ for joint training. This loss is obtained by adding weighting coefficients α, $1 - \alpha$ to the loss $\mathcal{L}_M$ based on the speech feature of the Mel spectrum and the speech loss $\mathcal{L}_S$ based on the speech variability contrast, respectively. The speech loss functions we choose are as follows:

If S, s, and N are represented as noisy speech, clean speech, and noise, the process of simulating noise can be represented by $S = s + N$. The optimization purpose of the speech enhancement model we build is to exclude the interference of noise from S, synthesize the enhanced speech s to make it as close as possible to the clean speech $\hat{s}$, and set such mapping to f, then we have $\hat{s} = f(S) \longrightarrow s$.

Scale-Invariant SNR Loss. The scale-invariant SNR loss is not affected by signal variations compared to the classical SNR loss.

$$s_{target} = \frac{\langle \hat{s}, s \rangle s}{||s||^2}, \tag{1}$$

$$e_{noise} = \hat{s} - s_{target}, \tag{2}$$

$$\mathcal{L}_{SI-SNR} = 10 log_{10} \frac{||s_{target}||^2}{||e_{noise}||^2}. \tag{3}$$

where s_{target} denotes the projection of the enhanced speech to the pure speech and e_{noise} evaluates the noise within the enhanced speech.

STFT Loss. The single short-time Fourier transform loss is the sum of the spectral convergence loss and the logarithmic short-time Fourier change amplitude loss.

$$\mathcal{L}_{STFT} = \mathbb{E}_{z \sim p(z), x \sim p_{data}}[\mathcal{L}_{sc}(s, \hat{s}) + \mathcal{L}_{mag}(s, \hat{s})], \tag{4}$$

where $\hat{s}$ denotes the generated speech, $\mathcal{L}_{sc}$ denotes the spectral convergence (spectral convergence) loss, and $\mathcal{L}_{mag}$ denotes the log short-time Fourier transform magnitude (log STFT magnitude) loss, respectively, defined as follows.

$$\mathcal{L}_{sc}(s, \hat{s}) = \frac{|||STFT(s)| - |STFT(\hat{s})|||_F}{|||STFT(s)|||_F}, \tag{5}$$

$$\mathcal{L}_{mag}(s, \hat{s}) = \frac{1}{M}|| \log|STFT(s)| - \log|STFT(\hat{s})| ||_1. \tag{6}$$

where $|SFTF(.)|$ denotes the short-time Fourier transform magnitude value, M denotes the number of elements of the magnitude value, $||.||_F$ denotes the Frobenius parametrization, defined as the squared and re-squared matrix elements: $||A||_F = \sqrt{\sum_{i=1}^{m} \sum_{j=1}^{n} |a_{ij}|^2}$.

3   Experiments

3.1   Data Preparation

In the context of a native Chinese environment, with the aim of excluding the influence of accents, we employed the Chinese Standard Mandarin Speech Corpus (CSMSC) [1]. CSMSC is a comprehensive Chinese standard female voice database with a sampling rate of 48 kHz, 16-bit precision, and features 10,000 utterances spoken by women aged 20 to 30 years. This database encompasses standard Mandarin speech with consistent timbre and speech rates. The average utterance length is 16 words, contributing to a total duration of approximately 12 h.

To optimize the denoising model, we extended the speaker-specific speech database to a speaker-independent one [10] to simulate noise-contaminated speech. Our approach involved incorporating various non-speech noise sources, each with different signal-to-noise levels. We randomly selected clean speech from the training set and mixed it to generate a significantly expanded set of noise-laden speech data. Given that noise segments may be shorter than the clean speech, we adopted a round-robin strategy and randomly segmented the noise for mixing.

Subsequently, signal-to-noise ratios were randomly selected from a range of [−5 dB, 10 dB], and this noisy speech was combined with 10,000 audio samples from the CSMSC database to form a foundational dataset of 180,000 speech samples. We then selected four distinct noise types-DKITCHEN, OMEETING, PCAFETER, and TBUS-for matching experiments, comparing clean and noisy speech. For each noise type, 500 samples from each of the 10,000 female voices were extracted as a validation set, another 500 as a test set, and the remainder served as the training set. This process yielded a comprehensive dataset, comprising 36,000 training samples, 2,000 validation samples, and 2,000 test samples, with a cumulative speech duration of approximately 43 h.

In the case of the English speech dataset, we utilized the publicly available expanded dataset VoiceBank+DEMAND [22], which we downsampled to 16 kHz to align with the parameters used in our experiments.

3.2 Model Training

In accordance with the loss calculation selected for the joint architecture, $\mathcal{L} = \alpha\mathcal{L}_M + (1 - \alpha)\mathcal{L}_S$, we set the weight coefficient α to 1. This implies that the training exclusively relies on losses derived from the Mel spectrum and does not involve interconnecting the two modules.

Separate Training. Consequently, we initially conduct separate training for the two modules and subsequently integrate them for joint training.

A. Denoising Model Training: During the training of the denoising model, the model's input is extracted directly from the full Mel spectrum segment containing noise, without any upsampling. Before feeding the data into this denoising model, we include a feature extraction module, specifically the Mel frequency spectrum module from the audio samples. The model parameters are configured as follows: g is set to 16, and the number of hidden cells in the LSTM cell h is set to 512. For training, we utilize the L1 loss function to compute the loss based on the Mel spectrum, denoted as $\mathcal{L}_M$. This loss quantifies the difference between the pure Mel spectrum of the input and the enhanced Mel spectrum. We employ the Adam optimizer with a learning rate of 0.001 for training and conduct 200 training epochs on the extended dataset. This process enables us to acquire the initially trained denoising network, which learns the mapping relationship from noisy speech to clean speech.

Similarly, we optimize the training of the denoising model using the VoiceBank+DEMAND database as proposed in the data preparation phase. We segment the speech in the training set into 2-second segments while preserving the length of the speech in the test set. Consistent with the parameter settings of CMGAN, we employ the Adam optimizer for the initial training of the denoising model in the joint framework, conducting 200 training epochs.

B. Vocoder Training: For training the HiFi-GAN, we configure the generator with 512 upsampling channels and set the upsampling rates to 8, 8, 2, and 2, with corresponding kernel sizes of 16, 16, 4, and 4. We adjust the learning rate of the ResBlock to 0.0002, use a batch size of 32, and introduce a learning rate decay rate with a 0.1% decay rate. Additionally, we specify convolutional kernel sizes for the ResBlock as 3, 7, and 11, with each assigned an expansion rate of 1, 3, and 5.

Training a high-capacity generative model like HiFi-GAN from scratch can be time-intensive, often requiring over 10 days for completion, particularly when conducting joint training with both the vocoder and the denoising module. To expedite the process, we opted for transfer learning and utilized pre-trained vocoders. Specifically, we employed the English pre-trained HiFi-GAN model "vocoder-hifigan-universal" available on HUGGING FACE.

Joint Training. Following the separate training of the denoising model and the HiFi-GAN vocoder, where α was set to 1, we embarked on exploring the impact of joint training when these two components are combined. To investigate this, we introduced varying weight coefficients α set to 0 and 0.5, thereby enabling the integration of multiple speech-based loss functions and the collective fine-tuning of the entire network. During our exploration to identify the optimal configuration, we examined several loss functions proposed in recent research, including SI-SNR Loss and STFT Loss. It's important to note that, due to limited training time, we regarded the vocoder module as invariant, focusing our adjustments solely on the denoising module. For optimization, we retained the use of the Adam optimizer and set the learning rate to 1×10^{-5}.

3.3   Baseline Model

OMLSA. The Optimally Modified Log-Spectral Amplitude Estimation Algorithm (OMLSA) is a more traditional speech enhancement model. For the application of OMLSA, the sampling rate is first unified to 16 kHz, and then adjusted mainly for its signal processing-dependent Hamming window. Finally, in order to adapt the speech length in the database, the Hanning window (Hanning) length is set to 512 sample points long, and the frameshift is set to 128 sample points long. On the Voice Bank database, the sampling length was set to 400, and the frameshift to 100 samples in order to obtain a window length of about 25 ms.

SEGAN. It is a model that uses straight generative adversarial networks for monophonic speech enhancement. For the training of SEGAN, we chose the same training parameters as much as possible in order to minimize the impact of different hyperparameters. However, since the generator and discriminator in the original SEGAN use the same kernel size, we set it equal to the median size of the kernels used in HiFi-GAN: 7. In addition, the loss function used for training was changed to the STFT loss function.

Spectral Mask. There is a masking effect in the human ear's perception of sound. The spectral mask is reflected in the fact that strong pure tones mask the weak pure tones in their vicinity, and the closer the weak pure tones are to the strong pure tones, the more easily they are masked. This property can be applied to the speech enhancement model. Here we will use the frequency masking-based speech enhancement model provided by SpeechBrain [19] for comparison experiments. The parameter settings: random seed, sampling rate, and batch size are the same. The original MSE loss function is replaced with the STFT loss function.

DCCRN-E. The Deep Complex Convolution Recurrent Network for Phase-Aware Speech Enhancement(DCCRN) [9] is a model based on the CRN architecture. It has designed a new complex-valued speech enhancement network to combine DCUNET [3] and CRN. It is based on the structure of CRN and applies complex CNN and complex batch normalization layers in the encoder and decoder to ensure that the real and imaginary parts follow the complex multiplication rule despite separate inputs. Among the three models given in the paper, DCCRN-E obtains the estimated speech from the calculation on the polar coordinate level.

In addition, we compared our results with several recent models on the Voice Bank+DEMAND dataset, such as WaveNet and MetricGAN. Given the inability to complete the complex training for the plural spectrum of MetricGAN, the pre-trained model results of SpeechBrain [19] on the Voice Bank+DEMAND training set are used directly in the paper.

4 Experimental Results and Analysis

This section presents the enhanced speech quality of the speech enhancement model from the objective perspective. Figures 3 and 4 visually exemplify spectrograms comparing the performance of speech enhancement, with the baseline model being SEGAN.

4.1 Results on CSMSC

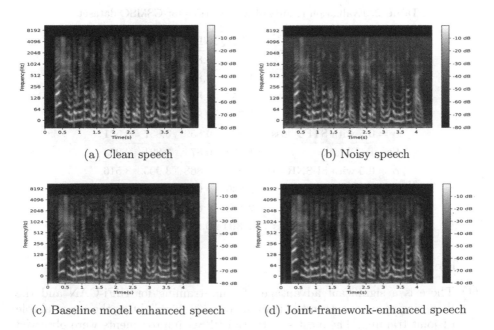

(a) Clean speech   (b) Noisy speech

(c) Baseline model enhanced speech   (d) Joint-framework-enhanced speech

Fig. 3. Spectrogram corresponding to the speech enhancement results of the speech samples on the CSMSC dataset

For objective evaluation, we adopted internationally recognized metrics for speech quality assessment. In our analysis, we employed four key metrics: PESQ, eSTOI, CSIG, and CBAK.

Table 1. PESQ evaluation results for the four noise categories on the CSMSC dataset

Model	KITCHEN	MEETING	CAFETER	BUS	AVG
Noisy	1.345	1.116	1.091	1.824	1.344
$\alpha = 1$	1.454	1.164	1.115	1.973	1.427
SEGAN	1.572	1.347	1.325	1.983	1.557
OMLSA	2.113	1.186	1.237	2.465	1.750
$\alpha = 0$, STFT loss	2.473	1.625	1.842	2.626	2.141
$\alpha = 0$, SI-SNR loss	**2.694**	**2.060**	**1.931**	**2.879**	**2.391**

To begin, Table 1 presents the PESQ evaluation results of the training experiments conducted on the Chinese speech dataset, CSMSC. The table categorizes the results into different columns based on the type of noise and provides an overview of noisy speech, results when α is set to 1 (indicating no joint training), and quality evaluations using the SI-SNR loss function and the STFT loss function with varying weighting factors. The last column represents the average scores, offering a comprehensive assessment of each model's overall performance.

Additionally, Table 2 displays the average scores achieved by various training models across the four evaluation metrics.

Table 2. Evaluation results of four metrics for CSMSC dataset

Model	PESQ	STOI	CSIG	CBAK
Noisy	1.344	0.815	3.021	2.107
$\alpha = 1$	1.427	0.821	3.032	2.127
SEGAN	1.557	0.827	3.162	2.138
OMLSA	1.750	0.832	3.719	2.608
$\alpha = 0.5$ with STFT loss	1.876	0.859	3.837	2.531
Spectral Mask	1.934	0.874	3.821	2.953
$\alpha = 0.5$ with SI-SNR loss	1.946	0.868	3.906	3.516
DCCRN-E	2.236	0.904	3.896	**3.712**
$\alpha = 0$ with STFT loss	2.141	0.878	3.914	3.572
$\alpha = 0$ with SI-SNR loss	**2.391**	**0.917**	**4.174**	3.697

The results reveal that:

(1) There is a significant advantage of joint training for HiFi-GAN and the denoising network over initial training ($\alpha = 1$), highlighting the essential role of joint training. The most substantial PESQ improvements were observed

in the KITCHEN and BUS noise classes. Compared to the OMLSA baseline, using $\alpha = 0$ with SI-SNR loss resulted in PESQ improvements of 1.35, 0.71, 0.69, and 0.41 for the four noise classes, with an average improvement of 0.64 points.

(2) The joint framework outperforms traditional speech enhancement methods across various metrics, including OMLSA, SEGAN, and spectral mask models. However, DCCRN-E, which utilizes a complex network, slightly outperforms our model in CBAK estimation.

(3) The SI-SNR loss function demonstrates a slight edge over the STFT loss function when utilized for $\mathcal{L}_S$. Looking at Table 2, the difference between evaluations with the two loss functions, using a weight coefficient of 0 (SI-SNR and STFT loss functions jointly), is 0.25, 0.02 (2%), 0.28, and 0.18. The emphasis on signal-to-noise ratio is apparent.

(4) Both speech-based loss function $\mathcal{L}_S$ and characteristic Mel spectrum-based loss function $\mathcal{L}_M$ yield positive results for model training. Notably, a weight coefficient of 0 performs the best in joint training. Comparing the results for weight coefficients α of 0, 0.5, and 1, a higher weight on the $\mathcal{L}_S$ loss function yields improvements across the metrics. This suggests a stronger correlation between the speech-based loss function and speech quality.

4.2   Results on Voice Bank+DEMAND

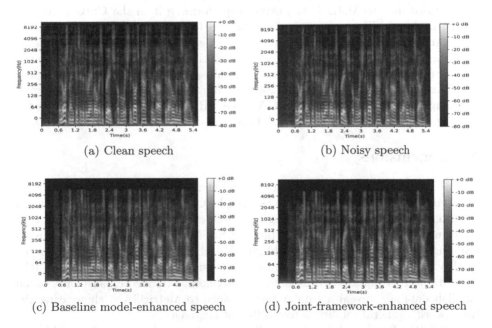

(a) Clean speech

(b) Noisy speech

(c) Baseline model-enhanced speech

(d) Joint-framework-enhanced speech

Fig. 4. Spectrum results corresponding to the speech enhancement model on the Voice Bank dataset

We calculated and compared the average scores of enhanced speech produced by each model using the same metrics as in previous papers [2, 32]. The results for MetricGAN are sourced from the original paper, and the SSNR metric was used, with higher values indicating lower signal-to-noise ratios. The model setup using the SI-SNR loss function demonstrated superior performance in previous experiments, as depicted in Table 3.

Table 3. Voice Bank+DEMAND dataset evaluation results for each indicator

Model	PESQ	CSIG	CBAK	COVL	SSNR	STOI
Noisy	1.97	3.35	2.44	2.63	1.68	0.91
SEGAN	2.16	3.48	2.94	2.80	7.73	0.92
Wiener	2.22	3.23	2.68	2.67	–	–
WaveNet	–	3.62	3.24	2.98	–	–
Spectral Mask	2.65	3.44	2.98	3.21	–	0.93
DCCRN-E	2.73	3.73	3.22	3.22	–	–
MetricGAN	**2.86**	3.99	**3.18**	**3.42**	–	–
Joint framework with SI-SNR loss	2.76	**4.01**	3.06	3.34	–	**0.93**

The above results reveal that:

(1) The framework surpasses SEGAN, Wiener, WaveNet, Spectral Mask-based models, and DCCRN-E on the English corpus. It demonstrates comparable performance to MetricGAN, even outperforming it in the CSIG metric to some degree. These comparisons emphasize the superiority of time-frequency domain models over time-domain models across various reference metrics. The framework's performance against MetricGAN underlines its potential for continuous improvement through architectural and training optimization.
(2) This speech enhancement framework exhibits strong generalizability across languages. It outperforms SEGAN, DCCRN-E, and Spectral mask-based algorithms on both the CSMSC dataset and the VoiceBank+DEMAND.

5   Conclusions

In this paper, we propose a speech enhancement method that leverages convolutional networks and neural network vocoders. The joint architecture features a denoising network and a GAN-based speech enhancement method. Extensive experiments explore the impact of popular SI-SNR and STFT loss functions on training outcomes. The results highlight the substantial improvement in speech quality achieved through joint training, surpassing OMLSA, SEGAN, DCCRN-E, and Spectral Mask-based models. Furthermore, extended experiments on an English database reveal a similar performance to MetricGAN, showcasing the framework's generalizability across different languages.

While the framework has been validated with mixed non-vocal noise, future work may expand its applicability to the removal of unseen speakers as noise. Additionally, exploring plural spectrum enhancement could enhance denoising effectiveness in diverse conditions.

Acknowledgements. I would like to express my deepest gratitude to my supervisor, Fei Wen, for his guidance throughout this project. This research was funded by Scientific and Technological Innovation 2030 under Grant 2021ZD0110900 and the Key Research and Development Program of Jiangsu Province under Grant BE2022059. This work was done in X-LANCE lab.

References

1. Baker, D.: Chinese standard mandarin speech copus. https://www.data-baker.com/open_source.html
2. Cao, R., Abdulatif, S., Yang, B.: CMGAN: conformer-based metric GAN for speech enhancement. arXiv preprint arXiv:2203.15149 (2022)
3. Choi, H.S., Kim, J.H., Huh, J., Kim, A., Ha, J.W., Lee, K.: Phase-aware speech enhancement with deep complex U-Net. In: International Conference on Learning Representations (2018)
4. Defossez, A., Synnaeve, G., Adi, Y.: Real time speech enhancement in the waveform domain (2020)
5. Du, Z., Zhang, X., Han, J.: A joint framework of denoising autoencoder and generative vocoder for monaural speech enhancement. IEEE/ACM Trans. Audio Speech Lang. Process. **28**, 1493–1505 (2020)
6. Fu, S.W., et al.: Boosting objective scores of a speech enhancement model by MetricGAN post-processing. In: 2020 Asia-Pacific Signal and Information Processing Association Annual Summit and Conference (APSIPA ASC), pp. 455–459. IEEE (2020)
7. Griffin, D., Lim, J.: Signal estimation from modified short-time Fourier transform. IEEE Trans. Acoust. Speech Signal Process. **32**(2), 236–243 (1984)
8. Hochreiter, S., Schmidhuber, J.: Long short-term memory. Neural Comput. **9**(8), 1735–1780 (1997)
9. Hu, Y., et al.: DCCRN: deep complex convolution recurrent network for phase-aware speech enhancement. arXiv preprint arXiv:2008.00264 (2020)
10. Jiang, W., Liu, Z., Yu, K., Wen, F.: Speech enhancement with neural homomorphic synthesis. In: 2022 IEEE International Conference on Acoustics, Speech and Signal Processing (ICASSP), ICASSP 2022, pp. 376–380. IEEE (2022)
11. Kawanaka, M., Koizumi, Y., Miyazaki, R., Yatabe, K.: Stable training of DNN for speech enhancement based on perceptually-motivated black-box cost function. In: 2020 IEEE International Conference on Acoustics, Speech and Signal Processing (ICASSP), ICASSP 2020, pp. 7524–7528. IEEE (2020)
12. Kong, J., Kim, J., Bae, J.: HiFi-GAN: generative adversarial networks for efficient and high fidelity speech synthesis. In: Advances in Neural Information Processing Systems, vol. 33, pp. 17022–17033 (2020)
13. Li, A., Zheng, C., Zhang, L., Li, X.: Glance and gaze: a collaborative learning framework for single-channel speech enhancement. Appl. Acoust. **187**, 108–499 (2022)
14. Li, H., Fu, S.W., Tsao, Y., Yamagishi, J.: iMetricGAN: intelligibility enhancement for speech-in-noise using generative adversarial network-based metric learning. arXiv preprint arXiv:2004.00932 (2020)
15. Luo, Y., Mesgarani, N.: Conv-TasNet: surpassing ideal time-frequency magnitude masking for speech separation. IEEE/ACM Trans. Audio Speech Lang. Process. **27**(8), 1256–1266 (2019)

16. van den Oord, A., et al.: WaveNet: a generative model for raw audio. arXiv preprint arXiv:1609.03499 (2016)
17. Pascual, S., Bonafonte, A., Serra, J.: SEGAN: speech enhancement generative adversarial network. arXiv preprint arXiv:1703.09452 (2017)
18. Prenger, R., Valle, R., Catanzaro, B.: WaveGlow: a flow-based generative network for speech synthesis. In: 2019 IEEE International Conference on Acoustics, Speech and Signal Processing (ICASSP), ICASSP 2019, pp. 3617–3621. IEEE (2019)
19. Ravanelli, M., et al.: SpeechBrain: a general-purpose speech toolkit. arXiv preprint arXiv:2106.04624
20. Rix, A.W., Beerends, J.G., Hollier, M.P., Hekstra, A.P.: Perceptual evaluation of speech quality (PESQ)-a new method for speech quality assessment of telephone networks and codecs. In: Proceedings of the 2001 IEEE International Conference on Acoustics, Speech, and Signal Processing (Cat. No. 01CH37221), pp. 749–752. IEEE (2001)
21. Taal, C.H., Hendriks, R.C., Heusdens, R., Jensen, J.: A short-time objective intelligibility measure for time-frequency weighted noisy speech. In: 2010 IEEE International Conference on Acoustics, Speech and Signal Processing, pp. 4214–4217. IEEE (2010)
22. Valentini-Botinhao, C., et al.: Noisy speech database for training speech enhancement algorithms and TTS models. University of Edinburgh. School of Informatics, Centre for Speech Technology Research (CSTR) (2017)
23. Wang, D., Chen, J.: Supervised speech separation based on deep learning: an overview. IEEE/ACM Trans. Audio Speech Lang. Process. **26**(10), 1702–1726 (2018)
24. Wang, P., Wang, D.: Enhanced spectral features for distortion-independent acoustic modeling. In: INTERSPEECH, pp. 476–480 (2019)
25. Wang, Y., et al.: Tacotron: towards end-to-end speech synthesis. arXiv preprint arXiv:1703.10135 (2017)
26. Wang, Z.Q., Wichern, G., Le Roux, J.: On the compensation between magnitude and phase in speech separation. IEEE Signal Process. Lett. **28**, 2018–2022 (2021)
27. Williamson, D.S., Wang, D.: Time-frequency masking in the complex domain for speech dereverberation and denoising. IEEE/ACM Trans. Audio Speech Lang. Process. **25**(7), 1492–1501 (2017)
28. Williamson, D.S., Wang, Y., Wang, D.: Complex ratio masking for monaural speech separation. IEEE/ACM Trans. Audio Speech Lang. Process. **24**(3), 483–492 (2015)
29. Williamson, D.S., Wang, Y., Wang, D.: Complex ratio masking for joint enhancement of magnitude and phase. In: 2016 IEEE International Conference on Acoustics, Speech and Signal Processing (ICASSP), pp. 5220–5224. IEEE (2016)
30. Xu, Y., Du, J., Dai, L.R., Lee, C.H.: A regression approach to speech enhancement based on deep neural networks. IEEE/ACM Trans. Audio Speech Lang. Process. **23**(1), 7–19 (2014)
31. Yu, G., Li, A., Zheng, C., Guo, Y., Wang, Y., Wang, H.: Dual-branch attention-in-attention transformer for single-channel speech enhancement. In: 2022 IEEE International Conference on Acoustics, Speech and Signal Processing (ICASSP), ICASSP 2022, pp. 7847–7851. IEEE (2022)
32. Zhou, L., Gao, Y., Wang, Z., Li, J., Zhang, W.: Complex spectral mapping with attention based convolution recurrent neural network for speech enhancement. arXiv preprint arXiv:2104.05267 (2021)

Accent-VITS: Accent Transfer for End-to-End TTS

Linhan Ma[1], Yongmao Zhang[1], Xinfa Zhu[1], Yi Lei[1], Ziqian Ning[1],
Pengcheng Zhu[2], and Lei Xie[1(✉)]

[1] Audio, Speech and Language Processing Group (ASLP@NPU),
School of Computer Science, Northwestern Polytechnical University, Xi'an, China
lxie@nwpu.edu.cn
[2] Fuxi AI Lab, NetEase Inc., Hangzhou, China

Abstract. Accent transfer aims to transfer an accent from a source speaker to synthetic speech in the target speaker's voice. The main challenge is how to effectively disentangle speaker timbre and accent which are entangled in speech. This paper presents a VITS-based [7] end-to-end accent transfer model named *Accent-VITS*. Based on the main structure of VITS, Accent-VITS makes substantial improvements to enable effective and stable accent transfer. We leverage a hierarchical CVAE structure to model accent pronunciation information and acoustic features, respectively, using bottleneck features and mel spectrums as constraints. Moreover, the text-to-wave mapping in VITS is decomposed into text-to-accent and accent-to-wave mappings in Accent-VITS. In this way, the disentanglement of accent and speaker timbre becomes be more stable and effective. Experiments on multi-accent and Mandarin datasets show that Accent-VITS achieves higher speaker similarity, accent similarity and speech naturalness as compared with a strong baseline (Demos: https://anonymous-accentvits.github.io/AccentVITS/).

Keywords: Text to speech · Accent transfer · Variational autoencoder · Hierarchical

1 Introduction

In recent years, there have been significant advancements in neural text-to-speech (TTS), which can generate human-like natural speech from input text. Accented speech is highly desired for a better user experience in many TTS applications. Cross-speaker accent transfer is a promising technology for accented speech synthesis, which aims to transfer an accent from a source speaker to the synthetic speech in the target speaker's voice. Accent transfer can promote cross-region communication and make a TTS system better adapt to diverse language environments and user needs.

An accent is usually reflected in the phoneme pronunciation pattern and prosody variations, both of which are key attributes of the accent rendering [12,14,15]. The segmental and suprasegmental structures may be in distinctive

© The Author(s), under exclusive license to Springer Nature Singapore Pte Ltd. 2024
J. Jia et al. (Eds.): NCMMSC 2023, CCIS 2006, pp. 203–214, 2024.
https://doi.org/10.1007/978-981-97-0601-3_17

pronunciation patterns for different accents and influence the listening perception of speaking accents [8, 25]. The prosody variations of accent are characterized by different pitch, energy, duration, and other prosodic appearance. To build an accent TTS system using accent transfer, the research problem can be treated as how to effectively *disentangle* speaker timbre and accent factors in speech. However, it is difficult to force the system to sufficiently disentangle the accent from the speaker timbre and content in speech since both pronunciation and prosody attributes are featured by local variations at the fine-grained level. And usually, each speaker has only one accent in the training phase which adds to the difficulty of disentangling.

Previous approaches attempting to disentangle accent attributes and speaker timbre are mainly based on Domain Adversarial Training (DAT) [5]. However, when the feature extraction function has a high capacity, DAT poses a weak constraint to the feature extraction function. Therefore, a single classifier with a gradient reversal layer in accent transfer TTS cannot disentangle the accent from the speaker's timbre, as the accent is varied in prosody and pronunciation. Additionally, gradient descent in domain adversarial training can violate the optimizer's asymptotic convergence guarantees, often hindering the transfer performance [1]. Applying DAT in accent transfer tasks, especially when each speaker has only one accent in the training phase, may result in inefficient and unstable feature disentanglement. Furthermore, there is a trade-off between speaker similarity and accent similarity, which means entirely removing speaker timbre hurts performance on preserving accent pronunciation [18].

Bottleneck (BN) features are recently used as an intermediate representation to supervise accent attribute modeling in TTS [24]. The BN feature, extracted from a well-trained neural ASR model, is considered to be noise-robust and speaker-independent [11, 19], which benefits speaker timbre and accent disentanglement. However, in the methods with BN as an intermediate representation [2, 13], models are often trained independently in multiple stages. This can lead to the issue of error accumulation and model mismatch between each stage, resulting in the degradation of synthesized speech quality and accent attributes.

In this paper, we propose an end-to-end accent transfer model, *Accent-VITS*, with a hierarchical conditional variational autoencoder (CVAE) [10] utilizing bottleneck features as a constraint to eliminate speaker timbre from the original signal. Specifically, we leverage the end-to-end speech synthesis framework, VITS [7], as the backbone of our model, since it achieves good audio quality and alleviates the error accumulation caused by the conventional two-stage TTS system consisting of an acoustic model and a vocoder. Based on the VITS structure, an additional CVAE is added to extract an accent-dependent latent distribution from the BN feature. The latent representation contains the accent and linguistic content and is modeled by the accented phoneme sequence input. The BN constraint factorizes the cross-speaker accent TTS into two joint-training processes, which are text-to-accent and accent-to-wave. The *text-to-accent* process takes the accented phoneme sequence as input to generate an accent-dependent distribution. The *accent-to-wave* process produces the speech distribution in the target accent and target speaker from the output accent distribution and is conditioned on speaker identity. This design enables more effective learning of accent attributes, leading to sufficient disentanglement and superior performance of accent transfer for the synthesized speech. Experimental results on Mandarin multi-accent datasets demonstrate the superiority of our proposed model.

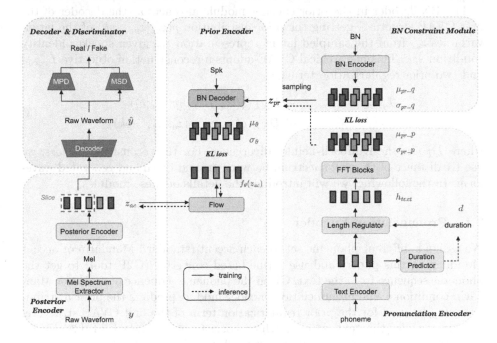

Fig. 1. Overview of Accent-VITS structure.

2 Method

This paper proposes a VITS-based end-to-end accent transfer model with a hierarchical conditional variational autoencoder (CVAE). As shown in Fig. 1, it mainly includes five parts: a posterior encoder, a decoder, a prior encoder, a pronunciation encoder, and a BN constraint module. The posterior encoder extracts the latent representation z_{ac} of acoustic feature from the waveform y, and the decoder reconstructs the waveform $\hat{y}$ according to z_{ac}:

$$z_{ac} = \text{PostEnc}(y) \sim q(z_{ac}|y) \tag{1}$$

$$\hat{y} = \text{Dec}(z_{ac}) \sim p(y|z_{ac}) \tag{2}$$

The prior encoder produces a prior distribution of z_{ac}. Since the prior encoder of CVAE in VITS cannot effectively disentangle accent from the speaker timbre and text content, we use a hierarchical CVAE structure to model accent information and acoustic features sequentially. We take the bottleneck feature (BN) that is extracted from the source wav by an ASR system as a constraint to improve the pronunciation information of accent in the latent space. The BN Encoder extracts the latent representation z_{pr} which contains accent pronunciation information from BN. The pronunciation encoder gets a prior distribution $p(z_{pr}|c)$ of the latent variables z_{pr} given accented phoneme sequence condition c. The BN decoder in the prior encoder module also acts as the decoder of the first CVAE structure, getting the prior distribution $p(z_{ac}|z_{pr}, spk)$ of the latent variables z_{ac} from the sampled latent representation z_{pr} given speaker identity condition spk. This hierarchical CVAE adopts a reconstruction objective L_{recon} and two prior regularization terms as

$$L_{cvae} = \alpha L_{recon} + D_{KL}(q(z_{pr}|BN)||p(z_{pr}|c)) \\ + D_{KL}(q(z_{ac}|y)||p(z_{ac}|z_{pr}, spk)) \tag{3}$$

where D_{KL} is the Kullback-Leibler divergence. For the reconstruction loss, we use L1 distance of the mel-spectrum between ground truth and generated waveform. In the following, we will introduce the details of these modules.

2.1   Pronunciation Encoder

We assign a different phoneme set to each accent (standard Mandarin or accent Mandarin in this paper) and use a rule-based converter (G2P tool) to get the phoneme sequence from the text. Given the phoneme sequence of accent or Mandarin condition c, the pronunciation encoder module predicts the prior distribution $p(z_{pr}|c)$ used for the prior regularization term of the first CVAE structure. In this module, the text encoder which consists of multiple FFT [20] blocks takes phoneme sequences as input and produces phoneme-level representation. Different from VITS, we use the length regulator (LR) in FastSpeech [16] to extend the phoneme-level representation to frame-level representation h_{text} [23]. The other multiple FFT blocks are used to extract a sequence of hidden vectors from the frame-level representation h_{text} and then generate the mean μ_{pr_p} and variance σ_{pr_p} of the prior normal distribution of latent variable z_{pr} by a linear projection.

$$p(z_{pr}|c) = N(z_{pr}; \mu_{pr\_p}(c), \sigma_{pr\_p}(c)) \tag{4}$$

2.2   BN Constraint Module

In this module, the BN encoder extracts the latent representation of pronunciation information z_{pr} from the BN feature and produces the posterior normal distribution $q(z_{pr}|BN)$ with the mean μ_{pr_q} and variance σ_{pr_q}. BN feature is usually the feature map of a neural network layer. Specifically, the BN adopted in this paper is the output of an ASR encoder, which is generally considered to contain only linguistic and prosodic information such as pronunciation, intonation, accent, and very limited speaker information [11]. The ASR model is usually trained with a large multi-speaker multi-condition dataset, and the BN feature extracted by it is also believed to be noise-robust and speaker-independent. The BN encoder consists of multiple layers of Conv1d, ReLU activation, Layer Normalization, Dropout, and a layer of linear projection to produce the mean and variance.

2.3 Prior Encoder

The BN decoder in the prior encoder module also acts as the decoder of the first CVAE structure. Given the speaker identity condition spk, the BN decoder extracts the latent representation of acoustic feature from sampled z_{pr} and generates the prior normal distribution with mean μ_θ and variance σ_θ of z_{ac}. Following VITS, a normalizing flow [4,17] f_θ is added to the prior encoder to improve the expressiveness of the prior distribution of the latent variable z_{ac}.

$$p(f_\theta(z_{ac})|z_{pr}, spk) = N(f_\theta(z_{ac}); \mu_\theta(z_{pr}, spk), \sigma_\theta(z_{pr}, spk)) \tag{5}$$

$$p(z_{ac}|z_{pr}, spk) = p(f_\theta(z_{ac})|z_{pr}, spk) \left| \det \frac{\partial f_\theta(z_{ac})}{\partial z_{ac}} \right| \tag{6}$$

2.4 Posterior Encoder

The posterior encoder module extracts the latent representation z_{ac} from the waveform y. The mel spectrum extractor in it is a fixed signal processing layer without updatable weights. The encoder firstly extracts the mel spectrum from the raw waveform through the signal processing layer. Unlike VITS, the posterior encoder takes the mel spectrum as input instead of the linear spectrum. We use multiple layers of Conv1d, ReLU activation, Layer Normalization, and Dropout to extract a sequence of hidden vector and then produces the mean and variance of the posterior distribution $q(z_{ac}|y)$ by a Conv1d layer. Then we can get the latent z_{ac} sampled from $q(z_{ac}|y)$ using the reparametrization trick.

2.5 Decoder

The decoder generates audio waveforms from the intermediate representation z_{ac}. We use HiFi-GAN generator G [9] as the decoder. For more efficient training, we only feed the sliced z_{ac} instead of the entire length into the decoder to generate the corresponding audio segment. We also use GAN-based [6] training to improve the quality of the synthesized speech. The discriminator D follows HiFiGAN's Multi-Period Discriminator (MPD) and Multi-Scale Discriminator (MSD) [9]. Specifically, the GAN losses for the generator G and discriminator D are defined as:

$$L_{adv}(G) = E_{(z_{ac})} \left[(D(G(z_{ac})) - 1)^2 \right] \tag{7}$$

$$L_{adv}(D) = E_{(y, z_{ac})} \left[(D(y) - 1)^2 + (D(G(z_{ac})))^2 \right] \tag{8}$$

Table 1. Experimental results in terms of subjective mean opinion score (MOS) with confidence intervals of 95% and two objective metrics. Note: the results of VITS-DAT are missing as the system cannot converge properly during training.

Accent	Model	Subjective Evaluation			Objective Evaluation	
		SMOS↑	NMOS↑	AMOS↑	Speaker cosine Similarity↑	Duration MAE↓
Shanghai	T2B2M	3.85 ± 0.03	3.68 ± 0.06	3.73 ± 0.05	0.76	3.51
	Accent-VITS	**3.86 ± 0.02**	**3.76 ± 0.02**	**3.76 ± 0.06**	**0.83**	**3.06**
Henan	T2B2M	3.79 ± 0.07	3.75 ± 0.03	3.59 ± 0.02	0.78	3.65
	Accent-VITS	**3.87 ± 0.02**	**3.92 ± 0.04**	**3.62 ± 0.04**	**0.81**	**3.21**
Dongbei	T2B2M	3.88 ± 0.05	3.87 ± 0.03	3.50 ± 0.03	0.79	3.49
	Accent-VITS	**4.01 ± 0.02**	**4.14 ± 0.05**	**3.88 ± 0.02**	**0.87**	**3.01**
Sichuan	T2B2M	3.94 ± 0.06	**3.82 ± 0.04**	3.69 ± 0.06	**0.84**	3.55
	Accent-VITS	**4.06 ± 0.04**	3.80 ± 0.02	**3.81 ± 0.06**	**0.84**	**3.14**
Average	T2B2M	3.87 ± 0.05	3.78 ± 0.04	3.63 ± 0.02	0.79	3.55
	Accent-VITS	**3.95 ± 0.04**	**3.91 ± 0.03**	**3.77 ± 0.06**	**0.84**	**3.11**

2.6   Duration Predictor

In the training process, the LR module expands the phoneme-level representation using the ground truth duration, denoted as d. In the inference process, the LR module expands the representation using the predicted duration, denoted as $\hat{d}$, obtained from the duration predictor. Unlike VITS, we utilize a duration predictor consisting of multiple layers of Conv1d, ReLU activation, Layer Normalization, and Dropout instead of a stochastic duration predictor due to the significant correlation between accent-specific pronunciation prosody and duration information [23]. For the duration loss L_{dur}, we use MSE loss between $\hat{d}$ and d.

2.7   Final Loss

With the above hierarchical CVAE and adversarial training, we optimize our proposed model with the full objective:

$$L = L_{adv}(G) + L_{fm}(G) + L_{cvae} + \lambda L_{dur} \tag{9}$$

$$L(D) = L_{adv}(D) \tag{10}$$

where $L_{adv}(G)$ and $L_{adv}(D)$ are the GAN loss of G and D respectively, and feature matching loss L_{fm} is added to improve the stability of the training. The L_{cvae} consists of the reconstruction loss and two KL losses.

3 Experiments

3.1 Datasets

The experimental data consists of high-quality standard Mandarin speech data and accent Mandarin speech data from four different regions: Sichuan, Dongbei (Northeast China), Henan, and Shanghai. Specifically, we use DB1[1] as the high-quality standard Mandarin data which contains 10,000 utterances recorded in a studio from a professional female anchor. The total duration is approximately 10.3 h. The accent data from the four regions were also recorded in a recording studio by speakers from these regions. Among them, Sichuan, Dongbei, and Shanghai each have two speakers, one male and one female respectively, while Henan has only one female speaker. In detail, Sichuan, Dongbei, Henan, and Shanghai accent data have 2794, 3947, 2049, and 4000 utterances, respectively. The duration of the accent data is approximately 13.7 h. So we have a total of 4 accents and 8 speakers in our training data.

All the audio recordings are downsampled to 16kHz. We utilize 80-dim mel-spectrograms with 50 ms frame length and 12.5 ms frame shift. Our ASR model is based on the WeNet U2++ model [21] trained on 10,000 h of data from the WenetSpeech corpus [22]. We use the Conformer-based encoder output as our BN feature with 512-dim. The BN feature is further interpolated to match the sequence length of the mel-spectrogram.

Table 2. Results of ablation studies.

Accent	Model	Subjective Evaluation			Objective Evaluation	
		SMOS↑	NMOS↑	AMOS↑	Speaker cosine Similarity↑	Duration MAE↓
Average	Accent-VITS	**3.95 ± 0.04**	**3.91 ± 0.03**	**3.77 ± 0.06**	**0.84**	**3.11**
	-BN encoder	3.82 ± 0.04	3.72 ± 0.01	3.61 ± 0.02	0.77	3.41
	-BN decoder	3.89 ± 0.06	3.86 ± 0.03	3.66 ± 0.03	0.82	3.13
	-BN (enc, dec)	3.05 ± 0.07	3.24 ± 0.04	2.93 ± 0.07	0.69	4.03

3.2 Model Configuration

We implemented the following three models for comparison.

- Text2BN2Mel (T2B2M) [2,24]: a three-stage accent transfer system composed of independently trained models for Text2BN, BN2Mel, and neural Vocoder. The Text2BN model predicts BN feature that contains accent pronunciation and content information from the input text. The BN2Mel model predicts mel-spectrogram based on the input BN feature and speaker identity. HiFi-GAN V1 is used as the vocoder.

---

[1] https://www.data-baker.com/open_source.html.

- VITS-DAT: an accent transfer model based on VITS and DAT. For disentangling accent and speaker timbre information, we add a DAT module composed of a gradient reversal layer and a speaker classifier to the output of the text encoder in VITS. Note that we also use a non-stochastic duration predictor and LR module instead of a stochastic duration predictor and MAS [7] in this model.
- Accent-VITS: the proposed accent transfer model in this paper.

In the text frontend processing, we assigned a different phoneme set for each accent or standard Mandarin. The ground truth phoneme-level duration of all datasets is extracted by force-alignment tools. The above comparison models are trained for 400k steps. The batch size of all the models is 24. The initial learning rate of all the models is 2e−4. The Adam optimizer with $\beta1 = 0.8$, $\beta2 = 0.99$ and $\epsilon = 10^{-9}$ is used to train all them.

3.3 Subjective Evaluation

The TTS test set consists of both short and long total of 30 sentences for each accent without overlap with the training set. Each test sentence was synthesized by combining all the speakers in the training data separately. The VITS-DAT system can not converge due to the instability of the combination of variational inference and DAT. Therefore we do not evaluate the results of VITS-DAT here. We randomly selected 20 synthetic utterances for each accent, resulting in a total of 80 utterances for subjective listening. We asked fifteen listeners for each accent to assess speaker similarity (SMOS) and speech naturalness (NMOS). There are twenty local accent listeners to evaluate the accent similarity (AMOS), with five listeners for each accent. Particularly for the SMOS test, we use target speakers' real recordings as reference. The results are summarized in Table 1.

Speaker Similarity. The results shown in Table 1 indicate that Accent-VITS can achieve the best performance in speaker similarity. Among the transfer results of all four accents, the SMOS score of Accent-VITS is better than that of T2B2M. This shows that our end-to-end model Accent-VITS effectively avoids the error accumulation and mismatch problems in the multi-stage model so that it can synthesize speech with more realistic target speaker timbre.

Speech Naturalness. In the NMOS test, we ask the listeners to pay more attention to the general prosody such as rhythm and expressiveness of the audio. From Table 1 we can see that the NMOS score of Accent-VITS is very close to T2B2M on the Sichuan accent and outperforms T2B2M on the other three accents. This indicates that the speech synthesized by the end-to-end model Accent-VITS is more natural than T2B2M on average.

Accent Similarity. In the AMOS test, we ask accent listeners to assess the similarity between synthesized speech and target accent, ignoring the naturalness of general prosody. The results in Table 1 show that Accent-VITS achieves a higher AMOS score than T2B2M, which indicates that Accent-VITS can model accent attribute information better than T2B2M thanks to its hierarchical modeling of accent information and acoustic features.

3.4   Objective Evaluation

Objective metrics, including speaker cosine similarity and duration mean absolute error (Duration MAE), are also calculated.

Speaker Cosine Similarity. We calculate the cosine similarity on the generated samples to further verify the speaker similarity. Specifically, we train an ECAPA-TDNN model [3] using 6000 h of speech from 18083 speakers to extract x-vectors. The cosine similarity to the target speaker audio is measured on all synthetic utterances. The results are also shown in Table 1. The speaker cosine similarity score of Accent-VITS is also higher than that of T2B2M on three accents except for the Sichuan accent. Compared with the T2B2M, Accent-VITS gets higher scores of speaker cosine similarity in Shanghai, Henan, and Dongbei accents. And in the Sichuan accent, both are equal. This further demonstrates that Accent-VITS is better than T2B2M in modeling the target speaker timbre.

Duration MAE. Prosody variations are key attributes of accent rendering, which is largely reflected in the perceived duration of pronunciation units. Therefore, we further calculate the duration mean absolute error between the predicted duration results of different models and the ground truth. The results in Table 1 show that Accent-VITS gets lower Duration MAE scores than T2B2M in transfer results of all four accents, which means that the transfer results of Accent-VITS are closer to the target accent in prosody than the transfer results of T2B2M.

3.5   Ablation Study

To investigate the importance of our proposed methods in Accent-VITS, three ablation systems were obtained by dropping the BN encoder and BN decoder respectively, and dropping both of them simultaneously, referred to as *-BN encoder*, *-BN decoder*, and *-BN (enc, dec)*. When dropping the BN encoder alone, the FFT blocks module directly predicts BN as an intermediate representation. We use MSE loss between the predicted BN and the ground truth BN as the constraint. The BN decoder module takes BN as input. When dropping the BN decoder alone, the distribution of z_{pr} directly as the prior distribution of z_{ac}. The flow module takes the sampled z_{pr} as input. When dropping both of them simultaneously, the FFT blocks module predicts the distribution of BN as the prior distribution of z_{ac}. We use MSE loss between the sampled BN and the ground truth BN as the constraint.

The results of ablation studies are shown in Table 2. As can be seen, dropping these methods brings performance degradation in terms of subjective evaluation and objective evaluation. Especially dropping both of them simultaneously leads to significantly performance degradation. This validates the effectiveness and importance of the hierarchical CVAE modeling structure in our proposed model.

4   Conclusions

In this paper, we propose Accent-VITS, a VITS-based end-to-end model with a hierarchical CVAE structure for accent transfer. The hierarchical CVAE respectively models accent pronunciation information with the constraint of BN and acoustic features with the constraint of mel-spectrum. Experiments on professional Mandarin data and accent data show that Accent-VITS significantly outperforms the Text2BN2Mel+Neural-Vocoder three-stage approach and the VITS-DAT approach.

References

1. Acuna, D., Law, M.T., Zhang, G., Fidler, S.: Domain adversarial training: a game perspective. In: The Tenth International Conference on Learning Representations, ICLR 2022, Virtual Event, 25–29 April 2022. OpenReview.net (2022)
2. Dai, D., et al.: Cloning one's voice using very limited data in the wild. In: IEEE International Conference on Acoustics, Speech and Signal Processing, ICASSP 2022, Virtual and Singapore, 23–27 May 2022, pp. 8322–8326. IEEE (2022)
3. Desplanques, B., Thienpondt, J., Demuynck, K.: ECAPA-TDNN: emphasized channel attention, propagation and aggregation in TDNN based speaker verification. In: Meng, H., Xu, B., Zheng, T.F. (eds.) Interspeech 2020, 21st Annual Conference of the International Speech Communication Association, Virtual Event, Shanghai, China, 25–29 October 2020, pp. 3830–3834. ISCA (2020)
4. Dinh, L., Sohl-Dickstein, J., Bengio, S.: Density estimation using real NVP. In: 5th International Conference on Learning Representations, ICLR 2017, Toulon, France, 24–26 April 2017, Conference Track Proceedings. OpenReview.net (2017)
5. Ganin, Y., et al.: Domain-adversarial training of neural networks. J. Mach. Learn. Res. **17**, 59:1–59:35 (2016)
6. Goodfellow, I.J., et al.: Generative adversarial nets. In: Ghahramani, Z., Welling, M., Cortes, C., Lawrence, N.D., Weinberger, K.Q. (eds.) Advances in Neural Information Processing Systems 27: Annual Conference on Neural Information Processing Systems 2014, Montreal, Quebec, Canada, 8–13 December 2014, pp. 2672–2680 (2014)
7. Kim, J., Kong, J., Son, J.: Conditional variational autoencoder with adversarial learning for end-to-end text-to-speech. In: Meila, M., Zhang, T. (eds.) Proceedings of the 38th International Conference on Machine Learning, ICML 2021, 18–24 July 2021, Virtual Event. Proceedings of Machine Learning Research, vol. 139, pp. 5530–5540. PMLR (2021)
8. Kolluru, B., Wan, V., Latorre, J., Yanagisawa, K., Gales, M.J.F.: Generating multiple-accent pronunciations for TTS using joint sequence model interpolation. In: Li, H., Meng, H.M., Ma, B., Chng, E., Xie, L. (eds.) INTERSPEECH 2014, 15th Annual Conference of the International Speech Communication Association, Singapore, 14–18 September 2014, pp. 1273–1277. ISCA (2014)

9. Kong, J., Kim, J., Bae, J.: HiFi-GAN: generative adversarial networks for efficient and high fidelity speech synthesis. In: Larochelle, H., Ranzato, M., Hadsell, R., Balcan, M., Lin, H. (eds.) Advances in Neural Information Processing Systems 33: Annual Conference on Neural Information Processing Systems 2020, NeurIPS 2020, 6–12 December 2020, virtual (2020)

10. Lee, S., Kim, S., Lee, J., Song, E., Hwang, M., Lee, S.: HierSpeech: bridging the gap between text and speech by hierarchical variational inference using self-supervised representations for speech synthesis. In: NeurIPS (2022)

11. Li, J., Deng, L., Gong, Y., Haeb-Umbach, R.: An overview of noise-robust automatic speech recognition. IEEE ACM Trans. Audio Speech Lang. Process. **22**(4), 745–777 (2014). https://doi.org/10.1109/TASLP.2014.2304637

12. Liu, R., Sisman, B., Gao, G., Li, H.: Controllable accented text-to-speech synthesis. CoRR abs/2209.10804 (2022). https://doi.org/10.48550/arXiv.2209.10804

13. Liu, S., Yang, S., Su, D., Yu, D.: Referee: towards reference-free cross-speaker style transfer with low-quality data for expressive speech synthesis. In: IEEE International Conference on Acoustics, Speech and Signal Processing, ICASSP 2022, Virtual and Singapore, 23–27 May 2022, pp. 6307–6311. IEEE (2022)

14. Loots, L., Niesler, T.: Automatic conversion between pronunciations of different English accents. Speech Commun. **53**(1), 75–84 (2011)

15. de Mareüil, P.B., Vieru-Dimulescu, B.: The contribution of prosody to the perception of foreign accent. Phonetica **63**(4), 247–267 (2006)

16. Ren, Y., et al.: FastSpeech: fast, robust and controllable text to speech. In: Wallach, H.M., Larochelle, H., Beygelzimer, A., d'Alché-Buc, F., Fox, E.B., Garnett, R. (eds.) Advances in Neural Information Processing Systems 32: Annual Conference on Neural Information Processing Systems 2019, NeurIPS 2019, Vancouver, BC, Canada, 8–14 December 2019, pp. 3165–3174 (2019)

17. Rezende, D.J., Mohamed, S.: Variational inference with normalizing flows. In: Bach, F.R., Blei, D.M. (eds.) Proceedings of the 32nd International Conference on Machine Learning, ICML 2015, Lille, France, 6–11 July 2015. JMLR Workshop and Conference Proceedings, vol. 37, pp. 1530–1538. JMLR.org (2015)

18. Shu, R., Bui, H.H., Narui, H., Ermon, S.: A DIRT-T approach to unsupervised domain adaptation. In: 6th International Conference on Learning Representations, ICLR 2018, Vancouver, BC, Canada, 30 April–3 May 2018, Conference Track Proceedings. OpenReview.net (2018)

19. Sun, L., Li, K., Wang, H., Kang, S., Meng, H.M.: Phonetic posteriorgrams for many-to-one voice conversion without parallel data training. In: IEEE International Conference on Multimedia and Expo, ICME 2016, Seattle, WA, USA, 11–15 July 2016, pp. 1–6. IEEE Computer Society (2016)

20. Vaswani, A., et al.: Attention is all you need. In: Guyon, I., et al. (eds.) Advances in Neural Information Processing Systems 30: Annual Conference on Neural Information Processing Systems 2017, Long Beach, CA, USA, 4–9 December 2017, pp. 5998–6008 (2017)

21. Yao, Z., et al.: WeNet: production oriented streaming and non-streaming end-to-end speech recognition toolkit. In: Hermansky, H., Cernocký, H., Burget, L., Lamel, L., Scharenborg, O., Motlícek, P. (eds.) Interspeech 2021, 22nd Annual Conference of the International Speech Communication Association, Brno, Czechia, 30 August–3 September 2021, pp. 4054–4058. ISCA (2021)

22. Zhang, B., et al.: WENETSPEECH: a 10000+ hours multi-domain mandarin corpus for speech recognition. In: IEEE International Conference on Acoustics, Speech and Signal Processing, ICASSP 2022, Virtual and Singapore, 23–27 May 2022, pp. 6182–6186. IEEE (2022)

23. Zhang, Y., Cong, J., Xue, H., Xie, L., Zhu, P., Bi, M.: Visinger: variational inference with adversarial learning for end-to-end singing voice synthesis. In: IEEE International Conference on Acoustics, Speech and Signal Processing, ICASSP 2022, Virtual and Singapore, 23–27 May 2022, pp. 7237–7241. IEEE (2022)

24. Zhang, Y., Wang, Z., Yang, P., Sun, H., Wang, Z., Xie, L.: AccentSpeech: learning accent from crowd-sourced data for target speaker TTS with accents. In: Lee, K.A., Lee, H., Lu, Y., Dong, M. (eds.) 13th International Symposium on Chinese Spoken Language Processing, ISCSLP 2022, Singapore, 11–14 December 2022, pp. 76–80. IEEE (2022)

25. Zhou, X., Zhang, M., Zhou, Y., Wu, Z., Li, H.: Accented text-to-speech synthesis with limited data. CoRR abs/2305.04816 (2023). https://doi.org/10.48550/arXiv.2305.04816

Multi-branch Network with Cross-Domain Feature Fusion for Anomalous Sound Detection

Wenjie Fang[1,2] (ID), Xin Fan[1,2] (ID), and Ying Hu[1,2(✉)] (ID)

[1] School of Computer Science and Technology, Xinjiang University, Urumqi, China
{fwj,fx}@stu.xju.edu.cn, huying@xju.edu.cn
[2] Key Laboratory of signal detection and processing in Xinjiang, Urumqi, China

Abstract. Anomalous sound detection (ASD) is a key technology to identify abnormal sounds in various industries. Self-supervised anomalous sound detection aims at detecting unknown machine anomalous sounds by learning the characteristics of the normal sounds using metainformation. In this paper, we propose a multi-branch network with cross-domain feature fusion (MBN-CFF) for self-supervised ASD task. The multi-branch network splits the complete feature representations and feeds them individually into classifiers to generate category predictions. The weighted loss, calculated by multiple predictions and the real labels, guides the model training process. We also design a cross-domain feature fusion (CFF) block for effectively fusing the time-domain and frequency-domain features and an attentive sandglass (AS) block for effectively extracting features. Experimental results on the DCASE2020 challenge task 2 show that our MBN-CFF network achieves the best performance with the AUC score of 94.73% and pAUC score of 88.74%, respectively, compared to the other five existing methods for anomalous sound detection. The results of ablation experiments show the effectiveness of CFF and AS blocks, multi-brach prediction (MBP).

Keywords: Anomalous Sound Detection · Self-supervised Learning · Machine Monitoring · Anomaly Detection

1 Introduction

The objective of Anomalous Sound Detection (ASD) is to determine whether a target object is normal or not based on the sound it generates, particularly in scenarios where only normal samples are available for training. Anomalous sounds often indicate malfunctions or potentially hazardous situations, so detecting anomalous sounds can help identify and remove the possible hazards as early as possible and avoid accidents. With the development of intelligent technology, ASD has been widely used in the fields of audio monitoring in industrial environments [4,20], industrial equipment monitoring [1,29] and others. In reality, the statistical characteristics of abnormal data are difficult to define, and it is

J. Jia et al. (Eds.): NCMMSC 2023, CCIS 2006, pp. 215–226, 2024.
https://doi.org/10.1007/978-981-97-0601-3_18

unfeasible to collect the abnormal data through a large number of destructive tests [26]. Therefore, an unsupervised anomalous sound detection method that learns the acoustic characteristics of normal sounds to distinguish between normal and anomalous sounds is better suited for this task [17].

Traditional reconstruction-based ASD methods have used autoencoder (AE), defining the reconstruction error between the input and reconstructed output of the model as the anomaly score and training the model to minimize the reconstruction error [15,24]. However, these methods may encounter the issue that, given an anomalous sample, the model can also reconstruct the anomalous parts of the spectrogram by the procedure resembling principal component analysis (PCA) [14]. Some generative models model the distribution of normal sounds and then identify the anomaly according to the deviation of test sample distribution from the normal sample one, such as IDNN [28] and Glow_Aff [7].

The abovementioned methods are trained without metadata from the original audio sample, such as the machine IDs and machine types. While on the DCASE 2020 dataset [17], many emerging self-supervised classification-based ASD methods [2,8,9,19,33,34,36] use the auxiliary machine IDs and perform better than AE-based unsupervised methods. These methods learn and make predictions based on the characteristics of 41 machine IDs across 6 machine types in the DCASE2020 dataset. However, the sound samples of different machine IDs within the same machine type have a large degree of distribution overlap, and the distinctions in these machine sounds tend to be relatively minor. These similarities impact the model classification performance and the ultimate detection effectiveness and make the classification task more difficult.

Most ASD methods use the log-mel spectrogram as the input feature of the network. However, some recent studies suggest that relying only on the log-mel spectrogram for ASD may not provide sufficient discrimination between normal and abnormal sounds [19,21,35]. By analyzing the spectrogram, Zeng et al. found that the distinguishable characteristics of certain machine types are primarily concentrated in the high-frequency range [35]. Whereas the log-mel spectrum, designed to emulate human hearing, filters out certain high-frequency components, which might hinder the model from capturing the effective features of machine sound. Therefore, they adopt the spectrogram followed by a high-pass filter. Liu et al. suggested that the log-mel spectrogram may filter out the high-frequency components of abnormal sounds that may have different characteristics, and thus supplemented the log-mel spectrogram with temporal information extracted from the original waveform [19]. Griffi et al. proposed the "high-frequency hypothesis" that anomaly detectors identify the anomalies using high-frequency components [21].

In this paper, we propose a self-supervised multi-branch ASD network with cross-domain feature fusion (MBN-CFF) for ASD task. The main contributions of this paper are as follows:

(1) We proposed a multi-branch network which divides the complete feature representation into the first half and second half along the time dimension, and

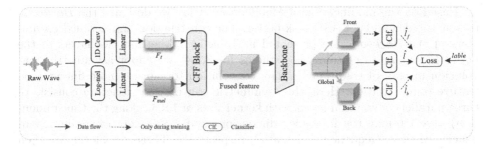

Fig. 1. The overall architecture of our proposed Multi-Branch Network with Cross-domain Feature Fusion (MBN-CFF).

feeds those two halves together with the complete feature representations into three classifiers, respectively.

(2) We design a Cross-domain Feature Fusion (CFF) block, introducing temporal feature compensation for the log-mel spectrogram.

(3) We designed an Attentive Sandglass (AS) block whose two branches are used to obtain detailed features and global dependencies that are valid for capturing short-term variations in non-stationary data samples and long-term trends in stationary data samples, respectively.

2   Proposed Method

In this section, we first introduce the comprehensive architecture of our proposed MBN-CFF network. Subsequently, we provide a detailed explanation of the cross-domain feature fusion (CFF), attentive sandglass (AS) blocks, and the ArcFace loss-based classifier.

2.1   Multi-branch Network Architecture

Inspired by a recent work for the person re-identification task [11], we proposed a multi-branch network which divides the complete feature representation into the first half and second half along the time dimension, and feeds those two halves together with the complete feature representations into three classifiers, respectively.

As depicted in Fig. 1, the raw waveform is first processed in two parallel pathways to extract the time-domain and log-mel frequency-domain features, respectively. These two types of features are subsequently integrated through a CFF block and passed into *Backbone*, whose output feature maps are split into front and back halves, together with the complete one, and fed into three classifiers. During the training phase, the predictions of these three classifiers are all used for calculating the loss, while in the testing phase, only the classifier input with the global feature maps generates the final prediction.

Based on the modification of MobileFaceNet [3], we designed the *Backbone* to suit the demands of ASD task better. The construction and detailed parameters of the *Backbone* are presented in Table 1. In the first two layers of the network, similar to certain prior studies [6,13,23], we adopt asymmetric convolution instead of traditional square convolution to capture more fine-grained feature representation. Here, the asymmetric convolution structure consists of three parallel convolution layers with kernel sizes of 1×3 (along the time dimension), 3×1 (along the frequency dimension) and 3×3, capturing the dependencies of time, frequency and spatial, respectively. Three AS blocks are then followed to capture both the local information and long-distance dependencies while maintaining a low computational load. After an AS block, the feature maps are downsampled 2 times along the time and frequency dimensions. The feature maps outputting from the final AS Block are split along the time dimension into the first (*Front*) and second (*Back*) half feature maps, which are further fed into three GDConv [3] and 1×1 convolutional layers to reduce feature channels and dimensions, together with the complete feature maps.

Table 1. The detailed structure of the *Backbone*. t, c and n represent the expansion factor, the channel number of outputs, and the number of repetitions. Only the first layer of each operator has a stride of s, while all others that of 1.

Operator	t	c	n	s
RHConv	–	64	1	2
RHConv	–	128	1	1
AS Block	2	256	2	2
AS Block	4	512	2	2
AS Block	2	256	2	2
GDConv	–	256	1	1
Conv2D	–	128	1	1

2.2 Cross-Domain Feature Fusion

As previously mentioned, the log-mel spectrogram mainly emphasizes non-high-frequency components, potentially leading to the omission of crucial information for distinguishing normal and abnormal machine sounds. Furthermore, based on empirical evidence, directly adding or concatenating different feature maps is inappropriate. To effectively utilize the high-frequency information and selectively emphasize more crucial regions of feature maps, we design a CFF Block for integrating the cross-domain features. The specific structure is illustrated in Fig. 2.

Given a single-channel normal/abnormal sound sample $x \in \mathbb{R}^{1 \times L}$ with length L. A short-time Fourier transform (STFT) initially processes the raw audio signal. Subsequently, the mel-scale filtering and then logarithmic operations to get the log-mel spectrogram $\boldsymbol{F} \in \mathbb{R}^{M \times N}$. Here, M represents the number of mel bins, and N is the number of time frames.

The time-domain feature $\boldsymbol{F}_t$ is acquired through a 1-D CNN-based network, which has the same structure as [19]. Subsequently, both the log-mel frequency feature $\boldsymbol{F}_{mel}$ and the time-domain feature $\boldsymbol{F}_t$ are further aligned to share the same feature representation dimension through individual linear layers.

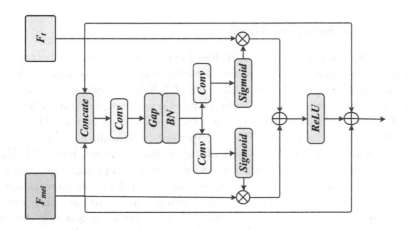

Fig. 2. Structure of Cross-domain Feature Fusion (CFF) Block.

As shown in Fig. 2, the aligned feature maps are firstly concatenated along the channel dimension. The concatenated feature map $\boldsymbol{F}_{cat}$ can be expressed as follows:

$$\boldsymbol{F}_{cat} = Concat\left(Linear(\boldsymbol{F}_{mel}), Linear(\boldsymbol{F}_t)\right) \tag{1}$$

where $Concat(\cdot)$ represents the concatenation operation and $Linear(\cdot)$ the linear operation.

To enhance the information complementarity among the feature maps across different channels and retain the high-frequency components of the spectrogram, the concatenated feature maps, $\boldsymbol{F}_{cat}$, are firstly passed through a convolutional layer with the kernel size of 1×1 followed by a global average pooling (GAP) and batch normalization (BN) operators. The resultant feature maps denoted as $\boldsymbol{F}_c$ can be obtained as follows:

$$\boldsymbol{F}_c = BN(f_{gap}(f_{conv}(\boldsymbol{F}_{cat}))) \tag{2}$$

Here, $f_{conv}(\cdot)$ and $f_{gap}(\cdot)$ represent the convolutional and global average pooling operations, and $BN(\cdot)$ is the batch normalization operation. $\boldsymbol{F}_c$ is fed into two fully connected layers followed by a *sigmoid* activation operation σ, separately, to generate two weights p_i.

$$p_i = \sigma(\omega_i \boldsymbol{F}_c), i = 1, 2 \tag{3}$$

where ω_i represents the weights matrix of a fully connection layer.

For the final output, we employed a residual operation that combined the original features with the weighted channel features. The final output F_{out}, can be obtained using the following formula:

$$F_{out} = \Theta((p_1 \otimes F_t) + (p_2 \otimes F_{mel})) + F_t + F_{mel} \tag{4}$$

where Θ denotes a rectified linear unit (ReLU) activation operation.

2.3 Attentive Sandglass Block

More recently, attention-based work has focused on performing better frame-level predictions for anomalous sound detection using autoencoder models with attention-based mechanisms [10,22,34]. The CNN-baed Inverse Residuals Block (IRB) has been regarded as a lightweight mobile network structure. Zhang et al. took a fresh perspective by reevaluating the combination of the IRB structure in MobileNetV2 [27] and the MHSA component in Transformer [30], aiming to integrate their respective advantages at the structure design level [37]. Hou et al. decomposed channel attention into sensing feature mappings along two directions, capturing long-range dependencies and precise location information [12]. Meanwhile, Wang et al. introduced axial attention to mitigate the computational complexity associated with global self-attention while ignoring the loss of local details caused by pooling or averaging operations [32]. Wan et al. devised a universal attention block for mobile semantic segmentation, characterized by squeezing axial and detail enhancement, which can be employed to construct a more cost-effective backbone architecture [31].

The original inverse residuals block performs expansion at first and then reduction. However, Zhou et al. proposed to flip this structure by introducing the sandglass block, which conducts identity mapping and spatial transformation at higher dimensions and thus alleviates information loss and gradient confusion. It has been verified that such a bottleneck structure is more beneficial than the inverted ones [38].

Based on the analysis mentioned above, we designed an attentive sandglass (AS) block. As depicted in Fig. 3, our proposed AS block consists of two branches: a sandglass branch for extracting the feature maps and an axial attention branch for capturing the attention weights.

The *query*, *key*, and *value* of multi-head attention, Q, K, and V, are derived through linear mapping of the input feature maps $\mathbf{X}$ with reduction. Q, K, and $V \in \mathbb{R}^{\frac{C}{r} \times F \times T}$ can be calculated as:

$$Q, K, V = \mathbf{X}W_q, \mathbf{X}W_k, \mathbf{X}W_v \tag{5}$$

where W_q, W_k, and W_v are learnable parameter matrices. Then, Q, K, and V are all passed through two 1-D pooling operations, $Fgap$ and $Tgap$, conducting along the time and frequency dimensions to obtain Q_f, K_f, $V_f \in \mathbb{R}^{\frac{C}{r} \times F \times 1}$ and Q_t, K_t, $V_t \in \mathbb{R}^{\frac{C}{r} \times 1 \times T}$, respectively. Passed through two Multi-Head Attention

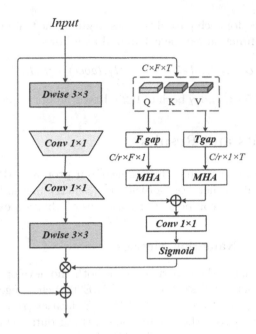

Fig. 3. Illustration of Attentive Sandglass (AS) Block. $Fgap$ and $Tgap$ denote 1-D average pooling operation along the frequency and time dimensions, respectively. $Dwise$ means the depth-wise convolution operation.

(MHA), we obtain two axial attention. Followed by a broadcast summation operation, we can get the spatial attention:

$$A(\mathbf{X}) = softmax(\frac{Q_f \cdot K_f{}^\top}{\sqrt{d}}) \cdot V_f + softmax(\frac{Q_t \cdot K_t{}^\top}{\sqrt{d}}) \cdot V_t, \qquad (6)$$

where $\top$ signifies the transpose operation, and d corresponds to the dimension of the $query$ or key.

Finally, through a 1×1 convolution and $sigmoid$ operation, the spatial coordinate information can be obtained. Ultimately, the detail enhancement features yielded by the sandglass block are multiplied with the spatial attention weights to accentuate the interested representations.

2.4 ArcFace Loss-Based Classifier

The output feature maps of $Backbone$ are split into the front and back halves, which are further passed into three classifiers together with the complete feature maps, respectively. Those three classifiers all adopt the ArcFace loss [5], which focuses on the angle between the features and the weights so as to improve the performance of classification.

The predictions $(\hat{l}_f, \hat{l}, \hat{l}_b)$ for machine IDs obtained from three classifiers are all used to calculate the loss with the true labels, respectively. Different weights

(α, β, γ) are assigned for each part of the loss to guide the training of the network. The loss objective function can be calculated as follows:

$$Loss = \alpha \cdot L_1(\hat{l}, label) + \beta \cdot L_2(\hat{l}_f, label) + \gamma \cdot L_3(\hat{l}_b, label) \qquad (7)$$

Here, we set α to 0.5, β, and γ both to 0.25, based on our experimental findings.

3   Experiments and Results

In this section, we evaluate the performance of our proposed MBN-CFF network by comparing it with other state-of-the-art (SOTA) methods and verify the effectiveness of three key components by a series of ablation experiments.

3.1   Dataset and Evaluation Metrics

Dataset. We conducted the experiments on both the development dataset and the additional dataset for Task 2 of the DCASE2020 challenge [17], which consists of the MIMII dataset and [25] ToyADMOS dataset [18]. The development dataset contains two parts: the training and testing data, while the additional dataset is only used for training. The MIMII dataset consists of four different machine types (Fan, Pump, Slider, and Valve), each of which contains seven machines with unique IDs. The ToyADMOS dataset contains two machine types (ToyCar and ToyConveyor) with seven and six different machine IDs, respectively. The complete datasets contain 41 machine IDs across 6 machine types, and each dataset contains both normal and abnormal sounds from toy/real machines. The abnormal sounds result from the machines that have been intentionally damaged. Each recording is a single-channel audio with a length of approximately 10 s, encompassing the machine's operational sounds and ambient noise. All recordings are sampled at a rate of 16 kHz.

In the experiments, the real normal sounds with machine IDs were used for training, while both the real normal and abnormal sounds for testing. The training set contains the training data from both the development and additional datasets, and the test data from the development dataset is used for evaluation.

Note that we did not use the DCASE 2021 challenge task 2 dataset because it was designed to investigate the domain shifts problem, which is outside the scope of our current study and will be investigated in our later work.

Evaluation Metrics. We evaluate the performance of the ASD network using the area under the receiver operating characteristic (ROC) curve (AUC) and the partial area under the ROC curve (pAUC). In our implementation, pAUC is calculated as the AUC over a low false positive rate range $[0, p]$, where p is set to 0.1.

3.2   Experimental Setting

We adopted the frame length of 1024 points, the hop length of 512, and the number of Mel filter bands of 128. Adam optimizer [16] was used with an initial learning rate of 0.0001 for model optimization, and the cosine annealing scheme was adopted as the learning rate decay strategy. The training process spanned 200 epochs with a batch size of 128. For the self-supervised classification task, we utilized the ArcFace loss with hyperparameters m and s set to 0.7 and 30, respectively.

3.3   Comparison with the State-of-Art Methods

In order to show the performance of the proposed MBN-CFF, we compare our proposed method with the other five state-of-the-art methods with the same data division and experimental setting on the DCASE2020 challenge task 2 dataset.

Table 2. Comparison with State-of-the-Art anomalous sound detection methods in terms of AUC (%) and pAUC (%) for different machine types.

Methods	Fan		Pump		Slider		Valve		ToyCar		ToyConveyor		Average	
	AUC	pAUC	AUC	pAUC	AUC	pAUC	AUC	pAUC	AUC	pAUC	AUC	pAUC	AUC	pAUC
MobileNetV2 [8]	80.19	74.40	82.53	76.50	95.27	85.22	88.65	87.98	87.66	85.92	69.71	56.43	84.34	77.74
Glow_Aff [7]	74.90	65.30	83.40	73.80	94.60	82.80	91.40	75.00	92.20	84.10	71.50	59.00	85.20	73.90
STgram-MFN [19]	94.04	88.97	91.94	81.75	99.55	97.61	99.64	98.44	94.44	87.68	74.57	63.60	92.36	86.34
SW-WaveNet [2]	**97.53**	91.54	87.27	82.68	98.96	94.58	99.01	97.26	95.49	90.20	81.20	68.20	93.25	87.41
CLP-SCF [9]	96.98	**93.23**	94.97	87.39	99.57	97.73	**99.89**	**99.51**	95.85	90.19	75.21	62.79	93.75	88.48
Ours	94.18	88.52	**95.03**	**88.66**	**99.58**	**97.80**	99.39	97.06	**97.01**	**90.89**	**83.18**	**69.52**	**94.73**	**88.74**

As shown in Table 2, our proposed MBN-CFF network achieves the best scores of metrics on four machine types, including Pump, Slider, ToyCar, and ToyConveyor, while only slightly lower than other methods on the Fan and Valve. Our MBN-CFF achieves the best overall performance among all compared models, compared with CLP-SCF [9], having the overall AUC increased by 1.02% and pAUC 0.26%. This proves the superiority of our model.

3.4   Ablation Study

For the task of ASD, we conducted a set of experiments on both the development dataset and additional dataset for Task 2 of the DCASE2020 challenge to verify the contribution of each key component in the proposed MBN-CFF network.

w/o CFF in Table 3 means the MBN-CFF network removes the CFF block, replacing it with the simple concatenation operation, and *w/o AS block* means that it removes the AS block replacing it with the original inverse residual block. *w/o MBP* means that the network only contains one branch where the complete feature representation is used for classification.

Table 3. Ablation results of three key components in our proposed MBN-CFF.

Methods	CFF	AS Block	MBP	AUC	pAUC
w/o CFF	–	✓	✓	94.08	87.92
w/o AS Block	✓	–	✓	93.84	88.30
w/o MBP*	✓	✓	–	93.91	88.11
MBN-CFF	✓	✓	✓	**94.73**	**88.74**

*: The process of multi-branch prediction

As can be seen from Table 3, the scores of AUC and pAUC both decrease to a certain extent when each key component is removed. When removing the CFF and AS blocks, respectively, the scores of AUC and pAUC of the networks decrease by 0.65% and 0.82%, 0.89% and 0.44%. When the multi-branch classifier structure is removed, and only global features are used, that of AUC and pAUC decrease by 0.82% and 0.63%. These indicate that each of the key components in our proposed MBN-CFF network is effective for ASD task.

4   Conclusion

This paper proposes a multi-branch network with cross-domain feature fusion (MBN-CFF) for self-supervised ASD task. We design a cross-domain feature fusion (CFF) block and an attentive sandglass (AS) block. The results of ablation experiments show the effectiveness of CFF and AS blocks, multi-branch prediction (MBP). Compared with five ASD methods, our proposed MBN-CFF network achieves the best performance on the whole.

Acknowledgements. This work is supported by the Multi-lingual Information Technology Research Center of Xinjiang (ZDI145-21).

References

1. Altinors, A., Yol, F., Yaman, O.: A sound based method for fault detection with statistical feature extraction in UAV motors. Appl. Acoust. **183**, 108325 (2021)
2. Chen, H., Ran, L., Sun, X., Cai, C.: SW-WAVENET: learning representation from spectrogram and WaveGram using WaveNet for anomalous sound detection. In: ICASSP 2023 - 2023 IEEE International Conference on Acoustics, Speech and Signal Processing (ICASSP), pp. 1–5 (2023)
3. Chen, S., Liu, Y., Gao, X., Han, Z.: MobileFaceNets: efficient CNNs for accurate real-time face verification on mobile devices. In: Zhou, J., et al. (eds.) CCBR 2018. LNCS, vol. 10996, pp. 428–438. Springer, Cham (2018). https://doi.org/10.1007/978-3-319-97909-0_46
4. Crocco, M., Cristani, M., Trucco, A., Murino, V.: Audio surveillance: a systematic review. ACM Comput. Surv. (CSUR) **48**(4), 1–46 (2016)
5. Deng, J., Guo, J., Xue, N., Zafeiriou, S.: ArcFace: additive angular margin loss for deep face recognition. In: Proceedings of the IEEE/CVF Conference on Computer Vision and Pattern Recognition (CVPR) (2019)

6. Ding, X., Guo, Y., Ding, G., Han, J.: ACNet: strengthening the kernel skeletons for powerful CNN via asymmetric convolution blocks. In: Proceedings of the IEEE/CVF International Conference on Computer Vision (ICCV) (2019)
7. Dohi, K., Endo, T., Purohit, H., Tanabe, R., Kawaguchi, Y.: Flow-based self-supervised density estimation for anomalous sound detection. In: ICASSP 2021 - 2021 IEEE International Conference on Acoustics, Speech and Signal Processing (ICASSP), pp. 336–340 (2021)
8. Giri, R., Tenneti, S.V., Cheng, F., Helwani, K., Isik, U., Krishnaswamy, A.: Self-supervised classification for detecting anomalous sounds. In: Detection and Classification of Acoustic Scenes and Events Workshop 2020 (2020)
9. Guan, J., Xiao, F., Liu, Y., Zhu, Q., Wang, W.: Anomalous sound detection using audio representation with machine id based contrastive learning pretraining. In: ICASSP 2023 - 2023 IEEE International Conference on Acoustics, Speech and Signal Processing (ICASSP), pp. 1–5 (2023)
10. Hayashi, T., Yoshimura, T., Adachi, Y.: Conformer-based id-aware autoencoder for unsupervised anomalous sound detection. DCASE2020 Challenge, Technical report (2020)
11. He, T., Shen, L., Guo, Y., Ding, G., Guo, Z.: SECRET: self-consistent pseudo label refinement for unsupervised domain adaptive person re-identification. In: Proceedings of the AAAI Conference on Artificial Intelligence, vol. 36, no. 1, pp. 879–887 (2022)
12. Hou, Q., Zhou, D., Feng, J.: Coordinate attention for efficient mobile network design. In: Proceedings of the IEEE/CVF Conference on Computer Vision and Pattern Recognition (CVPR), pp. 13713–13722 (2021)
13. Hu, Y., Zhu, X., Li, Y., Huang, H., He, L.: A multi-grained based attention network for semi-supervised sound event detection. arXiv preprint arXiv:2206.10175 (2022)
14. Jiang, A., Zhang, W.Q., Deng, Y., Fan, P., Liu, J.: Unsupervised anomaly detection and localization of machine audio: a GAN-based approach. In: ICASSP 2023 - 2023 IEEE International Conference on Acoustics, Speech and Signal Processing (ICASSP), pp. 1–5 (2023). https://doi.org/10.1109/ICASSP49357.2023.10096813
15. Kapka, S.: ID-conditioned auto-encoder for unsupervised anomaly detection. arXiv preprint arXiv:2007.05314 (2020)
16. Kingma, D.P., Ba, J.: Adam: a method for stochastic optimization. arXiv preprint arXiv:1412.6980 (2014)
17. Koizumi, Y., et al.: Description and discussion on DCASE2020 challenge task2: unsupervised anomalous sound detection for machine condition monitoring. arXiv preprint arXiv:2006.05822 (2020)
18. Koizumi, Y., Saito, S., Uematsu, H., Harada, N., Imoto, K.: ToyADMOS: a dataset of miniature-machine operating sounds for anomalous sound detection. In: 2019 IEEE Workshop on Applications of Signal Processing to Audio and Acoustics (WASPAA), pp. 313–317 (2019)
19. Liu, Y., Guan, J., Zhu, Q., Wang, W.: Anomalous sound detection using spectral-temporal information fusion. In: ICASSP 2022 - 2022 IEEE International Conference on Acoustics, Speech and Signal Processing (ICASSP), pp. 816–820 (2022)
20. Lojka, M., Pleva, M., Kiktová, E., Juhár, J., Čižmár, A.: Efficient acoustic detector of gunshots and glass breaking. Multimed. Tools Appl. 75, 10441–10469 (2016)
21. Mai, K.T., Davies, T., Griffin, L.D., Benetos, E.: Explaining the decision of anomalous sound detectors. In: Proceedings of the 7th Detection and Classification of Acoustic Scenes and Events 2022 Workshop (DCASE2022), Nancy, France (2022)

22. Mori, H., Tamura, S., Hayamizu, S.: Anomalous sound detection based on attention mechanism. In: 2021 29th European Signal Processing Conference (EUSIPCO), pp. 581–585 (2021)

23. Peng, C., Zhang, X., Yu, G., Luo, G., Sun, J.: Large kernel matters - improve semantic segmentation by global convolutional network. In: Proceedings of the IEEE Conference on Computer Vision and Pattern Recognition (CVPR) (2017)

24. Principi, E., Vesperini, F., Squartini, S., Piazza, F.: Acoustic novelty detection with adversarial autoencoders. In: 2017 International Joint Conference on Neural Networks (IJCNN), pp. 3324–3330 (2017)

25. Purohit, H., et al.: MIMII dataset: sound dataset for malfunctioning industrial machine investigation and inspection. arXiv preprint arXiv:1909.09347 (2019)

26. Ruff, L., et al.: A unifying review of deep and shallow anomaly detection. Proc. IEEE **109**(5), 756–795 (2021)

27. Sandler, M., Howard, A., Zhu, M., Zhmoginov, A., Chen, L.C.: MobileNetv 2: inverted residuals and linear bottlenecks. In: Proceedings of the IEEE Conference on Computer Vision and Pattern Recognition (CVPR) (2018)

28. Suefusa, K., Nishida, T., Purohit, H., Tanabe, R., Endo, T., Kawaguchi, Y.: Anomalous sound detection based on interpolation deep neural network. In: ICASSP 2020 - 2020 IEEE International Conference on Acoustics, Speech and Signal Processing (ICASSP), pp. 271–275 (2020)

29. Suman, A., Kumar, C., Suman, P.: Early detection of mechanical malfunctions in vehicles using sound signal processing. Appl. Acoust. **188**, 108578 (2022)

30. Vaswani, A., et al.: Attention is all you need. In: Guyon, I., et al. (eds.) Advances in Neural Information Processing Systems, vol. 30. Curran Associates, Inc. (2017)

31. Wan, Q., Huang, Z., Lu, J., Yu, G., Zhang, L.: SeaFormer: Squeeze-enhanced axial transformer for mobile semantic segmentation. arXiv preprint arXiv:2301.13156 (2023)

32. Wang, H., Zhu, Y., Green, B., Adam, H., Yuille, A., Chen, L.-C.: Axial-DeepLab: stand-alone axial-attention for panoptic segmentation. In: Vedaldi, A., Bischof, H., Brox, T., Frahm, J.-M. (eds.) ECCV 2020. LNCS, vol. 12349, pp. 108–126. Springer, Cham (2020). https://doi.org/10.1007/978-3-030-58548-8_7

33. Wu, J., Yang, F., Hu, W.: Unsupervised anomalous sound detection for industrial monitoring based on ArcFace classifier and gaussian mixture model. Appl. Acoust. **203**, 109188 (2023)

34. Zeng, X.M., et al.: Joint generative-contrastive representation learning for anomalous sound detection. In: ICASSP 2023 - 2023 IEEE International Conference on Acoustics, Speech and Signal Processing (ICASSP), pp. 1–5 (2023)

35. Zeng, Y., Liu, H., Xu, L., Zhou, Y., Gan, L.: Robust anomaly sound detection framework for machine condition monitoring. Technical report, DCASE2022 Challenge (2022)

36. Zhang, H., Guan, J., Zhu, Q., Xiao, F., Liu, Y.: Anomalous sound detection using self-attention-based frequency pattern analysis of machine sounds. arXiv preprint arXiv:2308.14063 (2023)

37. Zhang, J., et al.: Rethinking mobile block for efficient attention-based models. In: Proceedings of the IEEE/CVF International Conference on Computer Vision (ICCV), pp. 1389–1400 (2023)

38. Zhou, D., Hou, Q., Chen, Y., Feng, J., Yan, S.: Rethinking bottleneck structure for efficient mobile network design. In: Vedaldi, A., Bischof, H., Brox, T., Frahm, J.-M. (eds.) ECCV 2020. LNCS, vol. 12348, pp. 680–697. Springer, Cham (2020). https://doi.org/10.1007/978-3-030-58580-8_40

A Packet Loss Concealment Method Based on the Demucs Network Structure

Wenwen Li and Changchun Bao[✉]

Institute of Speech and Audio Information Processing, Faculty of Information Technology,
Beijing University of Technology, Beijing 100124, China
Liwnwn@emails.bjut.edu.cn, baochch@bjut.edu.cn

Abstract. Under the constrained real-time condition of network, packet loss often occurs in voice communications like voice over internet protocol (VoIP). Packet loss concealment (PLC) techniques often use the previous packets to recover the lost packet for improving the quality of speech communication. In this paper, a novel deep PLC approach is proposed, which uses a called Demucs network structure, i.e., a deep U-Net architecture with a long-short time memory (LSTM) network, to predict the lost packet in the time domain. Firstly, by combing the convolutions with gated linear unit (GLU), the encoder of network can systematically extract the high-level feature of each speech frame. Secondly, the LSTM layers are used to learn the long-term dependencies of speech frames. Finally, the U-Net architecture of the network is used to improve the gradient of information flow by using skip connections, which enhances the decoder's ability of reconstructing the lost speech frames. Additionally, the proposed architecture is optimized by utilizing multiple loss functions in the time and frequency domains. The experimental results show that the proposed method has better performance in perceptual evaluation of speech quality (PESQ) and short-term objective intelligibility (STOI).

Keywords: Packet loss concealment · LSTM · U-Net · Gated linear unit · Multiple loss functions

1   Introduction

In this interconnected world, voice communication plays an important role in real-time human interaction. Therefore, it is necessary to maintain the high-quality of voice communication. However, the packet loss is inevitable due to network congestion or latency in voice communication [1]. In addition, the packet loss causes the interrupt of speech communication, which degrades the overall communication experience. The process of recovering the lost packet is called as packet loss concealment [2]. By utilizing the correlation of the speech packets, the PLC techniques can estimate the lost packets that are consistent with the contents of the speech context.

Foundation item: the National Natural Science Foundation of China (No. 61831019).

A variety of traditional techniques have been proposed in attempting to address the issue of packet loss, for example, linear predictive (LP)-based algorithm and hidden Markov models (HMM)-based PLC algorithm [3]. However, these conventional PLC methods may tend to produce annoying artifact, especially consecutive packet loss. Additionally, these methods may obtain unsatisfactory performance under the high packet loss rates [4].

With the development of deep neural networks and its wide application in the field of speech signal processing, a large number of the PLC methods based on deep learning have been proposed. We call these approaches as deep PLC. Compared with traditional PLC algorithms, deep PLC methods have better performance, especially in the complex scenarios, e.g., consecutive packet loss and high packet loss rates. Deep PLC algorithms can be mainly divided into online and offline processing [5]. In the online methods, the system is required to make real-time prediction for the lost frames only based on previous content. Generally, online PLC systems benefit from low latency (typically no more than a frame). A deep neural network (DNN) was employed to individually map the amplitude and phase information of the lost frame in [4]. One of the characteristics of this approach is that the magnitude spectrum of lost frame can be estimated well, and the lost frame could be recovered. However, this approach has an unavoidable limitation which is the difficult to estimate the unstructured phase by the neural network. The above limitation was overcome in [5] by using convolutional recurrent network (CRN) on raw audio frames directly to estimate the succeeding frame. The CRN can capture the long-term dependencies of speech frames by using gated recurrent cells like LSTM. This has been shown to be very effective in language modelling. Furthermore, the CRN in [5] incorporates online training into the framework, allowing every non-lost frame to contribute to the model for PLC.

In contrast, offline methods handle larger segments of audio that include lost packets and leverage the context information of speech frames in both forward and backward directions. In [6], the auto-encoding neural network was utilized to map the amplitude information and the conventional algorithm was employed to estimate the phase information for improving PLC. Similar to the approach introduced above, a U-Net architecture [7] was introduced for PLC by way of spectrograms. Furthermore, some generative adversarial network (GAN) based approaches [8, 9] were proposed for PLC. However, these approaches prioritize to improve speech quality at the cost of the increased latency.

Inspired by the above analysis, we proposed a time-domain packet loss concealment method by using Demucs of the architecture [10, 11]. It consists of a causal model based on a series of causal convolutions and LSTM layers. Furthermore, our model works from waveform to waveform through hierarchical generation (using U-Net like skip-connections). We optimized the model to directly output the "prediction" version of the speech signal while minimizing a regression loss function (L1 loss) complemented with a frequency domain loss [12, 13].

The rest parts of this paper are organized as follows: The signal model and the packet loss simulator are given in Sect. 2. The structure of the Demucs and the loss function of the model are detailed in Sect. 3. The experimental results are shown and discussed in Sect. 4. The conclusions are drawn in Sect. 5.

2 Modeling of Packet Loss

2.1 Signal Model

According to the traditional PLC method [2], we assume that the audio sequence is decoded from a single packet just containing one frame. During the process of audio transmission, there is an operation of pre-processing which splits the continuous audio $s \in \mathbb{R}^{1 \times L}$ into the short-time frames. Therefore, the jth audio frame $s_j \in \mathbb{R}^{1 \times N}$ with a length of N can be represented as:

$$s_j(n) = s(N \cdot (j-1) + n), n = 0, \ldots, N-1 \tag{1}$$

where L is the length of the input audio and n is the time index of the jth audio frame. If the jth audio frame is lost, it becomes silence, and the sample values within this frame are set to zeros. Additionally, we further assume that a received frame may be clean and contain its packet loss state. With this definition, we have $\tilde{s}_j = I_j s_j$, where $I_j = 1$ when the jth frame is not lost, otherwise $I_j = 0$. In essence, the deep PLC algorithm is required to map the lost signal $\tilde{s}(n)$ into the original audio $s(n)$.

2.2 Packet Loss Simulator

In an online system, the modeling process of packet loss is important for the PLC. However, it is difficult to obtain the data of packet loss from the real network environment. Thus, a Gilbert-Elliott channel (GEC) model [14] is employed for packet loss simulation in this paper. This model can effectively simulate the packet loss with different packet loss rates (PLRs).

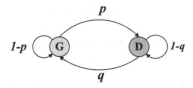

Fig. 1. Gilbert-Elloitt Channel Model

As shown in Fig. 1, the GEC model employing a two-state Markov model consists of two states, the "good" state (G) and the "bad" state (B). The transition probabilities between these two states are denoted as p and q, respectively. The packet loss rate in this model is represented as

$$P_{PLR} = \frac{q}{p+q} * P_G + \frac{p}{p+q} * P_B \tag{2}$$

where P_G and P_B are the packet loss probabilities in states G and B, respectively. We define $\lambda = 1 - (p + q)$, which indicates the burst or random characteristics of the channel. Once the packet loss rate P_{PLR} is given, the transition probabilities p and q could be derived as

$$p = (1 - \lambda) * (1 - \frac{P_B - P_{PLR}}{P_B - P_G}) \tag{3}$$

$$q = (1 - \lambda) * \frac{P_B - P_{PLR}}{P_B - P_G} \tag{4}$$

3   Proposed Method

3.1   U-Net Architecture

As shown in Fig. 2, the Demucs architecture forms an encoder-decoder structure with skip U-Net connections comprises the convolutional encoders, LSTM layers and the convolutional decoders. The architecture is defined by its number of encoder layer (L), initial number of hidden channels (H), layer kernel size (K), and stride (S).

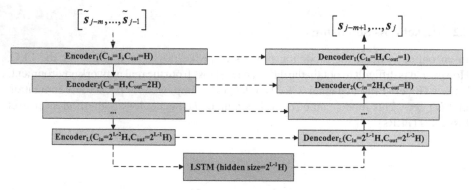

Fig. 2. The Architecture of Demucs

Firstly, we can see that the encoder blocks take the speech frames as input and produce a latent representation as output from Fig. 2. Each layer of the encoder blocks consists in a convolution layer with a kernel size of K_1 and stride of S_1 with $2^{i-1}H$ output channels, followed by a ReLU activation, a "1×1" convolution with $2^i H$ output channels and finally a GLU activation that converts back the number of channels to $2^{i-1}H$, please see Fig. 3(a) for a visual description.

Next, a recurrent neural network (RNN) module takes the latent representation of the last encoder layer as its input, and it outputs a non-linear transformation of the same size. As a variant of RNNs, the LSTM can effectively capture the long-term dependencies of speech signal and has a low probability of gradient vanishing, thus we choose LSTM to model the lost frame. To obtain a causal model, the unidirectional LSTM layers are utilized, which contain $2^{L-1}H$ hidden units.

Lastly, the output of the LSTM and the last layer of the encoder blocks are fed into the first layer of the decoder blocks. From Fig. 3(b), it can be seen that the ith layer of the

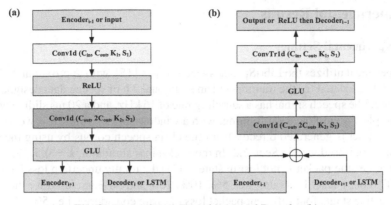

Fig. 3. View of Each Encoder (a) and Decoder (b)

decoder modules takes $2^{i-1}H$ channels as the input, and performs a "1×1" convolution on $2^i H$ channels, followed by a GLU activation function to produce $2^i H$ channels as the outputs. Subsequently, a transposed convolution is employed with a kernel size of K_1, stride of S_1, and $2^i H$ output channels, accompanied by a ReLU function. Notably, the last layer of the decoder modules produces the output of single-channel without ReLU, and there exists a skip connection linking the output of the ith encoder layer to the input of the ith decoder layer.

3.2 Loss Function

In the process of reconstructing the lost frames, if we focus on capturing the time-domain information, some frequency details may be neglected. In this paper, we combine the L1 loss on time waveform [12] with the Short-Time Fourier Transform (STFT) loss [13] on spectral magnitude. Formally, we define the STFT loss as the sum of spectral convergence (SC) loss and the magnitude (mag) loss as shown:

$$L_{stft}(s, \hat{s}) = L_{sc}(s, \hat{s}) + L_{mag}(s, \hat{s}) \tag{5}$$

where

$$L_{sc}(s, \hat{s}) = \frac{\big\| |STFT(s)| - |STFT(\hat{s})| \big\|_F}{\| |STFT(s)| \|_F} \tag{6}$$

$$L_{mag}(s, \hat{s}) = \frac{1}{T} \big\| \log|STFT(s)| - \log|STFT(\hat{s})| \big\|_1 \tag{7}$$

where $\|\cdot\|_F$ and $\|\cdot\|_1$ are the Frobenius norm and the L$_1$ norm, respectively. T indicates the sample points of STFT. Finally, we can optimize the network by minimizing the following function:

$$L_{sum}(s, \hat{s}) = \frac{1}{N} \| s - \hat{s} \|_1 + L_{stft}(s, \hat{s}) \tag{8}$$

4 Experimental Results

4.1 Experimental Settings

The experiment utilizes the LibriSpeech speech corpus [15], with approximately 20 h of speech data allocated for the training and an additional 2 h of speech data designated for the testing. The speech signal has a sampling rate of 16 kHz, and a 20 ms sliding window (320 samples) is employed for framing, with a window shift of 10ms. It is worth noting that the dataset is generated directly from the clean speech corpus by using the packet loss simulator introduced in Sect. 2.2. In the packet loss simulator, $\lambda = 0.5$, $P_G = 0$ and $P_B = 0.5$, i.e., the packet is not lost in state "G" and a totally uncertain loss happens in state "B". Four packet loss rates, i.e., 5%, 10%, 20% and 30%, are used for the testing. For the testing stage, four different packet loss rates are considered, i.e., 5%, 10%, 20%, and 30%. For the Demucs, the kernel sizes K_1 and K_2 are set to 8 and 1, respectively, strides S_1 and S_2 are set to 2 and 1, respectively, the number L of encoder blocks is set to 5 and the hidden channel H is set to 64. The Adam is used for the network optimization, and the learning rate is 0.0002.

4.2 Evaluation and Analysis

The performance evaluation is conducted using the perceptual evaluation of speech quality (PESQ) [16] and the short-term objective intelligibility (STOI) [17]. The proposed method is compared to various existing approaches, including the zero-filling approach [2], the DNN-based method [4], the residual network (ResNet) based method [18] and the CRN-based method [5]. The results for PESQ and STOI are presented in Table 1 and Table 2, respectively.

Table 1. PESQ Results

Methods	PESQ			
PLRs	5%	10%	20%	30%
Zero-fillings	3.16	2.68	2.09	1.65
DNN	3.29	2.91	2.54	2.22
ResNet	3.47	3.11	2.68	2.28
CRN	3.65	3.48	3.29	2.54
Demucs	**3.74**	**3.59**	**3.40**	**2.64**

From the results of PESQ and STOI, we can see that the deep PLC shown in the table obtain significant improvements compared with the zero-fillings-based method. However, when compared to other deep PLC methods in Table 1 and Table 2, the DNN-based approach presents a modest constraint in modeling the lost speech frames. This limitation arises from its utilization of fully connected networks to predict the lost frame, and it is

Table 2. STOI Results

Methods	STOI			
PLRs	5%	10%	20%	30%
Zero-fillings	0.9554	0.9121	0.8519	0.7939
DNN	0.9685	0.9433	0.8940	0.8405
ResNet	0.9779	0.9557	0.9156	0.8645
CRN	0.9902	0.9878	0.9753	0.9453
Demucs	**0.9908**	**0.9890**	**0.9779**	**0.9498**

slightly satisfactory for the unstructured phase spectrum predicted by DNN. The ResNet-based approach employs multiple residual-connected convolutional layers to recover the lost speech frames in time domain, which demonstrate better performance compared to the DNN-based method. Furthermore, the CRN-based method leverages convolutional operations for feature extraction, followed by RNN layers, which results in a significant improvement. The reason is the RNN layers can better model the correlation of the speech frames by using the previous information. Moreover, the Demucs-based method shows better performance than the CRN-FC based methods, this can be attributed to the use of GLU in both the encoder and decoder of Demucs, which strictly preserves feature information of temporal positions, thus improving the performance of feature extraction. Besides, the prediction accuracy of the network has been improved by utilizing a hybrid loss function that optimizes both time domain and frequency domain. Overall, the proposed Demucs method obtains the best PESQ and STOI results under different packet loss rates shown in Table 1 and Table 2. This indicates that the proposed method has better performance in the PLC compared with the reference methods.

5   Conclusions

This paper demonstrated how Demucs, an architecture originally designed for music source separation in the waveform domain, can be adapted into a causal packet loss concealment system. The U-Net architecture and the hybrid loss function enhanced prediction accuracy by effectively modeling the inter-frame correlation of the speech signal. The experimental results show that the proposed method has better performance in the task of packet loss concealment.

References

1. Wah, B.W., Xiao, S., Dong, L.: A survey of error-concealment schemes for real-time audio and video transmissions over the Internet. In: Proceedings International Symposium on Multimedia Software Engineering, Taipei, Taiwan, pp.17–24 (2000). https://doi.org/10.1109/MMSE.2000.897185
2. Suzuki, J., Taka, M.: Missing packet recovery techniques for low-bit-rate coded speech. IEEE J. Sel. Areas Commun. **7**, 707–717 (1989). https://doi.org/10.1109/49.32334

3. Rodbro, C.A., Murthi, M.N., Andersen, S.V., Jensen, S.H.: Hidden Markov model-based packet loss concealment for voice over IP. IEEE Trans. Audio Speech Lang. Process. **14**, 1609–1623 (2006). https://doi.org/10.1109/TSA.2005.858561

4. Lee, B.K., Chang, J.H.: Packet loss concealment based on deep neural networks for digital speech transmission. IEEE/ACM Trans. Audio Speech Lang. Process. **24**, 378–387 (2016). https://doi.org/10.1109/TASLP.2015.2509780

5. Lin, J., Wang, Y., Kalgaonkar, K., Keren, G., Zhang, D., Fuegen, C.: A time-domain convolutional recurrent network for packet loss concealment. In: IEEE International Conference on Acoustics, Speech and Signal Processing (ICASSP), Toronto, Canada, pp. 7148–7152 (2021). https://doi.org/10.1109/ICASSP39728.2021.9413595

6. Marafioti, A., Perraudin, N., Holighaus, N., Majdak, P.: A context encoder for audio inpainting. IEEE/ACM Trans. Audio Speech Lang. Process. **27**, 2362–2372 (2019). https://doi.org/10.1109/TASLP.2019.2947232

7. Kegler, M., Beckmann, P., Cernak, M.: Deep speech inpainting of time-frequency masks. arXiv preprint arXiv:1910.09058 (2019)

8. Shi, Y., Zheng, N., Kang, Y., Rong, W.: Speech loss compensation by generative adversarial networks. In: Asia-Pacific Signal and Information Processing Association Annual Summit and Conference (APSIPA ASC), Lanzhou, China, pp. 347–351 (2019). https://doi.org/10.1109/APSIPAASC47483.2019.9023132

9. Guan, Y., Yu, G., Li, A., Zheng, C., Wang, J.: TMGAN-PLC: audio packet loss concealment using temporal memory generative adversarial network. arXiv preprint arXiv:2207.01255 (2022)

10. Défossez A., Usunier N., Bottou L.: Music source separation in the waveform domain. arXiv preprint arXiv:1911.13254 (2019)

11. Defossez, A., Gabriel, S., Yossi, A.: Real time speech enhancement in the waveform domain. arXiv preprint arXiv:2006.12847 (2020)

12. Yamamoto, R., Song, E., Kim, J.M.: Parallel WaveGAN: a fast waveform generation model based on generative adversarial networks with multi-resolution spectrogram. In: IEEE International Conference on Acoustics, Speech and Signal Processing (ICASSP), Barcelona, Spain, pp. 6199–6203 (2020). https://doi.org/10.1109/ICASSP40776.2020.9053795

13. Yamamoto R., Song E., Kim J.M.: Probability density distillation with generative adversarial networks for high-quality parallel waveform generation. arXiv preprint arXiv:1904.04472 (2019)

14. Gilbert, E.N.: Capacity of a burst-noise channel. Bell Syst. Tech. J. **39**, 1253–1265 (1960). https://doi.org/10.1002/j.1538-73051960.tb03959.x

15. Panayotov, V., Chen, G., Povey, D., Khudanpur, S.: LibriSpeech: an ASR corpus based on public domain audio books. In: IEEE International Conference on Acoustics, Speech and Signal Processing (ICASSP), South Brisbane, QLD, Australia, pp. 5206–5210 (2015). https://doi.org/10.1109/ICASSP.2015.7178964

16. Rix, A.W., Beerends, J.G., Hollier, M.P., Hekstra, A.P.: Perceptual evaluation of speech quality (PESQ)-a new method for speech quality assessment of telephone networks and codecs. In: IEEE International Conference on Acoustics, Speech, and Signal Processing, Salt Lake City, UT, USA, pp. 749–752 (2001). https://doi.org/10.1109/ICASSP.2001.941023

17. Taal, C.H., Hendriks, R.C., Heusdens, R., Jensen, J.: An algorithm for intelligibility prediction of time-frequency weighted noisy speech. IEEE Trans. Audio Speech Lang. Process. **19**, 2125–2136 (2019). https://doi.org/10.1109/TASL.2011.2114881

18. Zhu, J., Bao, C., Huang, J.: Packet loss concealment method based on the simplified residual network. In: the 16th IEEE International Conference on Signal Processing (ICSP), Beijing, China, pp. 51–55 (2022). https://doi.org/10.1109/ICSP56322.2022.9964486

Improving Speech Perceptual Quality and Intelligibility Through Sub-band Temporal Envelope Characteristics

Ruilin Wu[1]📷, Zhihua Huang[1,2](✉)📷, Jingyi Song[1]📷, and Xiaoming Liang[1]📷

[1] School of Computer Science and Technology, Xinjiang University, Urumqi, China
{wuruilin,sjylily510677,107552201321}@stu.xju.edu.cn, zhhuang@xju.edu.cn
[2] Key Laboratory of Signal Detection and Processing in Xinjiang, Urumqi, China

Abstract. In the speech enhancement (SE) model, using auxiliary loss based on acoustic parameters can improve enhancement effects. However, currently used acoustic parameters focus on frequency domain information, neglecting the importance of time domain information. This study explores the effects of training the SE model utilizing temporal envelope parameters as auxiliary losses on perceptual quality and intelligibility, primarily focusing on accurately extracting temporal envelope parameters and optimizing losses. Initially, the effectiveness and robustness of the temporal envelope parameters in optimizing the SE model were verified. Subsequently, a temporal envelope loss based on sub-band weighted (ENVLoss) is proposed based on the perceptual characteristics of human hearing. Then, a method for constructing a joint loss function is proposed, integrating temporal and spectral acoustic features to promote the enhancement effect. Experimental results show that temporal envelope characteristics improve both the time domain and the time-frequency domain SE models. Compared to other acoustic parameter losses, the SE model using a sub-band temporal envelope auxiliary loss shows improvement in the PESQ, STOI, and MOS estimation metrics.

Keywords: Speech enhancement · Speech intelligibility · Temporal envelope

1 Introduction

In real-world speech-related applications, speech signals suffer from environmental noise pollution, leading to functional damage to the application system. Over the past few decades, the deep neural network (DNN) model has demonstrated powerful performance in the field of SE. Unlike traditional enhancement methods, DNN-based approaches possess superior processing capabilities, particularly in low signal-to-noise ratio (SNR) scenarios and/or non-stationary noise environments.

The DNN-based SE model typically uses point-wise differences between the waveforms or spectrograms of enhanced speech and clean speech as the objective

J. Jia et al. (Eds.): NCMMSC 2023, CCIS 2006, pp. 235–247, 2024.
https://doi.org/10.1007/978-981-97-0601-3_20

function, such as L1 and L2 losses. However, numerous studies have indicated the limitations of loss functions used commonly, which can result in artifacts or poor perceptual quality in the enhanced speech [1–4].

A novel research direction in the field of SE is generated according to the above issues, aimed at optimizing the perceptual quality of speech signals. Common strategies for optimization can be categorized into two types. The first approach involves introducing auxiliary losses to optimize the model, typically involving the computation of the difference in specific parameters between clean speech and the corresponding enhanced speech. The commonly used feature parameters in this approach include differentiable estimators of existing perceptual metrics [5] and speech representations produced by acoustic models [6–8]. However, the correlation of these parameters with human perception is somewhat limited, and the improvements provided by optimizing these parameters are restricted. It is worth noting that subjective intelligibility tests have shown that using auxiliary losses based on objective perceptual metrics does not promote speech intelligibility [9].

Another optimization strategy is directly optimizing the enhancement model by combining domain knowledge. Common methods include using perceptually motivated filter banks to simulate the human auditory system's adjustments to signals [10,11] or directly optimizing the model based on perceptual theories [12,13]. Nevertheless, these methods only provide implicit supervision and do not target specific characteristics. Moreover, due to the involvement of domain-specific knowledge, the performance improvements and scope of application provided by these strategies are limited.

In recent years, some advancements have been achieved in optimizing the performance of DNN by combining acoustic parameters. Statistical features related to timbre, such as cepstral statistics and the standard deviation of the Mel-Frequency Cepstral Coefficients (MFCCs), were attempted to use as a perceptual loss function to optimize the SE model [14]. In [15], fundamental frequency (F0), energy contour, pitch contour, and speaker identity were applied to represent pitch, loudness, and timbre elements to optimize perceptual quality. However, the features used in these studies represent only a tiny subset of the task-considered features. To better utilize the capabilities of optimization strategies based on acoustic parameters, researchers [16–18] employed the eGeMAPS [19] feature set, which is a more comparative set of features currently used. Furthermore, this approach was evaluated using standard English datasets, further proving the effectiveness of the chosen acoustic parameter feature set in improving speech perceptual quality.

In the strategy of fine-tuning the SE model through the auxiliary loss based on acoustic parameters, the current feature set being employed still exhibits certain limitations. According to the phonetic-aligned acoustic parameter (PAAP) loss proposed in [17], the loss function obtains accurate estimates of the 25 acoustic parameters in eGeMAPS using a feature estimation network. Subsequently, it references the correlation between these acoustic parameters and phonemes to assign weights to the features. Upon analyzing the features involved in the

PAAPLoss, it becomes evident that the feature set focuses on frequency domain features but lacks features related to time domain and phase. Nevertheless, the role of temporal features in speech enhancement has been confirmed through ablative experiments in the time and frequency domains [15]. In other words, temporal and frequency domain information are complementary, and introducing a time-domain loss training model could enhance the phase component of speech. Thereby, the perceptual quality of speech could be improved [20].

The temporal envelope of sound plays a crucial role in auditory perception and significantly influences speech intelligibility [21,22]. Studies [23,24] have shown that even when speech is severely degraded in the frequency domain, accurate speech recognition can be achieved by preserving the low-frequency amplitude modulation, primarily the temporal envelope. [25] verified the importance of the temporal envelope in the supervised learning approach for speech perception. [26] used the temporal envelope of reference speech as a control signal to modulate the envelope of noisy speech has the potential to improve speech quality. In subsequent studies, [27] achieved favorable results by directly training an SE model based on one-dimensional CNNs using temporal envelope features. Alternatively, [28] proposed an objective evaluation metric, the Gammachirp envelope distortion index, based on the temporal envelope to predict enhanced speech intelligibility. However, using the temporal envelope to construct an auxiliary loss to fine-tuning the SE model has not been studied in the existing literature. Based on the studies mentioned above, the main contributions are as follows:

1. The effectiveness of using sub-band temporal envelopes as an auxiliary loss to optimize the SE model has been demonstrated through experimentation.
2. Stem from the importance theory of sub-bands in human auditory perception, the temporal envelopes of different sub-bands are weighted through a multi-channel attention mechanism. Higher weights are assigned to the temporal envelopes of important sub-bands, making the auxiliary loss more in line with the auditory perception of the human ear.
3. A joint loss function is designed by integrating features from both the time and frequency domains. This allows the flexible combination of the proposed auxiliary loss with the existing loss functions of the SE model. This strategy enables model fine-tuning and further enhances the perceptual quality of the output audio.

2   Method

2.1   Temporal Envelope Representation

Let $s(t)$ represent the reference speech signal, where t is the time index. When the auditory system receives a sound signal, all signals are filtered into a series of narrowband signals through a bank of bandpass filters (auditory filters). To simulate the above signal processing, a set of K Gabor filters [29] divides the input speech signal into K parallel, contiguous frequency bands (also referred to as sub-bands). The chosen frequency bands cover the range of 80 to 6000 Hz. The bandwidth of each sub-band is approximately equivalent to 4.55 ERBs (equivalent rectangular bandwidths) [30].

During the sub-band envelope feature extraction process [26], we denote $s_c(t, k)$ as the sub-band signal at the output of the kth filter, where $k = 1, \cdots, K$. After half-wave rectification and low-pass filtering, we obtain the full-scale envelope vector $s_{AN}(t, k)$. The energy of the input signal is preserved using weighted vectors ω_G, and the final feature vector $S_{env} \in \mathbb{R}^{N \times K}$ is obtained by integrating the envelope power within a short-time window $\mu(t; \tau)$. Here, $N = \frac{L}{\tau}$, where L is the signal length, and τ is the time constant.

$$s_{AN}(t, k) = max(0, s_c(t, k)) *_t h(t) \tag{1}$$

$$s_{env} = s_{AN} \cdot \omega_G^T \tag{2}$$

$$S_{env}(n, k) = (\int_0^\tau s_{env}^2(n \cdot \tau + t, k) \cdot \mu(t; \tau) \mathrm{d}t)^{\frac{1}{2}}, n \in \{0, \cdots, N\} \tag{3}$$

In the above equations, $*_t$ represents the convolution operation in the time domain, and $h(t)$ represents the impulse response of the low-pass filter. Let the enhanced speech signal be denoted as $\hat{s}(t)$, and the sub-band temporal envelope features of both $s(t)$ and $\hat{s}(t)$ are extracted, resulting in feature vectors S_{env} and $\hat{S}_{env}$. The SE model is fine-tuned by minimizing the Mean Absolute Error (MAE) loss of the sub-band temporal envelope characteristics.

$$\mathcal{L}_{env}(S_{env}, \hat{S}_{env}) = MAE(S_{env}, \hat{S}_{env}) \tag{4}$$

2.2   The Temporal Envelope Loss Based on Sub-band Weighted

[31] points out that different frequency components make distinct contributions to human speech perception. Through psychophysical studies, [32] found that although all frequency bands convey useful envelope information, the envelope cues from higher frequency bands (above 1.8kHz) convey more information. Based on the above studies, we propose a sub-band weighted temporal envelope loss called ENVLoss.

S_{env} and $\hat{S}_{env}$ represent the sub-band temporal envelopes without weighted. We employ a multiscale temporal channel attention (MulCA) module [33] to obtain frequency band weights W. As shown in Fig. 1, S_{env} and $\hat{S}_{env}$ are first parallel passed through convolution layers with different kernel sizes, followed by average pooling layers and ReLU activation functions to generate features with different time scales $C_s, C_m, C_l \in \mathbb{R}^K$. Multiple fully connected layers fuse features from different time scales and obtain the final weight vector W, where $W = [W_0, \cdots, W_k, \cdots, W_K]^T \in \mathbb{R}^K$. The features of the weighted sub-band envelopes E and $\hat{E}$ are obtained by taking the dot product of S_{env} and $\hat{S}_{env}$ with W.

We define ENVLoss as the MAE between the sub-band envelopes of clean speech and enhanced speech, with the sub-band envelopes being weighted:

$$\mathcal{L}_{ENV}(S_{env}, \hat{S}_{env}) = MAE(S_{env} \odot W, \hat{S}_{env} \odot W) \tag{5}$$

This loss is end-to-end differentiable and only takes waveform as input. Therefore, the auxiliary loss proposed in this paper can be utilized to optimize any SE model with clean reference speech.

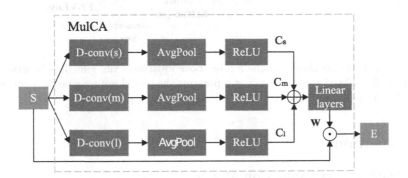

Fig. 1. In the structure diagram of MulCA, S represents the unweighted sub-band temporal envelope feature, and E represents the weighted sub-band temporal envelope.

2.3   Joint Loss Function

In this section, we employ a joint loss function composed of PAAPLoss[1] and the proposed ENVLoss to fine-tune the SE model. The workflow of the joint loss is illustrated in Fig. 2. Since the baseline model already possesses its inherent loss function, the final loss will be the original loss of each model combined with our proposed auxiliary loss.

$$\mathcal{L}_{joint} = \lambda \mathcal{L}_{ENV} + \gamma \mathcal{L}_{PAAP} \tag{6}$$

$$\mathcal{L}_{final} = \mathcal{L}_{original} + \mathcal{L}_{joint} \tag{7}$$

where λ and γ represent the weight hyperparameters for ENVLoss and PAAPLoss, respectively. $\mathcal{L}_{joint}$ represents the joint loss function. $\mathcal{L}_{original}$ is the original loss function of the enhancement model. $\mathcal{L}_{ENV}$ is the sub-band weighted temporal envelope loss, i.e., ENVLoss. $\mathcal{L}_{PAAP}$ is a phonetic-aligned acoustic parameter loss, i.e., PAAPLoss. During the backward propagation process, only the parameters of the enhancement model are optimized, while the parameters of the feature estimation network are frozen.

---

[1] https://github.com/muqiaoy/PAAP.

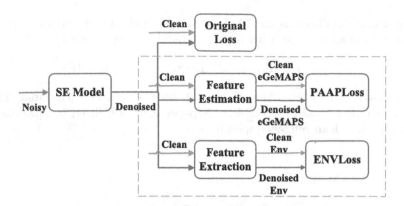

Fig. 2. The flowchart shown in the figure above illustrates the joint loss function process. The highlighted section in yellow represents the model improvements made in this study. (Color figure online)

3   Experiments

3.1   Dataset and Evaluation Metrics

The Interspeech 2020 DNS Challenge dataset [34] was used to prepare the required speech data for the experiment. The DNS dataset[2] comprises over 500 h of clean clips from 2150 speakers and over 180 h of noise clips from 150 categories. To train the model, we synthesized clean-noisy pairs using the noisy speech synthesis methods provided by the dataset. Each noisy speech clip lasts 30 s, with the SNR uniformly sampled between −10 dB and 5 dB.

The DNS dataset provides two non-blind test sets to evaluate the model: with-reverb and no-reverb. This study considers only the case without reverberation. Due to the dataset's SNR range not aligning with our requirements, we synthesized clean-noisy pairs for testing using clean speech clips from the no-reverb dataset and the noise clips from the DNS dataset. Each speech clip lasts 10 s, and the SNR ranges from −10 to 5 dB. It should be noted that the clean clips involved in the test set are independent of the training set.

We initially evaluated the model using classic objective metrics, including PESQ [35], STOI [36], SSNR [37], CSIG [38], CBAK [38], and COVL [38]. Considering that our goal is to improve speech's perceptual quality and intelligibility, human Mean Opinion Scores (MOS) is regarded as the gold standard for assessing the performance of the SE model. However, conducting MOS studies can be costly. Thus, we employ state-of-the-art estimation methods, such as DNS-MOS [39] and NORESQA-MOS [40], to estimate MOS values. For all metrics, higher values indicate better performance.

---

[2] https://github.com/microsoft/DNS-Challenge/tree/interspeech2020/master.

3.2 Experimental Results

Validity of Sub-band Temporal Envelope Auxiliary Loss. To validate the effectiveness of the sub-band temporal envelope auxiliary loss in the SE model, experiments were conducted by adding the auxiliary loss separately to the Demucs[3] and FullSubNet[4] Both of these SE models are open-source and provide pre-trained checkpoints. Demucs [41] is a mapping-based time domain model that directly takes noisy waveforms as inputs and estimated results for clean waveforms as outputs. FullSubNet [42] is a masking-based time-frequency domain fusion model. The model utilizes the spectral features of the full-band and the sub-band as input and outputs corresponding estimates for clean speech in both the full-band and sub-band.

Increasing the sub-band temporal envelope auxiliary loss function for fine-tuning the model, the experimental results are shown in Table 1. Most evaluation metrics have improved on the Demucs and FullSubNet. Show that using sub-band temporal envelope features as auxiliary loss effectively improves the enhanced model's performance. Meanwhile, both models exhibit a comparatively noticeable promotion in STOI, indicating that sub-band temporal envelopes are a crucial feature affecting speech intelligibility.

To determine the importance of the sub-band temporal envelope auxiliary loss, we set $\gamma = 0$ in Eq. (6). We adjusted the size of λ to observe the enhancement effect to obtain the optimal hyperparameter specification. The results in Table 2 show that as the weight of auxiliary loss changes, speech quality and intelligibility change in response. The model achieves optimal performance when $\lambda = 0.1$. Hence, we uniformly set the weight of the temporal envelope loss to 0.1 in subsequent experiments.

Table 1. The evaluation results of fine-tuning the SE model using temporal envelope parameters as an auxiliary loss on the synthetic test set are as follows, where $\mathcal{L}_{env}$ stands for sub-band temporal envelope loss.

	PESQ	STOI (%)	CSIG	CBAK	COVL	SSNR
Noisy	1.185	79.70	2.316	1.755	1.675	0.762
FullSubNet	2.063	90.49	3.476	2.896	2.747	7.934
FullSubNet + $\mathcal{L}_{env}$	**2.108**	**90.78**	**3.557**	**2.956**	**2.816**	**8.296**
Demucs	1.864	90.85	**3.35**	2.94	2.597	9.495
Demucs + $\mathcal{L}_{env}$	**1.878**	**91.15**	3.342	**2.951**	**2.601**	**9.617**

ENVLoss. To demonstrate the effectiveness of ENVLoss, we applied channel attention (MulCA) to weight the sub-band temporal envelope features in the FullSubNet. In contrast to previous FullSubNet-related papers, this study

---

[3] https://github.com/facebookresearch/denoiser.
[4] https://github.com/haoxiangsnr/FullSubNet.

Table 2. On FullSubNet, the optimal weight for the sub-band temporal envelope loss, where $\mathcal{L}_{env}$ stands for sub-band temporal envelope loss, and λ denotes the auxiliary loss weight.

FullSubNet $\mathcal{L}_{env}$ λ Ablation					
Weight	0.01	0.05	0.1	0.5	1
PESQ	2.092	2.101	**2.108**	2.104	2.104
STOI	90.78	90.74	**90.78**	90.73	90.77

applies channel attention to the sub-band temporal envelope features. MulCA considered the distinct sub-bands of features as separate channels, assigning them different weights. Table 3 shows that after fine-tuning the SE model with ENVLoss, the model achieved higher PESQ and STOI scores. Furthermore, a higher SSNR score indicates further improvement in the denoising performance of the model. The above results indicate that the importance of temporal envelope characteristics for speech perception varies across different frequency bands.

Table 3. On FullSubNet, the impact of sub-band importance on the temporal envelope loss, where $\mathcal{L}_{ENV}$ represents ENVLoss with an auxiliary loss weight of $\lambda = 0.1$.

	PESQ	STOI (%)	CSIG	CBAK	COVL	SSNR
Noisy	1.185	79.70	2.316	1.755	1.675	0.762
FullSubNet	2.063	90.49	3.476	2.896	2.747	7.934
FullSubNet + $\lambda\mathcal{L}_{env}$	2.108	90.78	**3.557**	2.956	**2.816**	8.296
FullSubNet + $\lambda\mathcal{L}_{ENV}$	**2.125**	**90.90**	3.54	**2.967**	2.814	**8.418**

The Effectiveness of the Joint Loss Function. In this section, the performance of the joint loss function is evaluated. Based on the [17] and the experiments mentioned above, the hyperparameters γ and λ values are set to 0.1.

The experimental results in Table 4 indicate that the joint loss function outperforms ENVLoss and PAAPLoss in all evaluation metrics. The qualitative analysis using spectrograms also supports the conclusion, as depicted in Fig. 3. Our interpretation of this result is that the joint loss function utilizes a more comprehensive set of feature parameters, using temporal information introduced by ENVLoss to complement the phase component lacking when fine-tuning with PAAPLoss.

To analyze the robustness of the auxiliary losses, we investigated the impact of various loss functions on the effect of the SE model at different SNR levels. We utilized clean speech and noise from the test set and synthesized 10-second test utterances at each fixed SNR, ranging from [−9 dB, −6 dB, −3 dB, 0 dB, 3 dB]. As shown in Fig. 4, the SE model fine-tuned with the joint loss exhibits higher improvements in PESQ and STOI at different SNR levels. Furthermore,

ENVLoss outperforms PAAPLoss in terms of evaluation metrics. The experimental result demonstrates that temporal envelope features are crucial to speech intelligibility [21].

Table 4. On FullSubNet, the effectiveness of the joint loss function, where $\mathcal{L}_{joint}$ represents the joint loss function with auxiliary loss weights $\lambda = \gamma = 0.1$.

	PESQ	STOI (%)	DNSMOS	NORESQA
Noisy	1.185	79.70	2.827	2.712
FullSubNet	2.063	90.49	3.608	3.749
FullSubNet + $\gamma\mathcal{L}_{PAAP}$	2.118	90.65	3.666	3.805
FullSubNet + $\lambda\mathcal{L}_{ENV}$	2.125	90.90	3.705	3.824
FullSubNet + $\mathcal{L}_{joint}$	**2.146**	**90.96**	**3.728**	**3.890**

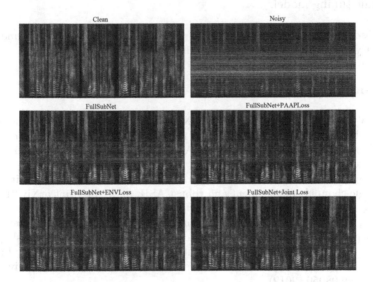

Fig. 3. On FullSubNet, a comparison of the spectrograms of enhanced speech for the FullSubNet with different loss functions.

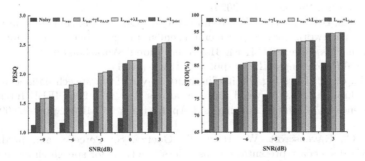

Fig. 4. The impact of different loss functions on the performance of FullSubNet under varying SNR conditions, with auxiliary loss weights $\lambda = \gamma = 0.1$.

4 Conclusion

This study investigated the role of auxiliary loss based on temporal envelope features in enhancing speech's perceptual quality and intelligibility. Firstly, we demonstrated on both the time domain and the time-frequency domain SE model that optimizing the model can be achieved by minimizing the difference in sub-band temporal envelope features. Secondly, we obtained a more perceptually relevant temporal envelope loss by weighting different frequency bands and constructed a joint loss function for fine-tuning the SE model. Experimental results show that the proposed loss functions are competitive across various evaluation metrics. Compared to other feature losses, the temporal envelope feature is differentiable and requires no additional training for estimation networks, significantly reducing model training time. However, the excessively long computation time for feature calculation remains an inevitable issue when employing the acoustic feature fine-tuning model.

Acknowledgements. This work is supported by the Natural Science Foundation of Xinjiang Uygur Autonomous Region of China (No. 2022D01C59).

References

1. Fu, S.W., Wang, T.W., Tsao, Y., Lu, X., Kawai, H.: End-to-end waveform utterance enhancement for direct evaluation metrics optimization by fully convolutional neural networks. IEEE/ACM Trans. Audio Speech Lang. Process. **26**(9), 1570–1584 (2018)
2. Plantinga, P., Bagchi, D., Fosler-Lussier, E.: Perceptual loss with recognition model for single-channel enhancement and robust ASR. arXiv preprint arXiv:2112.06068 (2021)
3. Turian, J., Henry, M.: I'm sorry for your loss: spectrally-based audio distances are bad at pitch. In: "I Can't Believe It's Not Better!" NeurIPS 2020 Workshop (2020)
4. Reddy, C.K., Beyrami, E., Pool, J., Cutler, R., Srinivasan, S., Gehrke, J.: A scalable noisy speech dataset and online subjective test framework. arXiv preprint arXiv:1909.08050 (2019)
5. Kolbæk, M., Tan, Z.H., Jensen, S.H., Jensen, J.: On loss functions for supervised monaural time-domain speech enhancement. IEEE/ACM Trans. Audio Speech Lang. Process. **28**, 825–838 (2020)
6. Guimarães, H.R., Beccaro, W., Ramírez, M.A.: Optimizing time domain fully convolutional networks for 3D speech enhancement in a reverberant environment using perceptual losses. In: 2021 IEEE 31st International Workshop on Machine Learning for Signal Processing (MLSP), pp. 1–6. IEEE (2021)
7. Sun, T., et al.: Boosting the intelligibility of waveform speech enhancement networks through self-supervised representations. In: 2021 20th IEEE International Conference on Machine Learning and Applications (ICMLA), pp. 992–997. IEEE (2021)
8. Close, G., Ravenscroft, W., Hain, T., Goetze, S.: Perceive and predict: self-supervised speech representation based loss functions for speech enhancement. In: ICASSP 2023 - 2023 IEEE International Conference on Acoustics, Speech and Signal Processing (ICASSP), pp. 1–5. IEEE (2023)

9. López-Espejo, I., Edraki, A., Chan, W.Y., Tan, Z.H., Jensen, J.: On the deficiency of intelligibility metrics as proxies for subjective intelligibility. Speech Commun. **150**, 9–22 (2023)

10. Rajeswari, M.R., Govind, D., Gangashetty, S.V., Dubey, A.K.: Improved epoch based prosody modification by zero frequency filtering of gabor filtered telephonic speech. In: 2023 National Conference on Communications (NCC), pp. 1–5. IEEE (2023)

11. Strauss, M., Torcoli, M., Edler, B.: Improved normalizing flow-based speech enhancement using an all-pole gammatone filterbank for conditional input representation. In: 2022 IEEE Spoken Language Technology Workshop (SLT), pp. 444–450. IEEE (2023)

12. Hsieh, T.A., Yu, C., Fu, S.W., Lu, X., Tsao, Y.: Improving perceptual quality by phone-fortified perceptual loss using wasserstein distance for speech enhancement. In: Proceedings of the Interspeech 2021, pp. 196–200 (2021). https://doi.org/10.21437/Interspeech.2021-582

13. Wang, T., Zhu, W., Gao, Y., Zhang, S., Feng, J.: Harmonic attention for monaural speech enhancement. IEEE/ACM Trans. Audio Speech Lang. Process. (2023)

14. Eng, N., Hioka, Y., Watson, C.I.: Using perceptual quality features in the design of the loss function for speech enhancement. In: 2022 Asia-Pacific Signal and Information Processing Association Annual Summit and Conference (APSIPA ASC), pp. 1904–1909. IEEE (2022)

15. Peng, C.J., Shen, Y.L., Chan, Y.J., Yu, C., Tsao, Y., Chi, T.S.: Perceptual characteristics based multi-objective model for speech enhancement. In: Proceedings of the Annual Conference of the International Speech Communication Association, INTERSPEECH, vol. 2022, pp. 211–215 (2022)

16. Yang, M., Konan, J., Bick, D., Kumar, A., Watanabe, S., Raj, B.: Improving speech enhancement through fine-grained speech characteristics. In: Proceedings of the Annual Conference of the International Speech Communication Association, INTERSPEECH, vol. 2022, pp. 2953–2957 (2022)

17. Yang, M., et al.: PAAPLoss: a phonetic-aligned acoustic parameter loss for speech enhancement. In: ICASSP 2023 - 2023 IEEE International Conference on Acoustics, Speech and Signal Processing (ICASSP), pp. 1–5. IEEE (2023)

18. Zeng, Y., et al.: TAPLoss: a temporal acoustic parameter loss for speech enhancement. In: ICASSP 2023 - 2023 IEEE International Conference on Acoustics, Speech and Signal Processing (ICASSP), pp. 1–5 (2023). https://doi.org/10.1109/ICASSP49357.2023.10094773

19. Eyben, F., et al.: The Geneva minimalistic acoustic parameter set (GeMAPS) for voice research and affective computing. IEEE Trans. Affect. Comput. **7**(2), 190–202 (2015)

20. Abdulatif, S., Armanious, K., Sajeev, J.T., Guirguis, K., Yang, B.: Investigating cross-domain losses for speech enhancement. In: 2021 29th European Signal Processing Conference (EUSIPCO), pp. 411–415. IEEE (2021)

21. Millman, R.E., Johnson, S.R., Prendergast, G.: The role of phase-locking to the temporal envelope of speech in auditory perception and speech intelligibility. J. Cogn. Neurosci. **27**(3), 533–545 (2015)

22. Moore, B.C.: The roles of temporal envelope and fine structure information in auditory perception. Acoust. Sci. Technol. **40**(2), 61–83 (2019)

23. Van Tasell, D.J., Soli, S.D., Kirby, V.M., Widin, G.P.: Speech waveform envelope cues for consonant recognition. J. Acoust. Soc. Am. **82**(4), 1152–1161 (1987)

24. Souza, P.E., Wright, R.A., Blackburn, M.C., Tatman, R., Gallun, F.J.: Individual sensitivity to spectral and temporal cues in listeners with hearing impairment. J. Speech Lang. Hear. Res. **58**(2), 520–534 (2015)
25. Thoidis, I., Vrysis, L., Markou, D., Papanikolaou, G.: Temporal auditory coding features for causal speech enhancement. Electronics **9**(10), 1698 (2020)
26. Soleymanpour, R., Brammer, A.J., Marquis, H., Heiney, E., Kim, I.: Enhancement of speech in noise using multi-channel, time-varying gains derived from the temporal envelope. Appl. Acoust. **190**, 108634 (2022)
27. Soleymanpour, R., Soleymanpour, M., Brammer, A.J., Johnson, M.T., Kim, I.: Speech enhancement algorithm based on a convolutional neural network reconstruction of the temporal envelope of speech in noisy environments. IEEE Access **11**, 5328–5336 (2023)
28. Yamamoto, K., Irino, T., Araki, S., Kinoshita, K., Nakatani, T.: GEDI: gammachirp envelope distortion index for predicting intelligibility of enhanced speech. Speech Commun. **123**, 43–58 (2020)
29. Moore, B.C., Glasberg, B.R.: A revision of Zwicker's loudness model. Acta Acust. Acust. **82**(2), 335–345 (1996)
30. Glasberg, B.R., Moore, B.C.: Derivation of auditory filter shapes from notched-noise data. Hear. Res. **47**(1–2), 103–138 (1990)
31. Chao, R., Yu, C., Wei Fu, S., Lu, X., Tsao, Y.: Perceptual contrast stretching on target feature for speech enhancement. In: Proceedings of the Interspeech 2022, pp. 5448–5452 (2022). https://doi.org/10.21437/Interspeech.2022-10478
32. Ardoint, M., Agus, T., Sheft, S., Lorenzi, C.: Importance of temporal-envelope speech cues in different spectral regions. J. Acoust. Soc. Am. **130**(2), EL115–EL121 (2011)
33. Chen, J., Wang, Z., Tuo, D., Wu, Z., Kang, S., Meng, H.: FullSubNet+: channel attention FullSubNet with complex spectrograms for speech enhancement. In: ICASSP 2022 - 2022 IEEE International Conference on Acoustics, Speech and Signal Processing (ICASSP), pp. 7857–7861. IEEE (2022)
34. Reddy, C.K., et al.: The INTERSPEECH 2020 deep noise suppression challenge: datasets, subjective testing framework, and challenge results. In: Proceedings of the Interspeech 2020, pp. 2492–2496 (2020). https://doi.org/10.21437/Interspeech.2020-3038
35. Rix, A.W., Beerends, J.G., Hollier, M.P., Hekstra, A.P.: Perceptual evaluation of speech quality (PESQ)-a new method for speech quality assessment of telephone networks and codecs. In: Proceedings of the 2001 IEEE International Conference on Acoustics, Speech, and Signal Processing (Cat. No. 01CH37221), vol. 2, pp. 749–752. IEEE (2001)
36. Taal, C.H., Hendriks, R.C., Heusdens, R., Jensen, J.: A short-time objective intelligibility measure for time-frequency weighted noisy speech. In: 2010 IEEE International Conference on Acoustics, Speech and Signal Processing, pp. 4214–4217. IEEE (2010)
37. Hansen, J.H., Pellom, B.L.: An effective quality evaluation protocol for speech enhancement algorithms. In: Fifth International Conference on Spoken Language Processing (1998)
38. Hu, Y., Loizou, P.C.: Evaluation of objective quality measures for speech enhancement. IEEE Trans. Audio Speech Lang. Process. **16**(1), 229–238 (2007)
39. Reddy, C.K., Gopal, V., Cutler, R.: DNSMOS: a non-intrusive perceptual objective speech quality metric to evaluate noise suppressors. In: ICASSP 2021 - 2021 IEEE International Conference on Acoustics, Speech and Signal Processing (ICASSP), pp. 6493–6497. IEEE (2021)

40. Manocha, P., Kumar, A.: Speech quality assessment through MOS using non-matching references. In: Proceedings of the Interspeech 2022, pp. 654–658 (2022). https://doi.org/10.21437/Interspeech.2022-407
41. Défossez, A., Synnaeve, G., Adi, Y.: Real time speech enhancement in the waveform domain. In: Proceedings of the Interspeech 2020, pp. 3291–3295 (2020). https://doi.org/10.21437/Interspeech.2020-2409
42. Hao, X., Su, X., Horaud, R., Li, X.: FullSubNet: a full-band and sub-band fusion model for real-time single-channel speech enhancement. In: ICASSP 2021 - 2021 IEEE International Conference on Acoustics, Speech and Signal Processing (ICASSP), pp. 6633–6637. IEEE (2021)

Adaptive Deep Graph Convolutional Network for Dialogical Speech Emotion Recognition

Jiaxing Liu, Sheng Wu, Longbiao Wang$^{(\boxtimes)}$, and Jianwu Dang

Tianjin Key Laboratory of Cognitive Computing and Application,
College of Intelligence and Computing, Tianjin University, Tianjin, China
longbiao_wang@tju.edu.cn

Abstract. With the increasing demand for humanization of human-computer interaction, dialogical speech emotion recognition (SER) has attracted the attention from researchers, and it is more aligned with actual scenarios. In this paper, we propose a dialogical SER approach that includes two modules, first module is a pre-trained model with an adapter, the other module is the proposed adaptive deep graph convolutional network (ADGCN). Since emotional speech data, especially dialogical data is scarce, it is difficult for models to extract enough information through limited data. A self-supervised pre-trained framework Data2vec is introduced. An adapter is designed to integrate pre-trained model to reduce the training cost while maintaining its extensive speech-related knowledge. The representations learned by adapter are from independent utterances (intra-utterance level), and ADGCN is proposed to model dialogical contextual information in one dialogue (inter-utterance level). Two residual mechanisms, adaptive residual and dynamic local residual are designed in the proposed ADGCN, which keep it from over-smoothing issues when increasing the number of layers to model global inter-utterance level contextual information. All experiments in this paper are conducted on IEMOCAP and achieved 76.49% in term of weighted accuracy.

Keywords: Dialogical Speech Emotion Recognition · Deep Graph Network · Adaptive Original Residual · Dynamic Local residual

1 Introduction

As a crucial part of the next generation of artificial intelligence (AI), affective computing can greatly improve cognitive efficiency, and enhance the experience of human-computer interaction (HCI). Emotion hidded in speech can help us to understand each other and improve the performance of HCI. Therefore, speech emotion recognition (SER) has a very realistic and scientific research value [1].

Emotional representation learning methods are the key to the success of the SER systems, and the introduced deep learning methods have overcomed the shortages of pre-defined limitations of traditional methods and achieved better

J. Jia et al. (Eds.): NCMMSC 2023, CCIS 2006, pp. 248–255, 2024.
https://doi.org/10.1007/978-981-97-0601-3_21

results. Among these deep learning models, CNN-based model [2] extracted features from the spectrogram, and introduced bidirectional long short-term memory (BLSTM) to model the contextual information from independent utterances (intra-utterance level). However, this most widely used baseline model ignored the importance of complete contextual modeling.

For inadequate contextual modeling, for example, sometimes one utterance has very few words, even only one word like 'ok'. Commonly using intra-utterance level model BLSTM cannot handle those situations well, and it is even difficult for humans to judge emotional states from just one utterance. Following the human cognitive process, placing utterances in dialogue scenes (inter-utterance) is a very effective and reasonable way. Dialogical (inter-utterance) contextual modeling has attracted the attention. The graph-based models [9,10] already got nice performance and could indeed model dialogue scenes. However, most of these existing graph-based models only contain one or two graph convolution layers, which limit their ability to extract information from high-order neighbors. For one dialogue, these shallow architectures are hard to capture dialogue global contextual information. So stacking more graph convolution layers are supposed for extracting global inter-utterance information. But it will bring over-smoothing problem, which degrades the performance of the model.

To study the contextual modeling problems, adaptive deep graph convolutional networks (ADGCN) for a dialogical (inter-utterance) SER system is proposed as shown in Fig. 1 which concludes an adapter and ADGCN. The simple structure adapter consists of two convolutional layers and one multilayer perceptron (MLP) layer is designed as the adapter of pre-trained Data2vec which could inject emotional knowledge to the pre-trained model and not flush original knowledge away. The ADGCN contains two residual mechanisms, adaptive original residual (AOR) and dynamic local residual (DLR) which make sure local inter-utterance information maintain, and global inter-utterance information be fully extracted.

2 Methodology

2.1 The Construction of the Dialogue Graph

The representations learned by designed adapter are intra-utterance level contextual information. For placing the utterances in one dialogue, the first is to construct one dialogue into graph format. Suppose $D = \{u_1, u_2, u_3, \cdots, u_n\}$ is a dialogue including n utterances with m dims features, and the defined dialogue graph (DiaG) is $G(U, A)$. Where, the $U \in \mathbb{R}^{n \times m}$ represents the features of nodes and u_i, represents the i-th row vector of U. $A \in \{0, 1\}^{n \times n}$ is an adjacent matrix denoting edges in DiaG, and if an edge exists from vertex u_i to vertex u_j, $A_{ij} = 1$ otherwise $A_{ij} = 0$.

For construction of edges in a DiaG, each node (utterance) is connected to itself and w future utterances in a dialogue, and all connections are directed from the past to the future. There are two reasons. First, in our daily life, the judgment about emotional states of current utterance is according to the past utterances

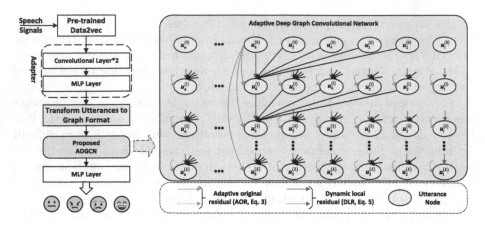

Fig. 1. The overall architecture of our speech emotion recognition system.

without the following sentences, which is why the edges in the constructed DiaG is directed. Second, if we hypothesized each node is connected to all the future utterances in a dialogue, the number of edges in a DiaG will reach $O(n^2)$, which is computationally very expensive for DiaGs with large number of utterances. So, the window size (w) is used to keep a reasonable number of edges in a DiaG.

2.2 Adaptive Deep Graph Convolutional Network

By some approximations on the polynomial of Laplacians, Kipf et al. [11] proposed the Graph Convolutional Network (GCN), used to encode the graph structure data. The layer-wise propagation rule can be represented as

$$H_{agg}^{(l+1)} = \tilde{D}^{-\frac{1}{2}} \tilde{A} \tilde{D}^{-\frac{1}{2}} H^{(l)}, \tag{1}$$

$$H^{(l+1)} = Relu\left(H_{agg}^{(l+1)} W^{(l)}\right). \tag{2}$$

where, $\tilde{A} \in \{0, 1\}^{n \times n}$ denotes adjacent matrix with self-loop, i.e. $\tilde{A} = A + I$; $\tilde{D} \in \mathbb{R}^{n \times n}$ represents the diagonal matrix consisted by degrees of nodes, i.e. $\tilde{D}_{ii} = \sum_i^n \tilde{A}_{i,:}$; $W^{(l)} \in \mathbb{R}^{d \times d}$ is a learnable matrix, and d is the hidden size; $H^{(l)} \in \mathbb{R}^{n \times d}$ denotes nodes embedding in the l-th layer.

As shown as Eq. (1) and Eq. (2), we split the GCN as two parts: *Laplacian transition* as Eq. (1) and *No-linear mapping* as Eq. (2). Then, according to the reference [12], *Laplacian transition* is actually a special form of Laplacian smoothing. As a result, when multiple GCN layers are stacked, the model will make representations of nodes tend to converge to a certain value and become indistinguishable, which is called over-smoothing [13]. So, most of the recent graph-based models, such as GCN [11] and GAT [14], can only work under 3 layers [15]. However, for a DiaG with large number of utterances, a shallow architectures will limit model ability to extract information from global inter-utterance contextual information.

So, in ADGCN, two optimization strategies, **Adaptive original residual (AOR)** and **Dynamic local residual (DLR)**, are proposed to extend GCN to deeper, which allowed model extract more comprehensively dialogical context information.

Adaptive Original Residual: Inspired by the ResNet model [16], we introduce a learnable adaptive residual connection to the original features of nodes in the Laplacian transition to alleviate over-smoothing problem. After adding Adaptive original residual, Eq. (1) can be rewritten as:

$$H_{agg}^{(l+1)} = (1 - s^{(l)})\tilde{D}^{-\frac{1}{2}}\tilde{A}\tilde{D}^{-\frac{1}{2}}H^{(l)} + s^{(l)}H^{(0)}, \tag{3}$$

$$s^{(l)} = Sigmoid\left(H^{(l)}q + b\right). \tag{4}$$

where, $s^{(l)} \in \mathbb{R}^{n \times 1}$ denotes proportions of connecting origin features in the l-th layer. In Eq. (4), the trainable $q \in \mathbb{R}^{d \times 1}$ and $b \in R$ consist a linear mapping function, and the normalization function $Sigmoid$ limits s in $(0, 1)$. According to s, we realize a adaptive residual that makes nodes original features maintain in node embedding by a adaptive scale.

Dynamic Local Residual: Similar to MLP, stacking too many *No-linear mapping* will arise vanishing gradient. So, a local dynamic residual module is added to Eq. (2) for each layer in order to avoid vanishing gradient. And the new *No-linear mapping* with local dynamic residual can be modified:

$$H^{(l+1)} = Relu\left(\alpha^{(l)}H_{agg}^{(l+1)}W^{(l)} + \left(1 - \alpha^{(l)}\right)H_{agg}^{(l+1)}\right), \tag{5}$$

where, $\alpha^{(l)} = 1/2l$ ensured the decay of the weight matrix dynamically increases as stacking more layers. According to Eq. (5), with the increasing depth, the coefficient of H_{agg} will adaptively increases, which reduced degradation of gradient in backpropagation.

3   Experiments and Analysis

3.1   Experimental Setup

The experiments are conducted on Interactive Emotional Dyadic Motion Capture database (IEMOCAP) [17] which consists of 5,531 utterances with four emotion categories: Neutrality (29%, N), Anger (20%, A), Sadness (20%, S) and Happiness (31%, H). The pre-trained model in this paper, we use Data2vec with one adapter which contains two 1D-CNN with channles (512, 256), strides (2, 1) and kernel width (5, 3), following one MLP with hidden size 512. For ADGCN, we set hidden size as 256. Adam is optimizer and ReLU is the activation. The batch size is set as 128 and we use leave-one-session-out to train the model. The CNN_BLSTM [2] and TFCAP [10] are spectrogram-based models which are selected as comparative model and DialogueGCN is introduced as dialogical SER model. DeepGCN is a model that directly adds GCN layers as deep GCN model. The † indicates only one layer of the proposed models.

3.2 Experiments Results and Analysis

Visualization. In this section, the t-distributed stochastic neighbor embedding (t-SNE) [18] is introduced to visualize the four emotional categories of the representations which are extracted by Data2vec with adapter (intra-utterance level) and GCN (inter-utterance level) [11]. Observing red dots (Happiness) in Fig. 2(a) which spreads all around the regions and which are greatly improved in Fig. 2(b). Some yellow dots in Fig. 2(b) are far away from their majority region and misclassify into the blue region (Neutrality). These phenomena indicate that intra- and inter- utterance level representations have their own characteristics which means considering the complementarity is necessary.

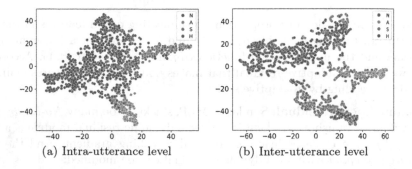

(a) Intra-utterance level   (b) Inter-utterance level

Fig. 2. The t-SNE visualizations of intra-and inter-utterance level representations

Relations Between Layers and Windows. The results of DeepGCN and proposed ADGCN are shown in Fig. 3 which is under the condition of same window size $(win = 4)$. First, observing results when both two these models are one layer, the result reflects the performance of extracting representations. The proposed ADGCN in one layer is better than traditional GCN and has absolute increases of 1.94% and 3.44% in terms of accuracy and F1-score. With the layers increasing, the inter-utterance level representations are gradually extracted. However, the DeepGCN quickly fall into the problem of over-smoothing. The proposed model ADGCN overcomes the problem of over-smoothing and achieve the best around 10 layers. A very interesting phenomenon should be noticed here is that the product of the numbers of layers and window-size approximates the average number of utterances in one dialogue. That also proves that our proposed model could learn the inter-utterance level information comprehensively and stably.

Classification Results of Comparative Experiments and the Ablation Study of Proposed ADGCN. In order to quantitatively evaluate the performance of the proposed ADGCN, the classification results of five comparative experiments are provided in Table 1. Also, the ablation study of proposed model

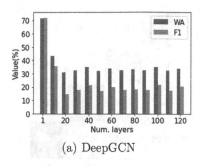 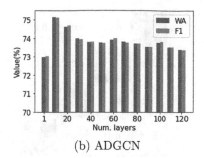

(a) DeepGCN   (b) ADGCN

Fig. 3. The results of DeepGCN and proposed ADGCN under the same window size ($w = 4$) in different depths

Table 1. The results of comparative experiments and the ablation study of proposed ADGCN. ADGCN† and DeepGCN† represent only the one layer of models

Model	WA(%)	UA(%)	F1(%)
CNN_BLSTM [19]	62.53	63.78	62.98
TFCAP [10]	66.50	64.85	65.60
Co-MLAI [20]	71.64	72.70	–
DeepGCN† [11]	71.15	69.22	70.14
DialogueGCN [9]	72.76	69.18	71.38
Data2vec	65.43	65.71	65.66
Data2vec_adapter	66.96	67.17	67.45
ADGCN†	73.09	73.19	73.58
ADGCN	**76.79**	**76.20**	**76.43**

is provided in the blow of Table 1. Our proposed global model achieves 76.79%, 76.20% and 76.43% in terms of weighted-accuracy (WA), unweighted-accuracy (UA) and F1-score (F1).

To further research and validate the performance when the layers increasing, two confusion matrices ADGCN† and ADGCN under the representation Data2vec_adapter are shown in Fig. 4. These two confusion matrices reflect that ADGCN† and ADGCN both show high sensitivity to all four emotion types. Compared with the ADGCN† (one layer), the proposed deep model ADGCN not only gets very balance results, but also the Happiness emotion achieves significant improvement.

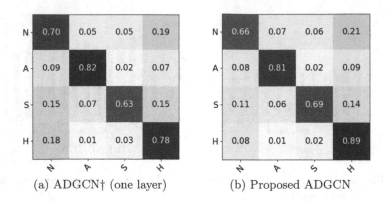

(a) ADGCN† (one layer)   (b) Proposed ADGCN

Fig. 4. The confusion matrices of ADGCN† and ADGCN

4   Conclusion

In this paper, we designed an adapter for the SSL-based pre-trained model Data2vec to extract significant intra-utterance information, and the visualization also proved the pre-trained model indeed bring a breakthrough. The extracted representations by the adapter intra-utterance level information which could not fit the need of actual needs. For the modeling of inter-utterance level contextual information, the ADGCN was proposed, and two residual mechanisms (AOR and DLR) were included. These two residual parts ensured that over-smoothing does not occur while fully extracting both local and global inter-utterance contextual information. Compared with the current intra-utterance methods and inter-utterance methods, our proposed model ADGCN got 76.79%, 76.20% and 76.43% with absolute increments more than 4.03%, 7.02% and 5.05% in terms of WA, UA and F1-score. The provided confusion matrices showed high sensitivity to all four emotion types. Based on the advantages of flexible expansion, in the future, we plan to investigate the performance of the proposed ADGCN on multimodal emotion fusion task.

References

1. Wani, T.M., Gunawan, T.S., Qadri, S.A.A., Kartiwi, M., Ambikairajah, E.: A comprehensive review of speech emotion recognition systems. IEEE Access **9**, 47795–47814 (2021)
2. Satt, A., Rozenberg, S., Hoory, R.: Efficient emotion recognition from speech using deep learning on spectrograms. In: Interspeech 2017, Stockholm, Sweden, pp. 1089–1093 (2017). https://doi.org/10.21437/Interspeech.2017-200
3. Devlin, J., Chang, M., Lee, K., Toutanova, K.: BERT: pre-training of deep bidirectional transformers for language understanding. arXiv preprint arXiv:1810.04805 (2018)
4. Liu, Y., et al.: RoBERTa: a robustly optimized BERT pretraining approach. arXiv preprint arXiv:1907.11692 (2019)

5. Grill, J., et al.: Bootstrap your own latent-a new approach to self-supervised learning. In: Advances in Neural Information Processing Systems, vol. 33, pp. 21271–21284 (2020)
6. Baevski, A., Zhou, Y., Mohamed, A., Auli, M.: Wav2vec 2.0: a framework for self-supervised learning of speech representations. In: Advances in Neural Information Processing Systems, vol. 33, pp. 12449–12460 (2020)
7. Hsu, W.N., Hubert Tsai, Y.H., Bolte, B., Salakhutdinov, R., Mohamed, A.: HuBERT: how much can a bad teacher benefit ASR pre-training?. In: ICASSP 2021–2021 IEEE International Conference on Acoustics, Speech and Signal Processing (ICASSP), Toronto, Ontario, Canada, pp. 6533–6537 (2021). https://doi.org/10.1109/ICASSP39728.2021.9414460
8. Baevski, A., Hsu, W.N., Xu, Q., Babu, A., Gu, J., Auli, M.: Data2vec: a general framework for self-supervised learning in speech, vision and language. arXiv preprint arXiv:2202.03555 (2022)
9. Ghosal, D., Majumder, N., Poria, S., Chhaya, N., Gelbukh, A.: DialogueGCN: a graph convolutional neural network for emotion recognition in conversation. arXiv preprint arXiv:1908.11540 (2019)
10. Liu, J., Song, Y., Wang, L., Dang, J., Yu, R.: Time-frequency representation learning with graph convolutional network for dialogue-level speech emotion recognition. In: Interspeech 2021, Brno, Czech Republic, pp. 4523–4527 (2021). https://doi.org/10.21437/Interspeech.2021-2067
11. Kipf, T.N., Welling, M.: Semi-supervised classification with graph convolutional networks. arXiv preprint arXiv:1609.02907 (2016)
12. Li, Q., Han, Z., Wu, X.: Deeper insights into graph convolutional networks for semi-supervised learning. In: 32nd AAAI Conference on Artificial Intelligence 2018, New Orleans, Louisiana, USA, pp. 3538–3545 (2018). https://doi.org/10.48550/arXiv.1801.07606
13. Liu, M., Gao, H., Ji, S.: Towards deeper graph neural networks. In: Proceedings of the 26th ACM SIGKDD International Conference on Knowledge Discovery & Data Mining, Virtual Event, CA, USA, pp. 338–348 (2020). https://doi.org/10.1145/3394486.3403076
14. Veličković, P., Cucurull, G., Casanova, A., Romero, A., Lio, P., Bengio, Y.: Graph attention networks. arXiv preprint arXiv:1710.10903 (2017)
15. Li, G., et al.: DeepGCNs: making GCNs go as deep as CNNs. arXiv preprint arXiv:1910.06849 (2021)
16. Targ, S., Almeida, D., Lyman, K.: ResNet in ResNet: generalizing residual architectures. arXiv preprint arXiv:1603.08029 (2016)
17. Busso, C., et al.: IEMOCAP: interactive emotional dyadic motion capture database. Lang. Resour. Eval. **42**, 335–359 (2008)
18. Belkina, A.C., Ciccolella, C.O., Anno, R., Halpert, R., Spidlen, J., Snyder-Cappione, J.E.: Automated optimized parameters for T-distributed stochastic neighbor embedding improve visualization and analysis of large datasets. Nat. Commun. **10**(1), 1–12 (2019)
19. Han, K., Yu, D., Tashev, I.: Speech emotion recognition using deep neural network and extreme learning machine. In: Fifteenth Annual Conference of the International Speech Communication Association, Singapore, Malaysia, pp. 223–227 (2014). https://doi.org/10.21437/Interspeech.2014-57
20. Guo, L.L., Wang, L.B., Dang, J.W.: A feature fusion method based on extreme learning machine for speech emotion recognition. In: 2018 IEEE International Conference on Acoustics, Speech and Signal Processing (ICASSP), Calgary, AB, Canada, pp. 2666–2670 (2018). https://doi.org/10.1109/ICASSP.2018.8462219

Iterative Noisy-Target Approach: Speech Enhancement Without Clean Speech

Yifan Zhang[1], Wenbin Jiang[2]($^{(\boxtimes)}$), Qing Zhuo[3], and Kai Yu[1]

[1] MoE Key Lab of Artificial Intelligence, X-LANCE Lab, Department of Computer Science and Engineering, Shanghai Jiao Tong University, Shanghai, China
{zhang_yifan,kai.yu}@sjtu.edu.cn
[2] School of Communication Engineering, Hangzhou Dianzi University, Hangzhou, China
wbjiang@hdu.edu.cn
[3] Department of Automation, Tsinghua University, Beijing, China
zhuoqing@tsinghua.edu.cn

Abstract. Traditional Deep Neural Network based speech enhancement usually requires clean speech as the target of training. However, limited access to ideal clean speech hinders its practical use. Meanwhile, existing self-supervised or unsupervised methods face both unsatisfactory performance and impractical source demand (e.g., various kinds of noises added to the same clean speech). Hence, there's a significant need to either release the restriction of training data or improve the performance. In this paper, we propose a training strategy that only requires noisy speech and noise waveform. It primarily consists of two phases: 1) With a pair of input and target constructed by adding noise to noisy speech itself for the training of DNN, the first round of training uses noisier speech (noise added to noisy speech) and noisy speech 2) For the following training, using the model trained last time to refine the noisy speech, construct new noisier-noisy pairs for next turn of training. Moreover, to accelerate the process, we apply the iteration into epochs. To evaluate the efficiency, we utilized a dataset including 10 types of real-world noises and made a comparison with two classic supervised and unsupervised methods.

Keywords: Speech Enhancement · Speech Denoising · Self-supervised · Deep Neural Network · Noise2Noise

1 Introduction

To obtain clean speech from noisy ones is the rudimentary purpose of speech enhancement (SE). The demand for audio and speech free from noises exists in various tasks, such as automatic speech recognition (ASR) [1], for enhanced speech sometimes leads to better criteria, which drives researchers to develop more effective SE methods. As an increasing number of high-quality speech and noise datasets [2–4] have become available, we are able to train Deep Neural Networks (DNN) with more efficiency by using clean speech in datasets as the training target. Experiments and practice in the recent decade have demonstrated its

J. Jia et al. (Eds.): NCMMSC 2023, CCIS 2006, pp. 256–264, 2024.
https://doi.org/10.1007/978-981-97-0601-3_22

ability to tackle challenges in both single-channel and multi-channel SE tasks [5–7]. However, these datasets are intentionally studio-recorded in a strictly silent environment thus it's hard to acquire as large a amount of data as other tasks in the audio domain (e.g., a classic SE dataset, VoiceBank-DEMAND [8], contains 300 h of recordings, while GigaSpeech [9], an ASR dataset, includes 10000 h of audio).

To surmount these challenges of supervised methods that use clean speech as a target (i.e., **Noise2Clean**), researchers keep exploring approaches to train a DNN model without clean data, in other words, unsupervised or self-supervised approaches. One strategy is **Noise2Noise**, which means the input and target are respectively mixtures of clean speech with noise to eliminate and another noise. This strategy was first proposed in the field of image denoising [10], and was illustrated as effective in speech enhancement [11,12] (see details in 2.2). In this strategy, clean speech is completely avoided. However, as mentioned in [13], unlike image in which it is simple to acquire an image with different types of noises, in speech, it's hard to observe multiple noisy signals with precisely the same speech due to time and space variance of audio signals. Now that another methodology was introduced.

To tackle this problem, researchers proposed **Noisy-target** training. Regarding the noisy speech as "clean", by simply adding the same type of noise to the noisy speech, we get a pair of noisy and "clean" speech, but actually, the "clean" one is relatively clean and the noisy one gets noisier.

In this paper, we proposed an improved self-supervised method. Inspired by the pioneering work [14], aiming to narrow the bias between Noisy-target method pairs and the noise-clean pairs, we introduced an iterative step in each round of training, using the currently trained model to modify the noisy speech continuously. We estimated that this method may have performance close to Noise2Noise. In terms of evaluation, we train DNNs on several types of noisy datasets through each method and calculate the score of speeches denoised by models. Then we will analyze the experimental results.

2   Related Work

2.1   Noise2Clean Training

This most classic and widely applied supervised training strategy [15] utilizes noisy speech $x = s + n$ as input and clean speech s as target, where n denotes a sample from a certain type of noise (the meaning of n is all the same in this paper). The prediction error (e.g. L_2 loss) to minimize is:

$$Loss_{n2c} = \mathbb{E}\{(f_\theta(s + n) - (s))^2\} \tag{1}$$

Thus, the additional noise n will be eliminated by the model f_θ. This supervised training method has realistic limitations in obtaining clean speech; therefore, several self-supervised systems have been developed.

2.2 Noise2Noise Training

In this self-supervised method, all that is needed is two noisy speech generated by mixing the same clean speech s with two different types of noise n_1, n_2, so that a denoising model could be achieved without clean training data. The model is trained using pairs $(x_1 = s + n_1, x_2 = s + n_2)$. Here are basic assumptions of noise [12]:

Condition 1. The additional noises are sampled from zero-mean distribution and uncorrelated to the clean (i.e., $\mathbb{E}\{n\} = 0$).

Condition 2. The correlation of different types of noises is close to zero.

The training goal is to minimize the prediction error between output $f_\theta(x_1)$ and target x_2, here L_2 is employed:

$$
\begin{aligned}
Loss_{n2n} &= \mathbb{E}\{(f_\theta(x_1) - x_2)^2\} \\
&= \mathbb{E}\{(f_\theta(x_1) - (s + n_2))^2\} \\
&= \mathbb{E}\{(f_\theta(x_1) - s)^2\} - \mathbb{E}\{2n_2(f_\theta(x_1) - s)\} + \mathbb{E}\{n_2^2\} \\
&= \mathbb{E}\{f_\theta(x_1) - s)^2\} - \mathbb{E}\{2n_2(f_\theta(x_1) - s)\} - Var(n_2) + \mathbb{E}\{n_2\}^2
\end{aligned}
\tag{2}
$$

According to **Condition 1**, $\mathbb{E}\{n_{1,2}\} = 0$. And **Condition 2** shows n_2 independent from all variables here, so the second term also equals 0. As for $Var(n_2)$, the variance of the sample distribution is equal to the variance of the population divided by the sampling size. So when the sample size is huge enough, we have:

$$
Loss_{n2n} = \mathbb{E}\{(f_\theta(x_1) - s)^2\} = Loss_{n2c}
\tag{3}
$$

Therefore, minimizing the discrimination between input and target will also minimize the error between noisy input and clean speech.

2.3 Noisy-Target Training

Noisy-target method [13] requires noisy speech $x = s + n$ and a sample from the same noise type n. By mixing x and n, we get noisier speech $y = x + n$ as input and have x as the target, forming the train pair (y, x). Therefore the DNN f_θ is trained to minimize the prediction error (e.g., L_2 loss):

$$
Loss_{NyT} = \mathbb{E}\{(f_\theta(x + n) - (x))^2\}
\tag{4}
$$

between output $f_\theta(x + n)$ and target x. It's obvious that this method is similar to Noise2Clean but does not use clean speech.

The methods discussed above lead to two findings. (1) We have confirmed that the model trained by datasets consisting of noisier and noisy speech in 3.3 could denoise the original noisy speech. (2) As results presented in [13] also indicate its gap away from the Noise2Clean method, we are now aware of the bias of data between noisier-noisy pairs and noisy-clean pairs could weaken the model in a visible scale. This inspired us to reduce the bias through an iterative method to improve the denoising performance.

3   Methodology

3.1   Basic Iterative Method

Our method is illustrated in Fig. 1. A model f is trained for the first epoch based on the same noisier-noisy pairs $(x+n, x)$ as the NoisyTarget method. Before the next epoch, we refine the dataset by utilizing the currently trained model f_1 to denoise the initial noisy speech $x = s + n$ and then add noise to the denoised speech. After that, we have a dataset D_1 consisting of new pair $(f_1(x)+n, f_1(x))$, whose bias is lower compared with the original noisier-noisy pair. In the next epoch, f will be trained on the modified dataset of $D_1 = \{(x_1 + n, x_1))\}$.

The following epochs are in the same way. To illustrate the process, let's consider the end of the i-th epoch. In this moment, we have the trained model f_i and dataset used in the i-th epoch $D_{i-1} = \{(x_{n-1} + n, x_{n-1})\}$. In the next epoch, the model will be continuously trained on the refined dataset D_i consisting of $(f_i(x_{n-1}) + n, f_i(x_{n-1}))$. It's obvious that in each epoch, the training pairs refined by the currently trained model are gradually closer to noisy-clean pairs $(s + n, s)$. Simultaneously, f_i is becoming closer to the model trained with the Noise2Clean method and is expected to perform better.

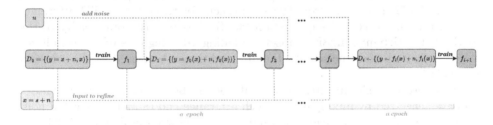

Fig. 1. An Overview of the iteration method

3.2   Iteration in Step

However, the method above is also time-consuming since we have to update the whole dataset after every epoch. For the purpose of accelerating the process, we propose another approach to refine the data.

As shown in Algorithm 1, in the first epoch, the noisier-noisy pairs are still used. Then the original data would be refined by the present trained model to construct new pairs in each epoch after that (the model is set to evaluation mode). Specifically, the newly built dataset won't be stored and used in the refinement in the next epoch.

Algorithm 1: Proposed iterative training algorithm

Data: noisy speech $x = s + n$, noise samples of a certain type n, number of epochs E

Result: model f_E

1 train f_1 on $(x + n, x)$;
2 **for** $i \leftarrow 2$ **to** E **do**
3 $\quad$ Copy f_{i-1} and fix its parameters;
4 $\quad$ Refine the data $x_i = f_{i-1}(x)$;
5 $\quad$ Add noise and construct pairs $(x_i + n, x_i)$;
6 $\quad$ Optimize f on pairs above;
7 **end**

---

4 Experiment Setups

4.1 Datasets

In experiments, we utilized the VoiceBank dataset as the source of clean speech (the clean speech will never appear in the entire training process of the unsupervised method and will only be used to synthesize noisy speech). The dataset contains 11572 utterances from the 28 speakers for training and 872 utterances from 2 unseen speakers for evaluation [8]. To illustrate the practicality of this method, the UrbanSound8K dataset, including 10 categories of noises, all collected from real-world noises like the air conditioner, was used as the source of noises [16].

The speech-noise mixtures of one category in both training and evaluation sets are generated by mixing each speech utterance and randomly selected noise utterance in the same category at random SNR between 0 dB and 10 dB. The mixing method is from [17].

4.2 Network Architecture

In experiments, we chose to train the 16-layer Deep Complex U-Net [18] (DCUNET-16) to test methods to be compared. This architecture has been proven effective in speech enhancement tasks, especially on the Voice-Bank+DEMAND dataset [6]. Details of modules in the deep complex network were described in [19]

The input of the network should be of the time-frequency domain. Time domain waveforms are first converted through Short Time Fourier Transform (STFT) with an FFT size of 512 (32 ms in 16 kHz sample rate), a hop length of 100, and a window length of 400.

The model will output a noise mask $\hat{M}_{t,f}$, and by multiplying $\hat{M}_{t,f}$ and the input spectrogram, we obtain the estimation spectrogram. The denoised waveform will be generated through the Inverse Short Time Fourier Transform (ISTFT) with the same parameters as STFT.

4.3   Training and Evaluation Details

To evaluate, we applied objective metrics, the wide-band Perceptual Evaluation of Speech Quality (PESQ) [20], and Short-Time Objective Intelligibility (STOI) [21] as the score of the model. The scores have been widely used in other speech tasks [22]. Since the metrics are not differentiable, in terms of the training loss, the weighted SDR loss function (denoted as $loss_{wSDR}$) is used, which is also applied in another speech enhancement task [12]. $loss_{wSDR}$ is defined as:

$$\alpha = \frac{||y||^2}{||y||^2 + ||x - y||^2} \tag{5}$$

$$loss_{wSDR} = -\alpha \frac{<y, \hat{y}>}{||y||||\hat{y}||} - (1 - \alpha) \frac{<x - y, x - \hat{y}>}{||x - y||||x - \hat{y}||} \tag{6}$$

$x, y, \hat{y}$ denote noisy speech, target speech and estimation respectively.

5   Results

Each training strategy's mean of PESQ scores is tabulated in Table 1. Each row corresponds to a noise category while N2C indicates the performance of the classic Noise2Clean approach, N2N suggests the performance of the Noise2Noise approach, NyT indicates the performance of Noisy-Target, and Ours presents the performance of our new iteration method.

The results illustrate that N2C was the most efficient for achieving the highest scores when processing all kinds of noisy speech. Among self-supervised methods, N2N performs slightly better than ours, while ours achieves a higher PESQ score in Siren and Street Music. Our method generally performs close to N2C and N2N. It's also clear that NyT only has a very tiny affection.

The minute inferior position of our method primarily derives from the ineffectiveness of the NyT method. The model trained in the first epoch could tremendously influence the successive training because if the refinement fails to denoise the original noisy speech x or even mitigate its quality, the following epochs will not be appropriately trained as expected. When NyT does not work, the disparity between the iterative method and the other two is much more significant, and this explanation accords with the results of Drilling and Jackhammer (The PESQ of 1.2 is close to that of the initial noisy speech).

Table 1. Average SNR, PESQ and STOI scores of denoising results of DCUNet-16 based on each method.

Noise Category		N2C	N2N	NyT	Ours
Air Conditioning(0)	SNR	13.70	13.73	10.66	13.50
	PESQ	1.868	1.837	1.465	1.813
	STOI	0.873	0.872	0.863	0.864
Car Horn(1)	SNR	15.80	15.67	11.85	15.18
	PESQ	2.045	2.038	1.604	2.023
	STOI	0.885	0.883	0.853	0.886
Children Playing(2)	SNR	13.40	13.48	10.58	12.25
	PESQ	1.738	1.724	1.412	1.685
	STOI	0.850	0.850	0.833	0.843
Dog Barking(3)	SNR	14.15	14.02	11.03	14.08
	PESQ	1.874	1.872	1.647	1.835
	STOI	0.880	0.879	0.863	0.869
Drilling(4)	SNR	14.37	14.36	10.68	14.14
	PESQ	1.735	1.724	1.272	1.668
	STOI	0.849	0.848	0.830	0.841
Engine Idling(5)	SNR	14.19	14.22	11.08	13.92
	PESQ	1.862	1.862	1.511	1.822
	STOI	0.876	0.877	0.865	0.866
Gunshot(6)	SNR	16.59	16.54	10.76	16.03
	PESQ	2.085	2.080	1.617	1.996
	STOI	0.881	0.880	0.857	0.874
Jackhammer(7)	SNR	13.97	13.95	10.66	13.17
	PESQ	1.731	1.718	1.260	1.633
	STOI	0.854	0.852	0.823	0.841
Siren(8)	SNR	15.31	15.19	11.28	14.94
	PESQ	1.943	1.923	1.489	1.934
	STOI	0.888	0.886	0.862	0.884
Street Music(9)	SNR	12.71	12.78	10.07	12.77
	PESQ	1.769	1.766	1.435	1.769
	STOI	0.858	0.859	0.840	0.854

6   Conclusion

In this work, we proposed a more applicable and effective method to train a DNN model without clean data. We improved the Noisy-target by using currently trained model. By training the DCUNET-16 to denoise 10 categories of real-world noise, we showed that our method (1) has the ability to train a model whose performance is close to the model from clean training target, (2) could work in situations where Noisy-Target method couldn't. Future works would involve using exploration on adding noise to a noisy speech with another type of noise to avoid the circumstances that NyT fails to take effect.

Acknowledgements. This research was funded by Scientific and Technological Innovation 2030 under Grant 2021ZD0110900 and the Key Research and Development Program of Jiangsu Province under Grant BE2022059.

References

1. Erdogan, H., Hershey, J.R., Watanabe, S., Le Roux, J.: Phase-sensitive and recognition-boosted speech separation using deep recurrent neural networks. In: 2015 IEEE International Conference on Acoustics, Speech and Signal Processing (ICASSP), pp. 708–712 (2015). https://doi.org/10.1109/ICASSP.2015.7178061
2. Garofolo, J., et al.: TIMIT Acoustic-Phonetic Continuous Speech Corpus (1993). 11272.1/AB2/SWVENO, https://hdl.handle.net/11272.1/AB2/SWVENO
3. Mesaros, A., Heittola, T., Virtanen, T.: Acoustic scene classification in DCASE 2019 challenge: closed and open set classification and data mismatch setups. In: DCASE (2019)
4. Heittola, T., Mesaros, A., Virtanen, T.: Acoustic scene classification in DCASE 2020 challenge: generalization across devices and low complexity solutions. In: Proceedings of the Fifth Workshop on Detection and Classification of Acoustic Scenes and Events (DCASE 2020), pp. 56–60 (2020). http://dcase.community/workshop2020/
5. Kawanaka, M., Koizumi, Y., Miyazaki, R., Yatabe, K.: Stable training of DNN for speech enhancement based on perceptually-motivated black-box cost function. In: ICASSP 2020–2020 IEEE International Conference on Acoustics, Speech and Signal Processing (ICASSP), pp. 7524–7528 (2020)
6. Hu, Y., et al.: DCCRN: deep complex convolution recurrent network for phase-aware speech enhancement. ArXiv abs/2008.00264 (2020)
7. Tzirakis, P., Kumar, A., Donley, J.: Multi-channel speech enhancement using graph neural networks. In: ICASSP 2021–2021 IEEE International Conference on Acoustics, Speech and Signal Processing (ICASSP), pp. 3415–3419 (2021). https://doi.org/10.1109/ICASSP39728.2021.9413955
8. Veaux, C., Yamagishi, J., King, S.: The voice bank corpus: design, collection and data analysis of a large regional accent speech database. In: Oriental COCOSDA held jointly with 2013 Conference on Asian Spoken Language Research and Evaluation (O-COCOSDA/CASLRE), 2013 International Conference. Institute of Electrical and Electronics Engineers (IEEE), United States, November 2013. https://doi.org/10.1109/ICSDA.2013.6709856
9. Chen, G., et al.: Gigaspeech: an evolving, multi-domain asr corpus with 10,000 hours of transcribed audio (2021). https://doi.org/10.48550/ARXIV.2106.06909, https://arxiv.org/abs/2106.06909
10. Lehtinen, J., et al.: Noise2noise: learning image restoration without clean data. ArXiv abs/1803.04189 (2018)
11. Alamdari, N., Azarang, A., Kehtarnavaz, N.: Improving deep speech denoising by Noisy2Noisy signal mapping. Appl. Acoust. **172**, 107631 (2021). https://doi.org/10.1016/j.apacoust.2020.107631. https://www.sciencedirect.com/science/article/pii/S0003682X20307350. ISSN 0003-682X

12. Kashyap, M.M., Tambwekar, A., Manohara, K., Natarajan, S.: Speech denoising without clean training data: a Noise2Noise approach. In: Proceedings of the Interspeech 2021, pp. 2716–2720 (2021). https://doi.org/10.21437/Interspeech. 2021–1130

13. Fujimura, T., Koizumi, Y., Yatabe, K., Miyazaki, R.: Noisy-target training: a training strategy for DNN-based speech enhancement without clean speech. In: 2021 29th European Signal Processing Conference (EUSIPCO), pp. 436–440 (2021). https://doi.org/10.23919/EUSIPCO54536.2021.9616166

14. Zhang, Y., Li, D., Law, K.L., Wang, X., Qin, H., Li, H.: IDR: self-supervised image denoising via iterative data refinement. In: 2022 IEEE/CVF Conference on Computer Vision and Pattern Recognition (CVPR), pp. 2088–2097 (2022)

15. Wang, D., Chen, J.: Supervised speech separation based on deep learning: an overview. IEEE/ACM Trans. Audio Speech Lang. Process. 26, 1702–1726 (2018)

16. Salamon, J., Jacoby, C., Bello, J.P.: A dataset and taxonomy for urban sound research. In: Proceedings of the 22nd ACM International Conference on Multimedia (2014)

17. Dubey, H., et al.: ICASSP 2022 deep noise suppression challenge. In: ICCASP (2022)

18. Choi, H.S., Kim, J.H., Huh, J., Kim, A., Ha, J.W., Lee, K.: Phase-aware speech enhancement with deep complex u-net. ArXiv abs/1903.03107 (2019)

19. Trabelsi, C., et al.: Deep complex networks. In: International Conference on Learning Representations (2018). https://openreview.net/forum?id=H1T2hmZAb

20. Rix, A.W., Beerends, J.G., Hollier, M., Hekstra, A.P.: Perceptual evaluation of speech quality (PESQ)-a new method for speech quality assessment of telephone networks and codecs. In: 2001 IEEE International Conference on Acoustics, Speech, and Signal Processing. Proceedings (Cat. No.01CH37221), vol. 2, pp. 749–752 (2001)

21. Taal, C.H., Hendriks, R.C., Heusdens, R., Jensen, J.: A short-time objective intelligibility measure for time-frequency weighted noisy speech. In: 2010 IEEE International Conference on Acoustics, Speech and Signal Processing, pp. 4214–4217 (2010). https://doi.org/10.1109/ICASSP.2010.5495701

22. Cao, R., Abdulatif, S., Yang, B.: CMGAN: conformer-based metric GAN for speech enhancement. In: INTERSPEECH (2022)

Joint Training or Not: An Exploration of Pre-trained Speech Models in Audio-Visual Speaker Diarization

Huan Zhao[1], Li Zhang[1], Yue Li[1], Yannan Wang[2], Hongji Wang[2], Wei Rao[2], Qing Wang[1], and Lei Xie[1(✉)]

[1] Audio, Speech and Language Processing Group (ASLP@NPU),
School of Computer Science, Northwestern Polytechnical University (NPU),
Xi'an, China
lxie@nwpu.edu.cn
[2] Tencent Corporation, Shenzhen, China

Abstract. The scarcity of labeled audio-visual datasets is a constraint for training superior audio-visual speaker diarization systems. To improve the performance of audio-visual speaker diarization, we leverage pre-trained supervised and self-supervised speech models for audio-visual speaker diarization. Specifically, we adopt supervised (ResNet and ECAPA-TDNN) and self-supervised pre-trained models (WavLM and HuBERT) as the speaker and audio embedding extractors in an end-to-end audio-visual speaker diarization (AVSD) system. Then we explore the effectiveness of different frameworks, including Transformer, Conformer, and cross-attention mechanism, in the audio-visual decoder. To mitigate the degradation of performance caused by separate training, we jointly train the audio encoder, speaker encoder, and audio-visual decoder in the AVSD system. Experiments on the MISP dataset demonstrate that the proposed method achieves superior performance and obtained third place in MISP Challenge 2022.

Keywords: audio-visual · speaker diarization · pre-trained model · joint traning

1 Introduction

Speaker diarization (SD) is the task of determining "who spoke when?" in an audio or video recording [20]. Its application is indispensable in multimedia information retrieval, speaker turn analysis, and multi-speaker speech recognition. With the emergence of wide and complex application scenarios of SD, single-modal (audio- or visual-based) SD encounters performance bottleneck [3,27,31]. Specifically, speech is intrinsically uncertain because it includes multiple speech signals produced by multi-speakers, interfered by echoes, other audio sources, and ambient noise. Likewise, identifying speakers from a single visual modal is highly complex, and it is restricted to detecting lip or facial movements from frontal close-up images of people. In more general scenarios, such as informal

J. Jia et al. (Eds.): NCMMSC 2023, CCIS 2006, pp. 265–275, 2024.
https://doi.org/10.1007/978-981-97-0601-3_23

social gatherings, people may not always face the cameras, which makes lip reading a difficult task [13]. To enhance the effectiveness of speaker diarization, it is crucial to utilize both visual and audio information.

Recently, several works have begun to leverage both visual and audio cues for speaker diarization. Sharma et al. [24] performed face clustering on the active speaker faces and showed superior speaker diarization performance compared to the state-of-the-art audio-based diarization methods. Sarafianos et al. [23] applied a semi-supervised version of Fisher Linear Discriminant analysis, both in the audio and the video signals to form a complete multimodal speaker diarization system. In addition to introducing the facial cue as the visual modal, several methods were proposed to leverage audio and lip cues for diarization motivated by the synergy between utterances and lip movements [7]. Wuerkaixi et al. [28] proposed a dynamic vision-guided speaker embedding (DyViSE) method in a multi-stage system to deal with the challenge in noisy or overlapped speech and off-screen speakers where visual features are missing. Tao et al. [25] proposed an audio-visual cross-attention mechanism for inter-modality interaction between audio and visual features to capture long-term speaking evidence. He et al. [15] proposed a novel end-to-end audio-visual speaker diarization (AVSD) method consisting of a lip encoder, a speaker encoder, an audio encoder, and an audio-visual speaker decoder. The end-to-end training strategy is similar to [25,28]. Although AVSD obtains a comparable improvement compared with single-modal speaker diarization, the separate optimization of speaker encoder and audio-visual speaker decoder is hard to get global optima. Furthermore, despite the demonstrated effectiveness of pre-trained speech models in various downstream speech tasks [5,17], there has been limited research investigating the impact of self-supervised pre-training models on AVSD.

To fill this gap, we employ supervised pre-trained speaker models (ResNet [14] and ECAPA-TDNN [10]) and self-supervised models (HuBERT [17] and WavLM [5]) as the speaker and audio embedding extractors in an end-to-end AVSD system. Subsequently, we evaluate the effectiveness of different frameworks, such as Transformer [26], Conformer [30], and cross-attention mechanism [25,28], in the audio-visual decoder module of the AVSD system. Furthermore, to mitigate the decline in diarization performance that may arise from separate training, we jointly train the pre-trained audio encoder and speaker encoder as well as the audio-visual decoder modules in the AVSD system. Finally, the experimental results on the MISP dataset [2] demonstrate joint training of self-supervised per-trained audio encoder, supervised pre-trained speaker encoder, and Transformer-based audio-visual decoder obtains considerable improvements compared with the results of AVSD [15].

2   Method

2.1   Overview

An overview of the improved AVSD framework is illustrated in Fig. 1, which consists of a lip encoder, an audio encoder, a speaker encoder, and an audio-visual

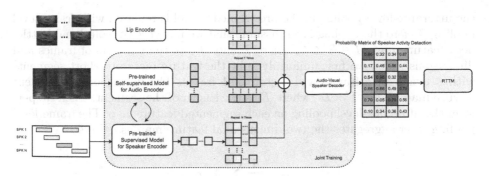

Fig. 1. Overview of the improved AVSD framework. Parts in oval boxes with gray backgrounds require joint training. Arrows represent the same model being capable of serving as both the audio encoder and the speaker encoder simultaneously. (Color figure online)

speaker decoder. Different from the AVSD framework in [15], we adopt the pre-trained self-supervised speech models (WavLM and Hubert) and pre-trained supervised speaker models (ResNet and ECAPA-TDNN) as our audio encoder and speaker encoder. Meanwhile, we jointly train the pre-trained speaker and audio encoders with the audio-visual speaker decoder to alleviate performance loss caused by separate training. Furthermore, we introduce different neural network frameworks (Transformer [26], Conformer [30], cross-attention [25]) in the audio-visual decoder to effectively fusion of audio and visual representations and decode out the probability of speaker activity detection. Finally, the probability matrix is converted to Rich Transcription Time Masked (RTTM).

2.2 Lip Encoder

The lip encoder is configured as [15], consisting of ResNet18 [14], Multi-Scale Temporal Convolutional Networks (MS-TCN) followed by three Conformer [30] and a BLSTM [16] layers. Firstly, we cut the lip regions of interest (ROIs) from the ground truth detection results from [2]. Assume that we have a sequence of lip ROIs $X_n = (X_n^1, X_n^2, \ldots, X_n^T) \in \mathbb{R}^{W \times H \times T}$, where n is the identity of speaker and T is the frame length. Then the ROIs are fed into the lip encoder. Then we stack embeddings of each speaker to get visual embeddings $E_V \in \mathbb{R}^{T \times N \times D^V}$. Moreover, the visual embeddings E_V are projected through a fully connected (FC) layer to speech activity probabilities $S = (S_1, S_2, \ldots, S_T) \in (0, 1)$. Therefore, the lipreading model can be considered as a visual speaker activity detection module, which has the capability to represent whether the state is speaking or not of a speaker. In the end, this module provides a visual representation of whether a speaker is speaking to the AVSD framework as the visual cues.

2.3 Supervised Pre-trained Models for Audio/Speaker Encoder

Supervised pre-trained speaker models trained with large-scale datasets have a super generalization ability [29]. In this paper, we use them as an audio encoder, the utterance-level pooling in the pre-trained model is replaced with frame-level pooling. Before the pooling layer, assume that we have the convolutional bottleneck feature map $M \in \mathbb{R}^{T \times D'}$, where T and D' are the numbers of frames and dimensions. Then we first uniformly split the feature map into short segments with a length of 5 frames and a shift of 1 frame. The dimensions of segment feature map M_i are $T' \times D'$, where T' is the length of the segment. Then we perform the utterance-level pooling on each segmented feature map. The frame-level pooling layer aggregates the two-dimensional feature map as follows:

$$\mu_i = \frac{1}{T'} \sum_{t=1}^{T'} M_{i,t}, \tag{1}$$

$$\sigma_i = \sqrt{\frac{1}{T'} \sum_{t=1}^{T'} (M_{i,t} - \mu_i)^2}, \tag{2}$$

where μ_i and σ_i are the mean and standard deviation vector of the i-th segment. Then, the concatenation of μ_i and σ_i represents i-th frame pooling result. Finally, we stack all frame pooling results as audio embeddings $E_A \in \mathbb{R}^{T \times D^A}$.

When the supervised pre-trained speaker models are used as a speaker encoder, the supervised models are applied to extract utterance-level speaker embeddings. In the training stage, we extract speaker embeddings $E_S \in \mathbb{R}^{D^A}$ with non-overlapped speech segments of each speaker in oracle labels. During the decoding stage, we compute speaker embeddings with the non-overlapped speech segments which are estimated by the result of visual speaker activity detection.

2.4   Self-supervised Pre-trained Models for Audio/Speaker Encoder

The pre-trained self-supervised models learned from large-scale datasets have shown super generalizability and discrimination, which shows consistent improvements across downstream speech tasks [1,11]. Inspired by this, we investigate two classical self-supervised pre-trained models (HuBERT [17] and WavLM [5]) in our AVSD system.

Similar to the utilization of supervised pre-trained speaker models as mentioned above, we conduct an investigation into the impacts of employing HuBERT and WavLM as the speaker encoder and the audio encoder within the end-to-end AVSD system. To harness the generalization and discriminative capabilities inherent in pre-trained models, we cross-validated the effect of self-supervised and supervised pre-trained speech models as speaker encoders and audio encoders in the AVSD system.

2.5   Audio-Visual Speaker Decoder

The audio-visual speaker decoder estimates the target-speaker voice activities by considering the correlations between each speaker. We explore Transformer [26], Conformer [30] and cross-attention [25] in the decoder module respectively. Given visual embeddings E_V, audio embeddings E_A, and speaker embeddings E_S, firstly audio embedding E_A are repeated N times and speaker embeddings are repeated T times, where N denotes the number of speakers and T is the length of frames. Then, we can obtain fusion audio-visual embeddings E_F by concatenating visual, audio, and speaker embeddings. For the Transformer, a speaker state detection module consisting of a two-layer Transformer encoder with six blocks is designed to extract the states of each speaker. Then, we apply another Transformer encoder with six blocks to estimate the correlations between each speaker. The utilization of Conformer is similar to that of Transformer. For the cross-attention decoder, we first perform cross-attention between lip and speaker embeddings extracted from the pre-trained speaker encoder. Then another cross-attention is further performed between the output of the first cross-attention and audio embeddings from the audio encoder. Finally, a linear layer is considered to obtain the outputs corresponding to the speech/non-speech probabilities for each of the N speakers respectively.

3   Experimental Setup

3.1   Dataset

We conduct experiments on the Multimodal Information Based Speech Processing (MISP) dataset [4]. MISP is a dataset for home TV scenarios, which provides over 100 h of audio and video recordings of 2–6 people in a living room interacting with smart speakers/TVs while watching and chatting. The supervised pre-trained model is trained on VoxCeleb [8,21] and CN-Celeb [12]. Then finetuned on the MISP training dataset. The self-supervised pre-trained model is trained on LibriSpeech dataset [22]. Table 1 details the specific data for training each module.

Table 1. The training data used for each module.

Module	Training data
Lip encoder	MISP
Supervised pre-trained model	CN-CELEB, VoxCeleb, MISP
Self-supervised pre-trained model	LibriSpeech
Audio-visual speaker decoder	MISP

3.2   Configurations

In this study, we employ ResNet34 [14] with channel configurations of 32, 64, 128, 256 and ECAPA-TDNN(1024) [10] as supervised pre-trained models. The embedding dimensions for ResNet34 and ECAPA-TDNN are set at 128 and

192, respectively. The loss function is additive angular margin softmax (AAM-softmax) [9] with a margin of 0.2 and a scale of 32. The speaker embedding models are trained with 80-dimensional log Mel-filter bank features with 25 ms window size and 10 ms window shift. We replace the utterance-level pooling of ResNet34 and ECAPA-TDNN with frame-level pooling to extract frame-level features. In addition, HuBERT Base and WavLM Base are utilized as self-supervised pre-trained models. The back-end fusion module consists of three layers of Transformer encoder. Each encoder layer has 6 blocks with the same config: 256-dim attentions with 2 heads and 1024-dim feed-forward layers. The Adam optimizer with a learning rate of 1e-4 is used for training. During training, the pre-trained model is frozen in the first 5 epochs and then unfrozen to joint training for another 10 epochs.

3.3   Comparison Methods

We compare the improved AVSD with other competitive multi-modal speaker diarization methods. They are listed in the following:

- AVSD [15]: The author introduces an end-to-end audio-visual speaker diarization method that employs CNN networks to extract audio embeddings and ivectors for speaker embeddings (CNN+ivector).
- WHU [6]: The author expands upon the Sequence-to-Sequence Target-Speaker Voice Activity Detection framework to concurrently identify the voice activities of multiple speakers from audio-visual signals.
- SJTU [19]: The author integrates Interchannel Phase Difference (IPD) to represent spatial characteristics and pre-trains a model based on ECAPA-TDNN to extract speaker embedding features.
- Cross-attention [25]: This work introduces a cross-attention mechanism [25] in the temporal dimension to effectively integrate audio-visual interactions.
- Conformer [30]: The authors present a method that combines convolutional neural networks (CNN) and Transformer to obtain both local and global information for sequences.

3.4   Evaluation Metric

We use Diarization Error Rate (DER) [18] as the evaluation metric, which computes the ratio of incorrectly attributed speaking segments or non-speech segments. DER consists of three components: False Alarm Rate (FA), Missing Speech Rate (MISS), and Speaker Error Rate (SPKERR). Lower values for all three metrics indicate superior performance of the speaker diarization system.

4   Results and Analysis

We present the results of the proposed audio-visual speaker diarization model in Table 2. The lip encoder is not always involved in training. For comparison, we show CNN-AE (joint) + ivector-SE (fixed) as a baseline and the corresponding DER is reported as 13.09%. When we use a pre-trained model, the effect improved by 9.82% compared to AVSD. Furthermore, when involving the audio encoder in joint training, DER further decreased by 28.86%. Training both audio and speaker encoders jointly leads to a 17.32% relative reduction on DER. In addition, we also compare the proposed method with other audio-visual methods. As shown in Table 2, the effect of our method is better than other audio-visual methods except WHU's.

Table 2. The results (%) of joint training with supervised and self-supervised model in AVSD. The audio encoder is abbreviated AE and the speaker encoder is abbreviated SE. (* represents joint training in the corresponding module).

Method	MISS	FA	SPKERR	DER
CNN-AE & ivector-SE [15]	4.01	5.86	3.22	13.09
WHU [6]	–	–	–	8.82
SJTU [19]	4.44	4.82	2.10	10.82
HuBERT-AE & ResNet-SE	3.72	5.66	2.53	11.92
HuBERT-AE* & ResNet-SE	**1.88**	**4.96**	**2.41**	**9.25**
HuBERT-AE* & ResNet-SE*	2.35	5.14	2.67	10.16

We also investigate the impact of different decoders on AVSD. Table 3 displays that the DER is 9.97% when using BLSTM as the decoder. When replacing BLSTM with the Conformer, the DER is reduced to 9.61%. Cross-attention method is also employed, leading to a further reduction in DER to 9.57%. Additionally, the best result of 9.54% is obtained when using Transformer as the decoder.

Table 3. DER (%) comparison among different audio-visual decoders in AVSD. Both the audio encoder and the speaker are ResNet.

Method	MISS	FA	SPKERR	DER
BLSTM	2.19	5.49	2.22	9.97
Conformer	2.01	5.50	2.10	9.61
Cross-Attention	1.35	6.26	1.95	9.57
Transformer	**1.36**	**6.23**	**1.92**	**9.54**

Next, we apply ablation experiments to assess the combined effect of supervised and self-supervised pre-trained models. Table 4 shows the experiments using only the supervised pre-trained model for training. Initially, after replacing the audio encoder with ResNet, DER decreased from 9.25% to 9.54%. Then, replacing the speaker encoder with ivectors led to a further decrease to 12.16%. Finally, replacing ResNet with CNN resulted in a DER reduction to 13.09%. We also performed a replacement with ECAPA-TDNN, which yielded results close to ResNet. The results indicate that the supervised model contributes more to the speaker encoder, which is related to the model's training objective.

Table 4. DERs (%) of only using supervised model in AVSD(* represents joint training the corresponding module).

Method	FA	MISS	SPKERR	DER
HuBERT-AE* & ResNet-SE	**1.88**	**4.96**	**2.41**	**9.25**
ResNet-AE* & ResNet-SE	1.36	6.32	1.92	9.54
ECAPA-AE* & ECAPA-SE	2.13	5.14	2.13	9.68
ResNet-AE* & ivector-SE	1.54	8.04	2.58	12.16
CNN-AE* & ivector-SE	4.01	5.86	3.22	13.09

Table 5 illustrates the experiments using only the self-supervised model. Firstly, after replacing the speaker encoder with HuBERT, the model's performance decreased from 9.25% to 10.14%. Then, replacing HuBERT with WavLM led to a further decrease in DER to 11.35%. Regardless of which self-supervised pre-trained model is used, the results are better than the baseline.

Table 5. DERs (%) of only using self-supervised model in AVSD(* represents joint training the corresponding module).

Method	MISS	FA	SPKERR	DER
HuBERT-AE* & ResNet-SE	**1.88**	**4.96**	**2.41**	**9.25**
HuBERT-AE* & HuBERT-SE	2.84	5.01	2.29	10.14
WavLM-AE* & WavLM-SE	2.03	5.84	3.48	11.35
CNN-AE* & ivector-SE	4.01	5.86	3.22	13.09

Figure 2 visualizes the RTTM on different methods. In comparison with the reference, it is observed that the baseline method exhibits numerous false alarms of other speakers. Then we replace the audio encoder and speaker encoder with ResNet, which can accurately predict the results. Furthermore, we perform a comparative analysis between the Transformer decoder and cross-attention methods. The latter demonstrated speaker errors around the 10th second. Moreover, the audio encoder is replaced with HuBERT, resulting in an improvement in detection.

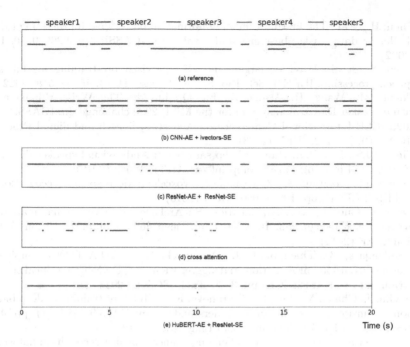

Fig. 2. Visualization of the difference between the results of different methods and the reference results.

5   Conclusion

In this paper, we use the pre-trained supervised and self-supervised speech models as speaker encoders and audio encoders to improve the AVSD framework. To mitigate the performance decline caused by separate learning, we jointly train the speaker encoder, audio encoder, and audio-visual speaker decoder. In addition, we explore different audio-visual speaker decoders in the AVSD framework, which are Transformers, Conformers, and cross-attention respectively. By applying HuBERT as the audio encoder and ResNet as the speaker encoder, we achieve the best performance. Compared with the AVSD, the improved AVSD has a considerable reduction in DER on the MISP dataset.

References

1. Baevski, A., Zhou, Y., Mohamed, A., Auli, M.: Wav2vec 2.0: a framework for self-supervised learning of speech representations. In: NeurIPS (2020)
2. Chen, H., Du, J., et al., D.Y.: Audio-visual speech recognition in misp2021 challenge: dataset release and deep analysis. In: Interspeech (2022)
3. Chen, H., et al.: Audio-visual speech recognition in misp2021 challenge: dataset release and deep analysis. In: Proceedings of the Annual Conference of the International Speech Communication Association, INTERSPEECH, vol. 2022, pp. 1766–1770 (2022)

4. Chen, H., et al.: The first multimodal information based speech processing (MISP) challenge: data, tasks, baselines and results. In: ICASSP, pp. 9266–9270. IEEE (2022)
5. Chen, S., et al.: WavLM: large-scale self-supervised pre-training for full stack speech processing. IEEE J. Sel. Top. Signal Process. **16**(6), 1505–1518 (2022)
6. Cheng, M., Wang, H., Wang, Z., Fu, Q., Li, M.: The WHU-Alibaba audio-visual speaker Diarization system for the MISP 2022 challenge. In: ICASSP 2023–2023 IEEE International Conference on Acoustics, Speech and Signal Processing (ICASSP), pp. 1–2. IEEE (2023)
7. Chung, J.S., Lee, B.J., Han, I.: Who said that?: Audio-visual speaker diarisation of real-world meetings. arXiv preprint arXiv:1906.10042 (2019)
8. Chung, J.S., Nagrani, A., Zisserman, A.: Voxceleb2: deep speaker recognition. In: INTERSPEECH, pp. 1086–1090. ISCA (2018)
9. Deng, J., Guo, J., Xue, N., Zafeiriou, S.: ArcFace: additive angular margin loss for deep face recognition. In: CVPR, pp. 4690–4699. Computer Vision Foundation/IEEE (2019)
10. Desplanques, B., Thienpondt, J., Demuynck, K.: ECAPA-TDNN: emphasized channel attention, propagation and aggregation in TDNN based speaker verification. In: INTERSPEECH, pp. 3830–3834. ISCA (2020)
11. Devlin, J., Chang, M., Lee, K., Toutanova, K.: BERT: pre-training of deep bidirectional transformers for language understanding. In: NAACL-HLT (1), pp. 4171–4186. Association for Computational Linguistics (2019)
12. Fan, Y., et al.: CN-CELEB: a challenging Chinese speaker recognition dataset. In: ICASSP, pp. 7604–7608. IEEE (2020)
13. Gebru, I.D., Ba, S., Li, X., Horaud, R.: Audio-visual speaker Diarization based on spatiotemporal Bayesian fusion. IEEE Trans. Pattern Anal. Mach. Intell. **40**(5), 1086–1099 (2017)
14. He, K., Zhang, X., Ren, S., Sun, J.: Deep residual learning for image recognition. In: CVPR, pp. 770–778. IEEE Computer Society (2016)
15. He, M.K., Du, J., Lee, C.H.: End-to-end audio-visual neural speaker diarization. In: Proceedings of the Interspeech 2022, pp. 1461–1465 (2022)
16. Hochreiter, S., Schmidhuber, J.: Long short-term memory. Neural Comput. **9**(8), 1735–1780 (1997)
17. Hsu, W., Bolte, B., Tsai, Y.H., Lakhotia, K., Salakhutdinov, R., Mohamed, A.: HuBERT: self-supervised speech representation learning by masked prediction of hidden units. IEEE ACM Trans. Audio Speech Lang. Process. **29**, 3451–3460 (2021)
18. Istrate, D., Fredouille, C., Meignier, S., Besacier, L., Bonastre, J.: NIST RT'05S evaluation: Pre-processing techniques and speaker Diarization on multiple microphone meetings. In: Renals, S., Bengio, S. (eds.) Machine Learning for Multimodal Interaction. MLMI 2005. LNCS, vol. 3869, pp. 428–439. Springer, Berlin, Heidelberg (2005). https://doi.org/10.1007/11677482_36
19. Liu, T., Chen, Z., Qian, Y., Yu, K.: Multi-speaker end-to-end multi-modal speaker diarization system for the MISP 2022 challenge. In: ICASSP 2023–2023 IEEE International Conference on Acoustics, Speech and Signal Processing (ICASSP), pp. 1–2. IEEE (2023)
20. Miró, X.A., Bozonnet, S., Evans, N.W.D., Fredouille, C., Friedland, G., Vinyals, O.: Speaker diarization: a review of recent research. IEEE Trans. Speech Audio Process. **20**(2), 356–370 (2012)
21. Nagrani, A., Chung, J.S., Zisserman, A.: VoxCeleb: a large-scale speaker identification dataset. In: INTERSPEECH, pp. 2616–2620. ISCA (2017)

22. Panayotov, V., Chen, G., Povey, D., Khudanpur, S.: Librispeech: an ASR corpus based on public domain audio books. In: ICASSP, pp. 5206–5210. IEEE (2015)

23. Sarafianos, N., Giannakopoulos, T., Petridis, S.: Audio-visual speaker diarization using fisher linear semi-discriminant analysis. Multimed. Tools Appl. **75**, 115–130 (2016)

24. Sharma, R., Narayanan, S.: Using active speaker faces for diarization in tv shows. arXiv preprint arXiv:2203.15961 (2022)

25. Tao, R., Pan, Z., Das, R.K., Qian, X., Shou, M.Z., Li, H.: Is someone speaking? Exploring long-term temporal features for audio-visual active speaker detection. In: Proceedings of the 29th ACM International Conference on Multimedia, pp. 3927–3935 (2021)

26. Vaswani, A., et al.: Attention is all you need. In: NIPS, pp. 5998–6008 (2017)

27. Watanabe, S., Mandel, M., et al.: Chime-6 challenge: tackling multispeaker speech recognition for unsegmented recordings. Interspeech (2020)

28. Wuerkaixi, A., Yan, K., Zhang, Y., Duan, Z., Zhang, C.: Dyvise: dynamic vision-guided speaker embedding for audio-visual speaker diarization. In: 2022 IEEE 24th International Workshop on Multimedia Signal Processing (MMSP), pp. 1–6. IEEE (2022)

29. Zhang, L., Li, Y., Wang, N., Liu, J., Xie, L.: NPU-HC speaker verification system for far-field speaker verification challenge 2022. INTERSPEECH (2022)

30. Zhang, Y., Lv, Z., et al.: MFA-Conformer: multi-scale feature aggregation conformer for automatic speaker verification. Interspeech (2022)

31. Zhou, H., et al.: Audio-visual wake word spotting in misp2021 challenge: dataset release and deep analysis. In: Proceedings of the Annual Conference of the International Speech Communication Association, INTERSPEECH, vol. 2022 (2022)

Zero-Shot Singing Voice Conversion Based on Timbre Space Modeling and Excitation Signal Control

Yuan Jiang[1,2], Yan-Nian Chen[1,2], Li-Juan Liu[2], Ya-Jun Hu[2], Xin Fang[2], and Zhen-Hua Ling[1(✉)]

[1] National Engineering Research Center for Speech and Language Information Processing, University of Science and Technology of China, Hefei 230026, China
zhling@ustc.edu.cn

[2] iFLYTEK Research, iFLYTEK Co., Ltd., Hefei 230088, China

Abstract. In recent years, singing voice conversion technology has rapidly advanced and is capable of generating high-quality singing voices. However, challenges persist, such as pitch fluctuations and significant differences in pitch ranges between source and target singers, leading to conflicts between the accuracy and similarity of the converted pitch. Additionally, the majority of current methods require some data from the target singer to train the model, and the performance and robustness of zero-shot singing voice conversion methods have not been explored. This paper introduces new modifications within the VITS framework to achieve zero-shot singing voice conversion: 1. Timbre space modeling based on Glow; 2. Incorporating excitation signal into the decoder for waveform generation to explicitly control the pitch; 3. Proposing dual-decoder to enhance the stability of 48 kHz waveform modeling for high-quality voice generation; 4. Proposing key shift based pitch mapping strategy for conversion stage. Experimental results demonstrate that Glow-based timbre space modeling enhances the similarity in zero-shot conversion. The excitation signal module and the dual-decoder contribute to naturalness and stability. The pitch mapping strategy effectively avoids out-of-tune problems without compromising similarity to the target singer.

Keywords: Singing Voice Conversion · Zero-shot · Timbre Space · Excitation Signal · Key Shift

1 Introduction

Singing plays an important role in human daily life, including information transmission, emotional expression, and entertainment. Singing voice conversion (SVC) aims to convert the singing voice of the source singer into the voice of the target singer while keeping the content and melody unchanged. Granting machines the ability to produce high-fidelity and expressive singing voices opens up new avenues for human-computer interaction. SVC is one of the possible approaches to achieving this goal.

J. Jia et al. (Eds.): NCMMSC 2023, CCIS 2006, pp. 276–286, 2024.
https://doi.org/10.1007/978-981-97-0601-3_24

In recent years, singing voice conversion using non-parallel data [1, 2] has garnered more attention because it does not rely on scarce parallel voice datasets. The key idea behind non-parallel SVC is to disentangle the linguistic content information and timbre information from the singing voice, and use the timbre embedding of the target singer to generate voice. Most recent SVC methods train a content encoder to extract content features from the source singing voice and a conversion model to transform content features into acoustic features or waveforms. The content encoder and the conversion model can be jointly trained as an auto-encoder. A WaveNet based auto-encoder has been used for unsupervised SVC [3] and can perform conversions among the singers that appear in the training set. Also, SVC methods can separately train the content encoder and the conversion model. These methods use a pre-trained automatic speech recognition (ASR) model as the content encoder to get speaker independent PPGs [4, 5]. The conversion model can directly generate waveforms from content features, for example, using generative adversarial networks (GANs) [6, 7]. Or it can transform content features into spectral features (e.g., mel spectrograms), conversion model architectures including GANs [8–10], variational auto-encoders (VAEs) [11], diffusion models [12]. And then it utilizes separately trained neural vocoders to generate waveforms. SVC methods has achieved success recently, capable of generating high-quality singing voices. However, problems such as pitch jitter and artifacts still exist, which will degrade the quality of converted speech. Source and target singers often have significantly different pitch range, which makes it hard to preserve melody form source singing voice without compromising similarity to the target singer. Besides, most existing methods require training data of the target singer to train the model, while the performance of zero-shot singing voice conversion have not been studied.

In the last two years, the end-to-end text-to-speech (TTS) model, VITS [13], has been proposed. It integrates the acoustic model and vocoder using a variational autoencoder (VAE) into a unified framework, enabling the direct synthesis of high-quality waveforms. While VITS can be used for voice conversion (VC), it cannot be directly applied to SVC.

This paper proposes a zero-shot singing voice conversion method based on timbre space modeling and excitation signal control, by adding novel modification to the VITS framework. The main contributions of this paper are as follows:

1. We propose a timbre space modeling method using Glow [14] model to obtain singer timbre embedding. The embedding is inserted into normalizing flow of VITS for timbre control, enabling zero-shot conversion ability.
2. We incorporate excitation signals into the decoder for waveform generation, to control the fundamental frequency of the generated speech.
3. We model 48 kHz waveforms to achieve high-fidelity speech, and we introduce the use of dual-decoder for modeling both 24 kHz and 48 kHz waveforms in a multi-task learning way, to ensure the stability of 48 kHz waveform modeling.
4. We propose key shift based pitch mapping strategy that maps the fundamental frequency of any input singing voice to the target singer's pitch range during conversion. This avoid out-of-tune problems without compromising similarity to the target singer.

Experiments conducted on an internal singing dataset show that the timbre space modeling method improves zero-shot conversion similarity compared to ECAPA-TDNN

[15] based speaker verification systems. The use of excitation signals enhances natural-ness and fundamental frequency stability. The proposed dual-decoder waveform mod-eling method improves stability and conversion naturalness compared to training with only one 48 kHz decoder.

The structure of this paper is as follows: Sect. 2 introduces the proposed method, while Sect. 3 and 4 cover experiments and conclusions, respectively.

2 Proposed Method

2.1 Overall Framework

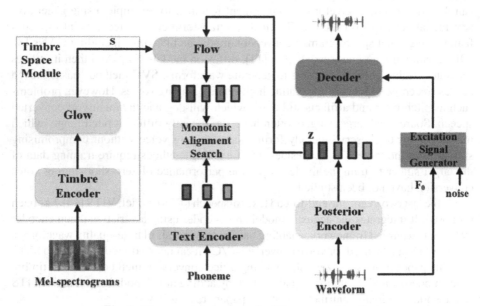

Fig. 1. The overview framework of the proposed method.

The proposed method is constructed based on VITS, and introduces some novel modification for zero-shot singing voice conversion. The entire framework is depicted in Fig. 1, which combines the acoustic model and vocoder using a conditional varia-tional auto-encoder, enabling end-to-end conversion. It comprises a posterior encoder, text encoder, decoder, excitation signal generator, normalizing flow, and timbre space module. The posterior encoder extracts latent variables z from the waveform y, and the decoder reconstructs the waveform $\hat{y}$ based on z:

$$z = PosEnc(y) \sim q(z|y) \tag{1}$$

$$\hat{y} = Dec(z) \sim p(y|z) \tag{2}$$

Given the phoneme c as a condition, the text encoder generates the prior distribution. The normalizing flow constructs the mapping from the prior distribution to z. The training process minimizes the loss L_{KL}, which is obtained by computing the KL divergence between the posterior distribution $q(z|y)$ and the prior distribution, where $z \sim q_\phi(z|y)$, and ϕ represents the model parameters. The training process also minimize the reconstruction loss L_{recon}, implemented as the L1 loss of the mel-spectrogram of the predicted waveform to that of the real waveform, which will be discussed in Sect. 2.5.

The structures of the text encoder, normalization flow, and posterior encoder are consistent with the original VITS. To address the problem of discontinuous and unstable pitch, we introduce an excitation signal generator (ESG) which produces the fundamental frequency information in the form of an excitation signal for the decoder, to explicitly control the pitch of generated voice. To enable zero-shot conversion ability, we propose to use timbre embedding vectors derived from Glow based timbre space modeling, and condition them on all affine coupling layers in the normalizing flow. Consequently, when training on multi-singer data, the normalizing flow learns to remove the singer's timbre information in the latent variable z while preserving content and melody information. Thus, we can use the timbre embedding to control timbre of converted voice.

2.2 Glow-Based Timbre Space Modeling

The current zero-shot voice conversion (VC) methods usually train a multi-speaker VC model, representing speaker information with a speaker-specific embedding vector. These embedding vectors can be either jointly trained in VC or obtained from a pretrained speaker recognition model. However, these methods do not explicitly model the distribution of speaker characteristics in a continuous space. This can lead to decreased voice quality, pronunciation issues and converted voices that might not accurately capture the target speaker's characteristics, when encountering new speakers during zero-shot conversion.

This paper employs a Glow model to model the distribution of speaker characteristics in a more compact and continuous timbre space. The proposed method first utilizes a timbre encoder module, implemented as a pre-trained speaker recognition model in this paper, to extract the raw speaker embedding vector h_s from the mel-spectrograms. Then h_s is modeled by a Glow model. The Glow model is a generative probabilistic model known for its ability to model complex probability distributions accurately [14]. It is a family of flows that can perform forward and inverse transformation, meaning it can project any complex probability distribution through the inverse transformation of the flow model into a standard normal distribution space. In this method, Glow is employed to fit and model the high-dimensional complex distribution of speaker characteristics accurately.

The Glow model is trained using a negative log-likelihood loss function L_{Glow}:

$$L_{Glow} = \log p(s; N(0, 1)) + \log \left(|\det(f_G)|^{-1} \right) \tag{3}$$

where s is the speaker timbre features extracted by the speaker encoder module and the Glow model, $\det(f_G)$ represents the determinant of the Jacobian matrix of the Glow model. By minimizing L_{Glow} and optimizing Glow model parameters until the loss converges, the distribution in the s-space is a standard normal distribution. Then we transform the raw speaker embedding distribution space to a standard normal distribution space, and treat it as the final timbre space distribution. Due to the smooth and continuous nature of the standard normal distribution, this approach yields more robust timbre embedding. Since the s-space is modeled to fit the overall speaker characteristics space, sampling from a standard normal distribution $N(s; 0, 1)$ yields a virtual speaker embedding vector $\tilde{s}$, which can be used to generate an extensive range of transformed singing voices with diverse speaker characteristics.

2.3 Decoder Incorporated with Excitation Signal

The decoder adopts the same model structure as HiFiGAN [16], which consists of multiple stacked upsampling layers and a multi-receptive field fusion (MRF) module. It gradually upsamples the input until it matches the sampling rate of the speech waveform.

To ensure better control and quality of the generated voice, this paper enhances the decoder with an excitation signal generator that explicitly controls the fundamental frequency when generating speech. The excitation signal generator takes the fundamental frequency and its harmonics as input and generates excitation signals, with voiced segments being a combination of sine signals and Gaussian noise, while unvoiced segments being Gaussian noise. These excitation signals, after undergoing multiple simple down-sampling and pre-processing network, are provided to the generator to predict the waveform. Specifically, the excitation signal generator generates a sine excitation signal $e_{1:T}$ based on the fundamental frequency $f_{1:T}$ and its harmonics. Assuming the instantaneous frequency of the t-th frame is f_t, for voiced frames $f_t > 0$, for unvoiced frames $f_t = 0$. The excitation signal corresponding to the H-th harmonic is:

$$
e_t^{<h>} = \begin{cases} \alpha \sin\left(\sum_{k=1}^{t} 2\pi \frac{(h+1) \times f_k}{f_s} + \phi\right) + n_t & \text{if } f_t > 0 \\ \frac{1}{3\sigma} n_t & \text{if } f_t = 0 \end{cases} \tag{4}
$$

where $n_t \sim \mathcal{N}(0, \sigma^2)$ is Gaussian noise, $\phi \in [-\pi, \pi]$ is the randomly generated initial phase during training and inference, and f_s is the waveform's sampling rate. The excitation signals corresponding to the fundamental frequency and H harmonics has the same resolution as the waveform. They are passed through a network composed of downsampling layers and MRF modules. This processing matches the resolution of the excitation signal with the upsampling resolution of the decoder. The outputs of each downsampling step and the corresponding upsampling step in the decoder, having the same resolution, are added as inputs to each layer of the MRF network in the decoder to predict the final waveform.

The model is trained by reconstruction loss L_{recon}, defined as:

$$
L_{\text{recon}} = E_{(y, \hat{y})} [\|M(y) - M(\hat{y})\|_1] \tag{5}
$$

where $M(.)$ is the function that transforms speech waveforms into Mel-spectrograms.

To produce high-fidelity speech, the model also uses an adversarial training method, which has a generator and discriminators. In this training method, the decoder is trained as generator (G) to generate speech which can deceive the discriminator, while the discriminator (D) is trained to distinguish between real speech and generator-generated speech. In this paper, a multi-period discriminator (MPD) and a multi-scale discriminator (MSD), same as those used in HiFiGAN, are employed to discriminate against the generated waveforms from the decoder. The loss functions for adversarial training are as follows:

$$L_{adv}(D) = E_{(y,\hat{y})}\left[(D(y) - 1)^2 + (D(\hat{y}))^2\right] \tag{6}$$

$$L_{adv}(G) = E_z\left[(D(\hat{y}) - 1)^2\right] \tag{7}$$

Additionally, the model utilizes the same feature matching loss as in HiFiGAN to enhance speech quality and training stability:

$$L_{fm}(G) = E_{(y,\hat{y})}\left[\sum_{l=1}^{T} \frac{1}{N_l}\left\|D^l(y) - D^l(\hat{y})\right\|_1\right] \tag{8}$$

where T represents the number of layers in the discriminator. D^l is the output of the l-th layer of the discriminator, and N_l represents the number of features in the l-th layer.

Thus, the total loss function for training the decoder is:

$$L = L_{adv}(G) + L_{fm}(G) + \lambda_{recon}L_{\text{recon}} \tag{9}$$

2.4 Dual-Decoder for High-Fidelity 48 kHz Waveform Modeling

The model is trained to generate high-quality speech at a 48 kHz sampling rate. Due to the high-frequency details in 48 kHz waveforms, it is challenging to model them directly. Solely modeling 48 kHz waveforms can lead to unstable training and affect its final performance. To address this, a dual-decoder approach is proposed. In this method, while only 48 kHz waveforms are used for extracting latent variables z, two decoders are employed to model both 24 kHz and 48 kHz waveforms, in a way of multi-task learning. Two decoders have the same structure, except for the upsampling factor. The loss function for training dual-decoder is defined as:

$$L_{dual} = \lambda_{24}L_{24k} + (1 - \lambda_{24k})L_{48k} \tag{10}$$

Initially, λ_{24} is set to 1, meaning only the 24 kHz decoder is trained, until convergence. Then, during the training process, λ_{24} is gradually reduced from 1 to 0.5, and this setting remains unchanged until convergence.

This dual-model approach is expected to enhance the stability of the training process and improves the overall naturalness and quality of the generated speech.

2.5 Key Shift Based Pitch Mapping Strategy for Conversion Stage

The conversion process need to extract the fundamental frequency from the source singer's voice, which can vary significantly from the target singer as different singers might have different pitch ranges. Directly copying the source singer's pitch would lead

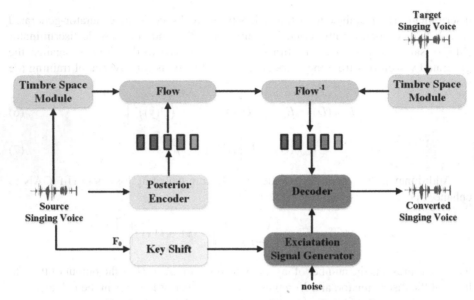

Fig. 2. The inference procedure of the proposed method.

to a converted voice that doesn't sound like the target singer's timbre. In traditional voice conversion tasks, Gaussian-based linear mean-variance transformation in log domain of fundamental frequency is applied. However, using such transformation in singing voice conversion can lead to out-of-tone. Therefore, a key shift based pitch mapping strategy is proposed.

First, we calculate the average fundamental frequency of the input source singing voice $F0_{src}^{<avg>}$ and the target one $F0_{trg}^{<avg>}$. This calculation is done in utterance level. Next we calculate the key difference between source and target singing voice and preserve the integer result as:

$$N_{key} = round\left(12 * \log_2\left(F0_{trg}^{<avg>}/F0_{src}^{<avg>}\right)\right) \tag{11}$$

Then, fundamental frequency of the source is shift by key difference as:

$$trans(f_o) = f_0 * 2^{N_{key}/12} \tag{12}$$

This pitch mapping strategy is expected to map the key of the source singing voice to the target singer's pitch range to prevent the out-of-tone problem, while does not degrades the speaker similarity in the conversion results.

The entire transformation process, as shown in Fig. 2, starts with extracting latent variables z from the source singer's voice by the posterior encoder. Then, the normalizing flow is used by the forward transformation to remove the source singer's timbre information of z given timbre embedding of the source singer. Afterward, the inverse transformation of the normalizing flow incorporates the target singer's timbre information given timbre embedding of the target singer. Finally, the decoder is used to generate the converted voice, in which fundamental frequency of source singer is transformed by above method and then used for generating excitation signal.

3 Experiments

3.1 Settings

In our experiments, we used an internal dataset of Chinese pop songs. Which included 120 singers, with each singing 20 popular songs. The total dataset comprised approximately 120 h of singing data, recorded at a sampling rate of 48 kHz and quantized with 16-bit precision. We also down-sampled the data to 24 kHz to train the 24 kHz decoder. We randomly selected 12 singers for testing, and the rest were used for training. The test set included 6 male and 6 female singers. We selected 3 male and 3 female singers as the source singers and the remaining singers as the target singers for conversion. From each source singer, 5 test speech samples were randomly chosen and converted to each target singer, resulting in 180 test samples.

In our model, the text encoder took phoneme labels as input (in Chinese, this usually corresponds to the consonants and vowels). We used the monotonic alignment search (MAS) algorithm for phoneme-to-latent-variable alignment. Since SVC does not require phoneme durations like TTS, we removed the duration predictor during training. The timbre encoder was built on the ECAPA-TDNN [15] structure, and pre-trained with an internal 60 kh speaker recognition dataset. The Glow module used for timbre space modeling followed the architecture of Glow-TTS [17]. Specifically, it is constructed with a stack of multiple blocks, each of which consists of an activation normalization layer, invertible 1×1 convolution layer, and affine coupling layer. We set the number of blocks as 6. The remaining hyper-parameters were consistent with VITS [13]: 192 dimensions for phoneme embedding vectors, 256 dimensions for latent variable z, and the text encoder included 6 FFTs with a self-attention mechanism using 2 heads. We used the AdamW [18] optimizer, with $\beta_1 = 0.9$, $\beta_2 = 0.99$, and weight decay $\lambda = 0.01$. The learning rate was decayed by a factor of $0.999^{1/8}$ in each training iteration, starting with an initial learning rate of 2×10^{-4}. Hyper-parameters for excitation signal generator are set as $\sigma = 0.003$, $\alpha = 0.1$, $H = 8$. All models were trained for 600,000 steps with a batch size of 64 on V100 GPUs.

We established 5 systems for comparison in our experiments:

1) **VITS:** Baseline system constructed using VITS with some small modifications: The logarithmic fundamental frequency values were concatenated with latent variable z as decoder input for controlling melody, and the decoder directly modeled 48 kHz waveforms. Speaker representations were extracted using the pre-trained speaker recognition system ECAPA-TDNN and inserted into the normalizing flow for timbre control. In coversion stage, Gaussian-based linear mean-variance transformations is used for fundamental frequency transformations.

2) **+Key Shift**: Same as the **VITS** baseline, but with the pitch conversion method changed to the key shift based method described in Sect. 2.5.

3) **+Ext**: Based on **+Key Shift**, and the decoder replaced pitch values with the excitation signal

4) **+24 kHz**: 24 kHz decoder was introduced for multi-task learning.

5) **+Glow(Proposed)**: The proposed method in Sect. 2, which used the timbre space module based on Glow as proposed in Sect. 2.2, replacing the pre-trained ECAPA-TDNN.

3.2   Results and Analysis

Naturalness and similarity are evaluated on the test set. Naturalness metric refers to whether or not the converted singing voice sound like singer singing, and listeners were asked to focus on melody and quality, while similarity refers to how closely the generated singing voices matched the timbre of the target singer. A total of 20 native Mandarin speakers participated in the listening test. Listeners were asked to rate each system using scores ranging from 1 to 5 for both naturalness and similarity.

Table 1. The naturalness and similarity mean opinion score (MOS) of all systems.

Methods	Naturalness	Similarity
VITS	2.96	3.38
+Key Shift	3.61	3.44
+Ext	3.75	3.46
+24 kHz	3.83	3.45
+Glow(Proposed)	**3.87**	**3.58**

Table 1 presents the mean opinion score (MOS) for each system. We can see that our proposed method achieved significantly higher naturalness and similarity than the **VITS** baseline.

Comparing **+Key Shift** to **VITS**, we can see that the proposed key shift based pitch mapping strategy improve naturalness significantly. We found that with traditional Gaussian-based linear mean-variance transformation, many of the converted singing speech was out-off-tune. Adding key shift also slightly improve similarity.

Comparing **+Ext** to **+Key Shift,** we can see that adding excitation signal improve naturalness. To measure the control of the excitation signal on the fundamental frequency of generated speech, STRAIGHT [19] was adopted to extract fundamental frequency from the convert speech, which was used for calculating root mean square error of F_0 (F_0 RMSE) with the given F_0. As shown in Table 2, adding excitation signal to decoder exhibits better control over F_0 than just concatenating F_0 value.

Table 2. F_0 RMSE results of **+F0Trans** and **+Ext**

Methods	F0 RMSE(Hz)
+Key shift	13.94
+Ext	7.05

Adding 24 kHz decoder for multi-task learning can slightly improve naturalness. However, we found the singing voice generated by model without 24 kHz decoder exists artifacts such as metallic noise sounds. Dual-decoder approach effectively removed the artifacts.

Comparing **Proposed** method with all other methods, similarity was improved significantly, which proved the effectiveness of the proposed timbre space modeling method.

4 Conclusion

This paper proposes a zero-shot singing voice conversion method based on timbre space modeling and excitation signal control. We add novel modification to the VITS framework for singing voice conversion: timbre space modeling based on Glow, incorporation of excitation signal for explicitly controlling the pitch, dual-decoder to enhance the stability of 48 kHz waveform modeling, and key shift based pitch mapping strategy for conversion. Experimental results demonstrate that Glow-based timbre space modeling enhances the similarity in zero-shot conversion. The excitation signal module and the dual-decoder contribute to naturalness and stability. The pitch mapping strategy effectively avoids out-of-tune problems without compromising similarity to the target singer.

References

1. Chen, X., Chu, W., Guo, J., Xu, N.: Singing voice conversion with non-parallel data. In: MIPR, pp. 292–296 (2019)
2. Lu, J., Zhou, K., Sisman, B., Li, H.: Vaw-gan for singing voice conversion with non-parallel training data. In: Asia-Pacific Signal and Information Processing Association Annual Summit and Conference, pp. 514–519 (2020)
3. Nachmani, E., Wolf, L.: Unsupervised singing voice conversion. In: Proceedings of the INTERSPEECH, pp. 2583–2587 (2019)
4. Li, Z., et al.: PPG-based singing voice conversion with adversarial representation learning. In: Proceedings of the ICASSP, pp. 7073–7077 (2021)
5. Wang, C., et al.: Towards high-fidelity singing voice conversion with acoustic reference and contrastive predictive coding. In: Proceedings of the INTERSPEECH, pp. 4287–4291 (2022)
6. Polyak, A., Wolf, L., Adi, Y., Taigman, Y.: Unsupervised cross-domain singing voice conversion. In: INTERSPEECH, pp. 801–805 (2020)
7. Liu, S., Cao, Y., Hu, N., Su, D., Meng, H.: FASTSVC: fast cross-domain singing voice conversion with feature-wise linear modulation. In: ICME, pp. 1–6 (2021)
8. Zhao, W., Wang, W., Sun, Y., Tang, T.: Singing voice conversion based on WD-GAN algorithm. In: Proceedings of the IAEAC, pp. 950–954 (2019)
9. Sisman, B., Vijayan, K., Dong, M., Li, H.: SINGAN: singing voice conversion with generative adversarial networks. In: Proceedings of the APSIPA, pp. 112–118 (2019)
10. Chandna, P., Blaauw, M., Bonada, J., Gomez, E.: WGANSing: a multi-voice singing voice synthesizer based on the wasserstein-gan. In: Proceedings of the 27th European Signal Processing Conference (2019)
11. Luo, Y., Hsu, C., Agres, K., Herremans, D.: Singing voice conversion with disentangled representations of singer and vocal technique using variational autoencoders. In: Proceedings of the ICASSP (2020)

12. Songxiang, L., Yuewen, C., Dan, S., Helen, M.: DIFFSVC: a diffusion probabilistic model for singing voice conversion. In: Proceedings of the ASRU, pp. 741–748 (2021)
13. Jaehyeon, K., Jungil, K., Juhee, S.: Conditional variational autoencoder with adversarial learning for end-to-end text-to-speech. In: Proceedings of the ICML, vol. 139, pp. 5530–5540 (2021)
14. Emiel, H., Rianne, V.D.B., Max, W.: Emerging convolutions for generative normalizing flows. In: International Conference on Machine Learning, pp. 2771– 2780 (2019)
15. Desplanques, B., Thienpondt, J., Demuynck, K.: ECAPA-TDNN: emphasized channel attention, propagation and aggregation in TDNN based speaker verification. In: INTERSPEECH (2020)
16. Kong, J., Kim, J., Bae, J.: HIFI-GAN: generative adversarial networks for efficient and high fidelity speech synthesis. Adv. Neural Inf. Process. Syst. **33** (2020)
17. Kim, J., Kim, S., Kong, J., Yoon, S.: Glow-TTS: a generative flow for text-to-speech via monotonic alignment search. Adv. Neural Inf. Process. Syst. (2020)
18. IIya, L., Frank, H.: Decoupled weight decay regularization. In: ICLR (2019)
19. Kawahara, H., Masuda-Katsuse, I., De Cheveigne, A.: Restructuring speech representations using a pitch-adaptive time-frequency smoothing and an instantaneous-frequency-based F0 extraction: possible role of a repetitive structure in sounds. Speech Commun.Commun. **27**(3–4), 187–207 (1999)

A Comparative Study of Pre-trained Audio and Speech Models for Heart Sound Detection

Yuxin Duan[1], Chenyu Yang[1], Zihan Zhao[1,2], Yiyang Jiang[2], Yanfeng Wang[1,2], and Yu Wang[1,2(✉)]

[1] Cooperative Medianet Innovation Center, Shanghai Jiao Tong University, Shanghai, China
yuwangsjtu@sjtu.edu.cn
[2] Shanghai AI Lab, Shanghai, China

Abstract. Cardiovascular disease screening is critically anchored in heart sound auscultation. As deep learning methodologies advance, the impetus toward automating heart sound detection grows, aiming to curtail reliance on specialized clinicians. However, the compilation and annotation of expansive high-fidelity datasets present challenges, attributing to both necessary expertise and environmental complexities. In this landscape, transfer learning, harnessing extensive pre-trained models, emerges as a potential solution. In our investigation, we rigorously assessed established audio and speech models—PANNs, SSAST, BEATs, HuBERT, and WavLM—using the PhysioNet/CinC 2016 dataset. Preliminary results showcased the pre-tuning BEATs model's superior performance, achieving an accuracy of approximately 90%. However, following optimization procedures, the PANN-V1 model surpassed its counterparts, registering an accuracy of 94.02%. Our study further delved into the models' robustness against various noise paradigms. Pink noise was observed to be more disruptive than white noise, with the PANN-V2 model demonstrating notable resilience across both noise spectra. Contrarily, impulse noise exhibited a minimal perturbative effect. In a more pragmatic setting, we evaluated the models using the CirCor DigiScope Dataset, emphasizing specific demographics such as pediatric and antenatal populations. It was discerned that these particular demographics, coupled with ambient clinical noise, can indeed modulate model performance. Within this context, the BEATs model retained commendable proficiency, achieving a 65.33% accuracy. This study provides insights into model selection and fine-tuning, fostering more informed decision in the selection of pre-trained models for heart sound processing and analysis.

Keywords: Heart sound detection · Pre-trained models · Noise resistance

© The Author(s), under exclusive license to Springer Nature Singapore Pte Ltd. 2024
J. Jia et al. (Eds.): NCMMSC 2023, CCIS 2006, pp. 287–301, 2024.
https://doi.org/10.1007/978-981-97-0601-3_25

1   Introduction

Heart sound auscultation, a non-invasive test, has an irreplaceable role in the early diagnosis of cardiovascular disease [11]. However, the analysis and interpretation of heart sounds requires a high degree of professional skill, which means that there are extremely high demands on the professional training of doctors [19].

In recent years, with the rapid development of deep learning technologies, which are expected to solve the problem of over-reliance on specialized physicians, automated heart sound detection has received a lot of attention [13]. Indeed, studies have utilized deep learning methods such as deep neural networks, convolutional neural networks, and recurrent neural networks to achieve encouraging results in heart sound detection [6].

However, in order to achieve efficient automated heart sound detection, researchers have faced multiple challenges, such as difficulties in data collection, expert annotation and environmental noise [26]. In this context, transfer learning using deep learning models becomes a promising solution. These models are expected to capture the intrinsic properties in audio without the need for expensive domain expert knowledge [27]. Although audio pre-trained models are preferred for the audio task of heart sound detection, Speech models, especially Self-Supervised Learning (SSL) speech models, are capable of capturing meaningful latent representations of speech audio segments [31]. SSL speech models, such as HuBERT [12], have also been shown to be effective in non-speech scenarios [18]. Thus the potential of speech models for heart sound tasks is worth pursuing.

Furthermore, in the actual acquisition of heart sound signals, interference from external noises such as medical equipment, the patient's own physiological activities, or the environmental background is unavoidable [22]. These noises can adversely affect the quality of the heart sound signal, thus reducing the accuracy and reliability of the heart sound detection task. It is crucial to evaluate the performance of the models in various noisy environments in order to ensure that they are reliable and robust in real scenarios.

The main contribution of this study is to evaluate and compare the performance of multiple pre-trained models in heart sound detection. First, on the PhysioNet/CinC 2016 dataset [17], we perform a comprehensive performance evaluation and analysis of several mainstream audio pre-trained models, PANNs [16], SSAST [10], and BEATs [4], as well as large-scale speech pre-trained models, HuBERT [12] and WavLM [3]. Based on this preliminary study, for the better performing model configurations among them, we further explore the noise immunity performance of these models. We analyze the effects of five different SNRs of white and pink noise, as well as impulse noise of different frequencies and durations on these models. In addition, in order to get closer to the real clinical environment and to evaluate the application potential of the models more comprehensively, we choose the CirCor DigiScope dataset [21] for an in-depth study and analysis of the better-performing model configurations. This dataset specifically focuses on pediatric and prenatal populations and contains a large amount of realistic environmental noises. These experiments provide insights and references on how to select and finetune pre-trained models and how to optimize the processing of heart sound data, hopefully providing a solid foundation for future clinical applications.

The rest of this paper is organized as follows: Sect. 2 describes the background of heart sound and five pre-trained models. Section 3 describes the dataset we used, Sect. 4 details our methodology and experimental settings, and Sect. 5 showcases the results and discussions. Finally, in Sect. 6, we summarize our findings and suggest future research directions.

2 Preliminaries

2.1 Heart Sound

In the human cardiac system, normal heart sounds are produced primarily by the vibration of the heart valves as they open and close during each cardiac cycle, and by the turbulence of the blood as it enters the arteries [21]. Two heart sounds are included: the first heart sound, S1, and the second heart sound, S2 [26]. Other sounds may be associated with physiologic or pathologic conditions. For example, the third heart sound, S3, may be a sign of heart failure, and a murmur may indicate a defective valve or a hole in the wall [1]. In clinical practice, murmurs include two types, systolic murmurs and diastolic murmurs [20].

2.2 PANNs

Pre-trained Audio Neural Networks (PANNs) [16] are deep learning models designed for audio pattern recognition tasks, which are pre-trained on the large-scale dataset called AudioSet [8], which contains over 5,000 h of audio recordings with 527 sound classes. Even with limited data, PANNs can be efficiently fine-tuned and perform significantly on new tasks. In addition, PANNs can be easily transferred to a variety of audio pattern recognition tasks and have outperformed several previous state-of-the-art(SOTA) systems.

2.3 SSAST

Transformer [28]-based models have shown promising performance and excellent suitability in different domains. Recently, [9] proposed the Audio Spectrogram Transformer (AST), which achieved SOTA results for audio classification tasks. However, the training of AST relies on a large amount of supervised training data, which is expensive and hard to collect. To extend AST into various downstream tasks, [10] further explores the Self-Supervised Audio Spectrogram Transformer (SSAST), which leverages unlabeled audio corpus from AudioSet [8] and Librispeech [23] to pretrain the AST model, thus reducing the needs of supervised training samples. SSAST leverages the Masked Spectrogram Patch Modeling (MSPM) technique during the pre-training stage to gain a better performance.

2.4 BEATs

In recent years, self-supervised learning (SSL) has advanced notably in language, vision, speech, and audio. SSL mainly employs acoustic feature reconstruction loss and discrete label prediction loss as pre-training objectives. Discrete label prediction is favored for its high-level semantic alignment with human audio perception, but the continuity and diversity of audio data hinder direct semantic segmentation. Therefore current audio SSL models still rely on reconstruction loss. To address this, [4] proposes Bidirectional Encoder representation from Audio Transformers(BEATs), an iterative framework optimizing both an acoustic tokenizer and an SSL model. BEATs is compatible with various masked audio prediction models and demonstrates superior performance on various audio and speech classification tasks, achieving new SOTA results on benchmarks like AudioSet-2M(AS-2M) [8] and ESC-50 [25].

2.5 HuBERT

Additionally, we consider speech-based pre-trained models. Hidden Unit BERT (HuBERT) [12] addresses three key challenges in self-supervised speech representation learning: multiple sound units in utterances, lack of a pre-training lexicon, and variable sound unit lengths without clear segmentation. It employs offline clustering to provide aligned target labels for prediction loss, applying this loss only to masked regions in partially masked continuous input. This encourages learning a combined acoustic and language model. HuBERT's effectiveness hinges on the consistency of unsupervised clustering, not the quality of cluster labels. HuBERT uses Librispeech 960 h [23] and Libri-Light 60k hours [14] datasets for unsupervised pre-training, which performs well on multiple evaluation subsets and is applicable to various downstream tasks.

2.6 WavLM

SSL excels in Automatic Speech Recognition(ASR) but less so in other speech tasks due to the complexity of speech signals, which encompass speaker identity, paralinguistics, etc. Addressing this, WavLM [3] is proposed to learn generic speech representations. It uniquely combines masked speech prediction with denoising, enhancing both speech content modeling and its applicability to non-ASR tasks, such as speaker diarization and speech enhancement. WavLM improves the Transformer architecture by introducing gated relative position bias for better sequence understanding, boosting performance with minimal parameter increase. To mitigate data discrepancies, WavLM uses up to 94k hours of unlabeled audio pre-training data, including Libri-Light [15], GigaSpeech [2], and VoxPopuli [30] datasets. These datasets contain examples from different scenarios, enabling WavLM to be more comprehensively adapted to various speech processing tasks.

3   Datasets

PhysioNet, the moniker of the Research Resource for Complex Physiologic Signals, was established in 1999 under the auspices of the National Institutes of Health (NIH)[1]. It provides free access to voluminous physiological and clinical data, and is one of the most commonly used open-source data platforms for AI in healthcare. PhysioNet, in cooperation with the annual Computing in Cardiology conference, hosts sequential challenges aimed at solving clinical and basic science problems. In this section, we present two currently sizable heart sound PCG datasets for subsequent validation.

3.1   Physionet/CinC 2016 Database

The PhysioNet/CinC 2016 Database, featuring public heart sounds, was assembled for the international competition PhysioNet/ Computing in Cardiology Challenge (CinC) 2016 [24]. This database comprised nine heart sound collections from global research teams, gathered in diverse clinical and non-clinical settings using various devices [17]. The recordings were categorized into normal and abnormal types, with abnormal ones from patients with confirmed heart disease, including heart valve and coronary artery disease. The training and test sets were unbalanced, favoring more normal than abnormal recordings. The lengths of these recordings varied, with minimum, maximum, and average durations of 5.3, 122.0, and 20.8 s, respectively. Each PCG signal was downsampled to 2000 Hz. This study uses data from the publicly available training set, including 3,240 instances with high signal quality. Among these, 2,141 instances were obtained from the Dalian University of Technology heart sounds database (DLUTHSDB), and the PCG signals were all collected from the mitral position at the chest.

3.2   CirCor DigiScope Dataset

The CirCor DigiScope Dataset, the largest pediatric heart sound dataset, was gathered during two cardiac screening campaigns in Pernambuco, Brazil. A total of 5,282 recordings were collected from 1,568 patients, including 63% children, 20% infants, and 8% pregnant women. Nearly all subjects had audio samples taken from the four standard auscultation points: PV (Pulmonary), TV (Tricuspid), AV (Aortic) and MV (Mitral). In this study, we utilized the public training set comprising 3,163 instances, evenly distributed between normal and abnormal data. All samples were collected in an outpatient setting and exhibited various background noises, including stethoscope friction, crying and laughter. Thus, the automatic analysis of CirCor dataset is a difficult task [21].

---

[1] https://physionet.org/.

4   Methodology and Experimental Settings

4.1   Comparison of Heart Sound Detection Performance Using Different Pre-trained Models

In the first experiment, we comprehensively evaluated the performance of multiple pre-trained models using relatively high quality data.

Model Configurations. We adopted several pre-trained models under different configurations and fine-tuning strategies, including Baseline, PANNs [16], SSAST [10], BEATs [4], HuBERT [12], and WavLM [3].

Baseline Model. We utilized the FBANK-TDNN as our baseline model, which is composed of an filter bank(FBANK) feature extraction module and a time-delayed neural network (TDNN) [29] layer that eventually links to a classifier. The FBANK extractor employs a frequency filter bank to capture energy distribution across frequency bands from the input signal. The TDNN layer adapts to different context sizes and dilation rates for analyzing audio at various temporal resolutions. Following feature extraction and transformation, a classifier with a Sigmoid activation function generates the final prediction. This baseline model offers a basic yet effective method for heart sound detection, serving as a foundation for future improvement and optimization.

PANNs. We adopt two architectures of PANNs, Wavegram-CNN and Wavegram-Logmel-CNN. The former takes the Wavegram only as the input, which is extracted from the waveform by a series of one-dimensional CNNs on the time domain, while the latter combines both Wavegram and log-mel spectrogram together, which utilizes extra information. Finally, the feature embeddings are fed into the two-dimensional Convolutional Neural Network (2D-CNN) layers.

SSAST. SSAST, an adaptation of the Audio Spectrogram Transformer (AST), employs a Transformer encoder with 768 embedding dimensions, 12 layers, and 12 attention heads. SSAST utilizes the MSPM technique during pre-training. This process entails dividing an audio spectrogram into patches or frames, randomly masking some, then sequencing and feeding them into the Transformer encoder. The encoder outputs patch or frame representations which are then processed using mean pooling to obtain an audio clip level representation. Finally, classification is executed using a linear head based on this representation. This study introduces three SSAST configurations: SSAST-patch-400, SSAST-frame-400, and SSAST-patch-250, which utilize patch-based and frame-based pre-training with 400 and 250 masked patches or frames, respectively. The term patch refers to a square portion of the spectrogram, while frame indicates a rectangular region of the spectrogram in the temporal order.

BEATs. BEATs combines an acoustic tokenizer and an audio SSL model in an iterative framework. It employs the Vision Transformer(ViT) [7] structure as a backbone and updates the tokenizer and SSL model parameters alternately. It starts with a random-projection acoustic tokenizer and laverages the latest SSL model in subsequent iterations to teach the tokenizer. In each iteration, the labels from tokenizer will be further used to train the SSL model. BEATs models have 12 Transformer encoder layers, 768-dimensional hidden states, and 8 attention heads. BEATs models were optimized in three iterations and we select the third iteration models for the first experiment.

HuBERT. HuBERT comprises a convolutional waveform encoder with 7 layers and 512 channels, a BERT [5] encoder with multiple identical Transformer blocks, a projection layer, and a code embedding layer. In pre-training, HuBERT uses a k-means algorithm and two clustering iterations to cluster MFCC features, creating discrete acoustic unit labels for speech frame assignments and model learning. In this study, we used the "HuBERT-base-ls960" configuration, featuring a BERT encoder with 12 layers, 8 attention heads and 768 embedding dimensions, pre-trained on the labeled Librispeech 960 h dataset.

WavLM. WavLM utilizes a masked speech prediction and denoising framework, simulating noisy or overlapping speech to predict pseudo-labels for masked regions. WavLMs have a Transformer encoder with gated relative position bias. All configurations are employed for this experiment. WavLM Base and Base plus use 12 Transformer layers, 768-dimensional states, and 8 attention heads. WavLM Large has 24 layers, 1024-dimensional states, and 12 heads. WavLM Base is pre-trained on LibriSpeech 960 h dataset. WavLM Base plus and WavLM Large are pre-trained on 94k large-scale diverse data.

For convenience, we use plural forms to refer to various configurations within a category of pre-trained models. For example, "SSASTs" represents multiple models with different configurations within the SSAST category. All models utilized and their abbreviations correspond as follows:

- **Baseline**: FBANK-TDNN with multi-layer perceptron.
- **PANN-{V1,V2}**[2]: Wavegram-CNN MLP and Wavegram-Logmel-CNN MLP, respectively
- **SSAST-{P400,P250,F400}**[3]: SSAST patch-based and frame-based pre-training with 400 masked patches, 250 masked patches and 400 masked frames, respectively
- **BEATs-{iter3, iter3-ft-AS2M-cpt1/2}**[4]: The third iteration model of BEATs and BEATs-iter3 finetuned on AS2M cpt1/2, respectively
- **BEATs-iter3+AS2M/AS20K**(see footnote 4): The third iteration model trained with additional AS2M or AS20K data.

---

[2] https://zenodo.org/records/3987831.
[3] https://github.com/YuanGongND/ssast.
[4] https://github.com/microsoft/unilm/tree/master/beats.

- **BEATs - iter3+AS2M/AS20K-ft-AS2M-cpt1/2**(see footnote 4):
  BEATs - iter3+AS2M/AS20K finetuned on AS2M cpt1/2.
- **HuBERT-Base**[5]: HuBERT-base-ls960.
- **WavLM-{Base, Base+, Large}**[6]: WavLM Base, Base plus and Large, respectively.

Experimental Data. The experimental data was obtained from the Physionet/CinC 2016 Database [17]. It comprised 3240 records, which were divided into a 9:1 ratio for training and validation. The original dataset contained 2575 records labeled as 0 (normal) and 665 records labeled as 1 (abnormal), accounting for approximately 79.48% and 20.52% of the dataset, respectively. To maintain data representativeness, we ensured that the label proportions in the training and validation sets matched those of the original dataset. To handle varying data lengths, we sliced and padded the data to ensure each segment lasted for a consistent 3 s.

Loss Function. The loss function used for all experiments in this study is the Binary Cross-Entropy(BCE) loss, defined as follows:

$$\text{BCELoss}(y, \hat{y}) = -\frac{1}{N} \sum_{i=1}^{N} \left(y_i \log\left(\hat{y}_i\right) + (1 - y_i) \log\left(1 - \hat{y}_i\right) \right) \tag{1}$$

where N is the number of instances, $\hat{y}_i$ is the model prediction output for i-th instances, and y_i is 1 for abnormal instances and 0 for normal instances.

Evaluation Metrics. We use four popular metrics for evaluation: Accuracy, F1 Score, Precision, and Recall [32]. All experiments in this study will use all or a part of them. These evaluation metrics provide a comprehensive, multi-dimensional view of the model's performance.

4.2   Noise Resistance Performance

The purpose of this experiment is to explore the effect of different noises on the heart sound detection task.

Model Configurations. We selected five top-performing models from the first experiment. They are PANN-V1, PANN-V2, BEATs - iter3+AS20K and BEATs - iter3+AS20K-ft-AS2M-cpt1/2.

Experimental Data. We introduced color and impulse noises to the Physionet/CinC 2016 data in order to simulate signals in different noise environments. For color noises, white and pink noise, we chose five different SNRs: –10 dB, –5 dB, 0 dB, 5 dB and 10 dB. Figure 1 (a)(c)(d) illustrates the comparison between the original signal and color noise-added signals.

---

[5] https://github.com/facebookresearch/fairseq/tree/main/examples/HuBERT.
[6] https://github.com/microsoft/unilm/tree/master/wavlm.

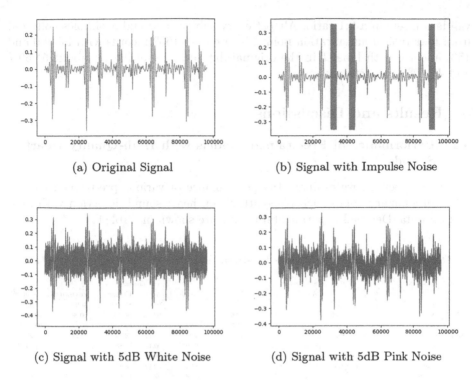

(a) Original Signal

(b) Signal with Impulse Noise

(c) Signal with 5dB White Noise

(d) Signal with 5dB Pink Noise

Fig. 1. Comparison of original signal and signals with different noise types. (Each impulse noise lasts 0.1 s per second in (b).)

For impulse noises, we applied three additional frequencies and durations. Frequencies consist of one piece of noise per audio clip, one piece and two pieces of noise per second. The durations include 0.1, 0.2 and 0.3 s, respectively. We utilized a random sequence with amplitudes in the range {−max signal, max signal} to randomly replace segments of the signal. We ensured that the noise did not overlap when the length of the signal allowed. Figure 1 (a)(b) illustrates a comparison between the original and impulse noise-added signal.

4.3 Robustness of Models in Real Scenarios

In this section we analyze the performance of the pre-trained models on the real pediatric outpatient settings.

Model Configurations. Eight models with excellent performance were selected based on the first experiment. They are PANN-V1/V2, SSAST-P250, BEATs - iter3+AS20K, BEATs - iter3+AS20K-ft-AS2M-cpt1/2, HuBERT-Base and WavLM-Base.

Experimental Data. The experimental data was taken from the CirCor DigiScope Dataset, covering 3163 data records, and was divided into training and validation sets in a 9:1 ratio. Almost every subject had audio samples collected from four typical auscultation points. There were 1632 records with label 0 and 1531 records with label 1 in the original dataset, accounting for about 51.60% and 48.40%, respectively.

5 Results and Discussions

5.1 Performance of Pre-trained Models with High-Quality Heart Sounds

In this subsection, we evaluate the performance of various pre-trained models in an preliminary comprehensive synthesis on heart sound detection with high quality data. Detailed experimental results are shown in Table 1.

Table 1. Performance of Various Pre-trained Models in Heart Sound Detection

Model	Model Name	Params	w/o finetune				with finetune			
			Acc (%)	F1 score (%)	Precision (%)	Recall (%)	Acc (%)	F1 score (%)	Precision (%)	Recall (%)
Baseline	FBANK-TDNN	2M	–	–	–	–	90.60	85.00	87.92	84.93
PANN	V1	81M	68.50	9.90	66.67	5.35	**94.02**	**90.75**	**90.86**	**90.64**
	V2		83.76	70.48	85.59	59.90	93.35	89.80	89.21	90.40
SSAST	F400	89M	72.55	38.12	70.49	26.12	85.14	76.14	79.24	73.27
	P250		82.30	70.39	76.76	65.01	90.13	84.83	84.38	85.30
	P400		81.32	66.90	78.43	58.32	90.09	85.00	83.31	86.76
BEATs	Iter3	90M	89.07	82.82	84.28	81.41	92.10	88.11	85.83	90.52
	Iter3-ft-AS2M-cpt1		89.50	82.76	87.82	78.86	92.14	87.86	87.76	87.97
	Iter3-ft-AS2M-cpt2		90.29	84.03	89.78	78.98	92.45	88.53	87.07	90.04
	Iter3+AS2M		90.33	84.31	88.72	80.32	92.17	88.00	87.32	88.70
	Iter3+AS20K		**90.48**	**84.86**	87.48	**82.38**	92.45	88.16	89.49	86.88
	Iter3+AS2M-ft-AS2M-cpt1		89.62	83.10	87.82	78.86	92.69	88.56	89.66	87.48
	Iter3+AS2M-ft-AS2M-cpt2		89.66	82.75	**89.89**	76.67	92.25	88.13	87.44	88.82
	Iter3+AS20K-ft-AS2M-cpt1		89.15	82.62	85.75	79.71	93.39	89.86	89.32	90.40
	Iter3+AS20K-ft-AS2M-cpt2		90.25	84.18	88.59	80.19	93.04	89.36	88.45	90.28
HuBERT	Base	90M	78.80	65.69	68.98	62.70	91.43	86.84	86.31	87.36
WavLM	Base	95M	78.61	65.44	68.58	62.58	92.69	88.46	90.37	86.63
	Base+		75.19	51.87	69.67	41.31	92.61	88.25	**90.86**	85.78
	Large	317M	71.06	26.98	73.51	16.52	91.98	87.30	89.53	85.18

Without fine-tuning, PANNs exhibit divergent performance. PANN-V1 model achieves 68.50% accuracy but its F1 score is notably low at 9.90%. SSASTs in different configurations also show performance differences, accuracy ranges from 72.55% to 82.30%, but F1 scores vary widely. Patch-based SSASTs performing better. SSAST-F400 displays lower F1 scores, indicating limitations in handling imbalanced data. HuBERT-Base and WavLMs perform moderately well. The performance of WavLM-Large model is comparatively poor, likely due to complexity and overfitting. BEATs outperform the others, with accuracies ranging from 89.07% to 90.48% and relatively high F1 scores. These results serve as an initial performance benchmark, suggesting the generalization ability and limitations of the pre-trained models in heart sound detection task without fine-tuning.

After fine-tuning, PANN-V1 achieves the highest accuracy and F1 score among all models, at 94.02% and 90.75%, respectively. Accuracy of SSASTs remains around 90%, falling short of expectations. The F400 model still performs relatively inferior, suggesting that square patches are more effective in capturing frequency structure features in heart sound detection. BEATs achieves accuracies between 92.10% and 93.39%, with improved F1 scores. HuBERT and WavLM demonstrate potential for applications in heart sound detection. HuBERT-Base improves its accuracy to 91.43% with an F1 score of 86.84%. While it has a similar number of parameters as BEATs, its performance is slightly inferior both before and after fine-tuning. WavLMs all achieve around 92% accuracy and significant F1 score improvements. Despite having several times more parameters, WavLM-Large does not outperform the other models. This may indicate that increasing model parameters and complexity may not necessarily lead to improved performance in heart sound detection.

In summary, BEATs have the best raw performance in heart sound detection task. After fine-tuning, PANN-V1 achieves the best performance in terms of both accuracy and F1 score.

5.2   Analysis of Model's Noise Resistance Performance

In this subsection, we evaluate the performance of the pre-trained model in various simulated noise environments.

Color Noise Results. The accuracy of different pre-trained models varies with SNRs as shown in Fig. 2, where (a) is the white noise condition and (b) is the pink noise condition.

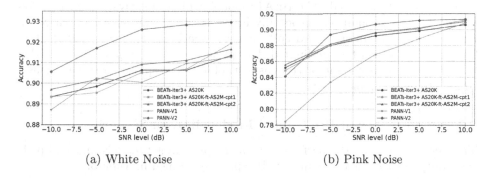

(a) White Noise   (b) Pink Noise

Fig. 2. Accuracy of various models across different SNR levels (from −10 dB to 10 dB) for heart sounds contaminated with color noises (white and pink). (Color figure online)

As SNRs increased gradually from −10 dB to 10 dB, all models exhibited improved accuracy. Under white noise conditions, the PANN-V2 model demonstrated an obvious advantage, with accuracy rising from 90.56% at −10 dB to 92.96% at 10 dB. This indicates that PANN-V2 model not only has commendable robustness at low SNR, but also enhances its performance as SNR increases. Among the BEATs models, BEATs-iter3+AS20K-ft-AS2M-cpt2 exhibited relatively excellent comprehensive noise immunity.

Pink noise has a more detrimental effect on model performance. This may be due to its stronger presence in the low-frequency spectrum, causing interference with heart sound signals in those frequency ranges. Despite some interference from pink noise, the PANN-V2 model maintains the highest accuracy across all SNRs except −10 dB. Its accuracy improves from 84.11% at −10 dB to 91.27% at 10 dB. This improvement can be attributed to the effective multi-scale feature extraction of PANN-V2, combining high and low frequency information. The BEATs model performs similarly overall and also exhibits high stability.

Impulse Noise Results. The experimental results of model accuracy as affected by impulse noise frequencies and durations are shown in Table 2.

Table 2. Accuracy(%) of Various Models with Impulse Noise

Model	Model Configurations	Impulse Noise (Frequencies - Durations)								
		1/3 s-0.1 s	1/3 s-0.2 s	1/3 s-0.3 s	1/1 s-0.1 s	1/1 s-0.2 s	1/1 s-0.3 s	2/1 s-0.1 s	2/1 s-0.2 s	2/1 s-0.3 s
PANNs	V1	92.17	92.29	92.41	91.70	91.39	91.39	91.45	91.31	90.44
	V2	92.49	93.04	92.55	93.04	92.55	92.17	92.37	91.07	89.89
BEATs	Iter3+AS20K	92.02	92.33	92.02	92.17	92.14	91.82	92.47	91.90	91.86
	Iter3+AS20K-ft-AS2M-cpt1	92.53	92.37	92.47	92.49	92.25	93.16	92.73	92.49	92.06
	Iter3+AS20K-ft-AS2M-cpt2	92.61	92.96	92.25	93.08	92.61	92.37	92.84	92.43	91.98

Although the performance broadly tends to decline with increased frequency and duration of impulse noise, it is not significantly disruptive. However, under the extreme noise conditions, such as 2/1 s-0.3 s, some disruption in model performance is still observed. The BEATs model exhibits relatively robust performance across all conditions, especially in the most extreme scenario, where the BEATs - iter3+AS20K-ft-AS2M-cpt2 achieves the highest accuracy at 91.98%. One possible reason for the limited disruption caused by impulse noise is that these impulses may enhance certain aspects of heartbeat pulses or serve as masks similar to data enhancement.

5.3   Performance of Pre-trained Models in Real Scenarios

This section analyzes the performance of the pre-trained models in a real pediatric outpatient clinic environment. Results are detailed in Table 3.

Table 3. Performance of Various Pre-trained Models (CirCor DigiScope Dataset)

Model	Model Name	Params	w/o finetune				with finetune			
			Acc (%)	F1 score (%)	Precision (%)	Recall (%)	Acc (%)	F1 score (%)	Precision (%)	Recall (%)
Baseline	FBANK-TDNN	2M	–	–	–	–	61.98	52.43	70.51	41.73
PANN	V1	81M	59.75	50.76	65.79	41.32	64.13	**63.68**	64.77	**62.63**
	V2		60.00	47.79	**69.33**	36.46	63.88	63.43	64.51	62.39
SSAST	P250	89M	56.36	50.56	58.63	44.44	58.72	50.57	63.40	42.06
BEATs	Iter3+AS20K	90M	61.20	51.97	68.65	41.81	64.63	61.85	67.44	57.12
	Iter3+AS20K-ft-AS2M-cpt1		**62.48**	**59.10**	65.27	**53.99**	**65.33**	59.25	**72.27**	50.21
	Iter3+AS20K-ft-AS2M-cpt2		61.61	53.71	68.06	44.36	65.21	61.83	68.82	56.13
HuBERT	Base	90M	57.40	50.46	60.62	43.21	62.07	53.50	69.57	43.46
WavLM	Base	95M	56.03	47.53	59.29	39.67	62.60	58.58	65.98	52.67

BEATs-Iter3+AS20K-ft-AS2M-cpt1 has the best overall performance, achieving the highest accuracy both before and after fine-tuning, reaching 62.48% and 65.33% respectively. Its precision also stands out after fine-tuning at 72.27%. PANN-V1 achieves an F1 score of 63.68% and a recall of 62.63% after fine-tuning.

Compared to the CirCor DigiScope Dataset, Models generally have a better performance on the Physionet/CinC 2016 Database. Several factors may contribute to this difference. Firstly, the Physionet/CinC 2016 Database has a larger proportion of normal data compared to abnormal data, making it more likely for models to learn features from normal data. In contrast, the CirCor DigiScope Dataset maintains a relatively balanced ratio of normal and abnormal data, requiring models to work harder to learn features from both types, potentially leading to lower performance.

Moreover, only 13% of the data in the Physionet/CinC 2016 Database used in this study had a clear indication of noise contamination, whereas all of the CirCor DigiScope Dataset was recorded in a real pediatric outpatient setting, introducing various sources of noise, from stethoscope friction to crying or laughter. The complex noise in real-world environments could adversely affect model performance.

Additionally, 66% of the Physionet/CinC 2016 Database data used in this study was collected from the mitral position on the chest, whereas the CirCor DigiScope Dataset includes audio samples from four typical auscultation points for almost every subjects. The diversity in auscultation locations might impact model performance. Specific reasons await further research.

6   Conclusion

In this study, we investigated the performance of audio pre-trained models (PANNs, SSAST, BEATs) and speech pre-trained models (HuBERT, WavLM) on heart sound detection at three levels. We first evaluated the models using Physionet/CinC 2016 Database, and found that the non-fine-tuned BEATs models achieved an accuracy of approximately 90%. The fine-tuned PANN-V1 model performed even better with an accuracy of 94.02%. We also examined the models' robustness against different noise paradigms. Pink noise was more disruptive than white noise, and the PANN-V2 model showed resilience across both noise spectra. Impulse noise had minimal effect. To further validate the models, we evaluated them using the CirCor DigiScope dataset, which included data from multiple listening points and specific demographics. The results showed that the models' performance was affected by the demographic features and clinical noise. The BEATs model achieved an accuracy of 65.33% in this case. Overall, the BEATs model performed well in all environments, while the speech pre-trained models did not excel in the heart sound task. These findings provide insights for optimizing the model and improving heart sound detection accuracy. Future work should focus on optimizing the model's robustness to real environmental noise and its ability to adapt to different listening positions. Additionally, research on the interpretability of heart sound auscultation is of great interest.

Acknowledgements. This work was supproted by National Natural Science Foundation of China (No. 62106140), Shanghai Science and Technology Committee (No. 21511101100), and State Key Laboratory of UHD Video and Audio Production and Presentation.

References

1. Ahlström, C.: Processing of the Phonocardiographic Signal: methods for the intelligent stethoscope. Ph.D. thesis, Institutionen för medicinsk teknik (2006)
2. Chen, G., Chai, S., Wang, G., Du, J., et al.: Gigaspeech: an evolving, multi-domain ASR corpus with 10,000 hours of transcribed audio. arXiv preprint arXiv:2106.06909 (2021)
3. Chen, S., Wang, C., Chen, Z., Wu, Y., et al.: WavLM: large-scale self-supervised pre-training for full stack speech processing. IEEE J. Sel. Top. Signal Process. **16**(6), 1505–1518 (2022)
4. Chen, S., Wu, Y., Wang, C., Liu, S., et al.: Beats: audio pre-training with acoustic tokenizers. arXiv preprint arXiv:2212.09058 (2022)
5. Devlin, J., Chang, M.W., Lee, K., Toutanova, K.: Bert: pre-training of deep bidirectional transformers for language understanding. arXiv preprint arXiv:1810.04805 (2018)
6. Dong, F., Schuller, B., Qian, K., Ren, Z., et al.: Machine listening for heart status monitoring: introducing and benchmarking HSS — the heart sounds Shenzhen corpus. IEEE J. Biomed. Health Inform. 1–13 (2019)
7. Dosovitskiy, A., Beyer, L., Kolesnikov, A., Weissenborn, D., et al.: An image is worth 16×16 words: transformers for image recognition at scale. arXiv preprint arXiv:2010.11929 (2020)
8. Gemmeke, J.F., Ellis, D.P., Freedman, D., Jansen, A., et al.: Audio set: an ontology and human-labeled dataset for audio events. In: IEEE International Conference on Acoustics, Speech and Signal Processing (ICASSP), pp. 776–780. IEEE (2017)
9. Gong, Y., Chung, Y.A., Glass, J.: AST: audio spectrogram transformer. arXiv preprint arXiv:2104.01778 (2021)
10. Gong, Y., Lai, C.I., Chung, Y.A., Glass, J.: SSAST: self-supervised audio spectrogram transformer. In: Proceedings of the AAAI Conference on Artificial Intelligence, vol. 36, pp. 10699–10709 (2022)
11. Hanna, I.R., Silverman, M.E.: A history of cardiac auscultation and some of its contributors. Am. J. Cardiol. **90**(3), 259–267 (2002)
12. Hsu, W.N., Bolte, B., Tsai, Y.H.H., Lakhotia, K., et al.: Hubert: self-supervised speech representation learning by masked prediction of hidden units. IEEE/ACM Trans. Audio Speech Lang. Process. **29**, 3451–3460 (2021)
13. Ismail, S., Siddiqi, I., Akram, U.: Localization and classification of heart beats in phonocardiography signals —a comprehensive review. EURASIP J. Adv. Signal Process. **2018**(1), 26 (2018)
14. Kahn, J., et al.: Libri-Light: a benchmark for ASR with limited or no supervision. In: ICASSP (2020)
15. Kahn, J., Rivière, M., Zheng, W., Kharitonov, E., et al.: Libri-light: a benchmark for ASR with limited or no supervision. In: IEEE International Conference on Acoustics, Speech and Signal Processing (ICASSP), pp. 7669–7673. IEEE (2020)
16. Kong, Q., Cao, Y., Iqbal, T., Wang, Y., et al.: PANNs: large-scale pretrained audio neural networks for audio pattern recognition. IEEE/ACM Trans. Audio Speech Lang. Process. **28**, 2880–2894 (2020)

17. Liu, C., Springer, D., Li, Q., et al.: An open access database for the evaluation of heart sound algorithms. Physiol. Meas. **37**(12), 2181–2213 (2016)
18. Ma, Y., et al.: On the effectiveness of speech self-supervised learning for music. arXiv preprint arXiv:2307.05161 (2023)
19. Mangione, S.: Cardiac auscultatory skills of physicians-in-training: a comparison of three English-speaking countries. Am. J. Med. **110**(3), 210–216 (2001)
20. Noor, A.M., Shadi, M.F.: The heart auscultation. From sound to graphical. J. Eng. Technol. **4**(2), 73–84 (2013)
21. Oliveira, J., Renna, F., Costa, P.D., Nogueira, M., et al.: The circor digiscope dataset: from murmur detection to murmur classification. IEEE J. Biomed. Health Inform. **26**(6), 2524–2535 (2022)
22. Panah, D.S., Hines, A., McKeever, S.: Exploring the impact of noise and degradations on heart sound classification models. Biomed. Signal Process. Control **85**, 104932 (2023)
23. Panayotov, V., Chen, G., Povey, D., Khudanpur, S.: Librispeech: an ASR corpus based on public domain audio books. In: IEEE International Conference on Acoustics, Speech and Signal Processing (ICASSP), pp. 5206–5210. IEEE (2015)
24. PhysioNet: The physionet cardiovascular signal toolbox (2016). https://physionet.org/content/challenge-2016/1.0.0/. Accessed 28 June 2023
25. Piczak, K.J.: ESC: dataset for environmental sound classification. In: Proceedings of the 23rd ACM International Conference on Multimedia, pp. 1015–1018 (2015)
26. Ren, Z., Chang, Y., Nguyen, T.T., Tan, Y., et al.: A comprehensive survey on heart sound analysis in the deep learning era. arXiv preprint arXiv:2301.09362 (2023)
27. Ren, Z., Cummins, N., Pandit, V., Han, J., et al.: Learning image-based representations for heart sound classification. In: Proceedings of the DH, pp. 143–147. Lyon, France (2018)
28. Vaswani, A., et al.: Attention is all you need. Adv. Neural Inf. Process. Syst. **30** (2017)
29. Waibel, A., Hanazawa, T., Hinton, G., Shikano, K., Lang, K.J.: Phoneme recognition using time-delay neural networks. In: Backpropagation, pp. 35–61. Psychology Press (2013)
30. Wang, C., Riviere, M., Lee, A., Wu, A., et al.: Voxpopuli: a large-scale multilingual speech corpus for representation learning, semi-supervised learning and interpretation. arXiv preprint arXiv:2101.00390 (2021)
31. Wu, T.Y., Hsu, T.Y., Li, C.A., Lin, T.H., Lee, H.V.: The efficacy of self-supervised speech models for audio representations. In: HEAR: Holistic Evaluation of Audio Representations, pp. 90–110. PMLR (2022)
32. Yuenyong, S., Nishihara, A., Kongprawechnon, W., Tungpimolrut, K.: A framework for automatic heart sound analysis without segmentation. Biomed. Eng. Online **10**, 1–23 (2011)

CAM-GUI: A Conversational Assistant on Mobile GUI

Zichen Zhu[1] , Liangtai Sun[1], Jingkai Yang[1], Yifan Peng[1], Weilin Zou[1],
Ziyuan Li[1], Wutao Li[1], Lu Chen[1(✉)], Yingzi Ma[2], Danyang Zhang[1],
Shuai Fan[3], and Kai Yu[1(✉)]

[1] X-LANCE Lab, Department of Computer Science and Engineering MoE Key Lab
of Artificial Intelligence, SJTU AI Institute, Shanghai Jiao Tong University,
Shanghai, China
{JamesZhutheThird,chenlusz,kai.yu}@sjtu.edu.cn
[2] Sichuan University, Chengdu, Sichuan, China
[3] AISpeech Ltd., Suzhou, China

Abstract. Smartphone assistants are becoming more and more popular
in our daily lives. These assistants mostly rely on the API-based Task-
Oreiented Dialogue (TOD) systems, which limits the generality of these
assistants, and the development of APIs would cost much labor and
time. In this paper, we develop a **C**onversational **A**ssistant on **M**obile
GUI (**CAM-GUI**), which can directly perform GUI operations on real
devices, without the need of TOD-related backend APIs. To evaluate the
performance of our assistant, we collect a dataset containing dialogues
and GUI operation traces. From the experiment demonstrations and user
studies, we show that the **CAM-GUI** reaches promising results. Our
system demonstration can be downloaded from this link. You can watch
our demo through this link.

Keywords: Human Computer Interaction · Interface Understanding ·
Natural Language Processing

1 Introduction

Smart assistants are widely used in mobile phones and computers. These assistants are mainly based on Task-Oriented Dialogue systems [1,13]. These assistants work in a similar pipeline. First, the assistants will obtain the user's commands through voice or text input, and extract useful information from it. With
the history information, the assistants will call the backend APIs to fetch information or perform some operations on the device. Finally, they will generate a
response based on the results.

This process most commonly relies on the APIs co-developed by electronic
device manufacturers and software developers, which are costly in terms of labor
and time for the huge amount of third-party applications and are difficult to cope
with the rapid iterative migration of software and systems. There are also a few
functions based on websites of third-party applications. The control progress is
similar to that with installed applications but more convenient.

© The Author(s), under exclusive license to Springer Nature Singapore Pte Ltd. 2024
J. Jia et al. (Eds.): NCMMSC 2023, CCIS 2006, pp. 302–315, 2024.
https://doi.org/10.1007/978-981-97-0601-3_26

In order to provide wider support to visually impaired groups and elderly users, screen reader services like TalkBack[1] and Narrator[2] can read out contents displayed on the screen. Specifically, the screen reader applications on the Android system including TalkBack and smart assistants mentioned before comprehend the existing content through Accessibility Service[3] or Android Debug Bridge (ADB)[4]. They collect the layout information (View-Hierarchy, VH) and the screenshot of the current interface through several APIs, analyze the categories, functions, and meanings of different components, and guide users to understand the content on the interface. Since information like component descriptions obtained from VH are filled in by the software developers, it is not guaranteed that these descriptions are accurate and consistent, and there are even missing cases, for which visual information from screenshots is needed to assist in the progress.

Graphical User Interface (GUI) is currently the most important tool for people to get information and interact with smart devices like mobile phones and computers. Massive information from applications and websites is presented on the screen through images, text, and animations. Users decide the next command based on this known information, and execute the next operation by operations, such as clicking, long pressing, and dragging.

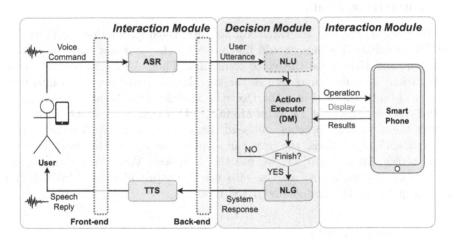

Fig. 1. The system architecture.

If the agents can simulate how human users get and understand the information, and interact with the device in the same way as users, the smartphone

---

[1] https://github.com/google/talkback.

[2] http://aka.ms/narratorguide.

[3] https://developer.android.com/reference/android/accessibilityservice/ AccessibilityService.

[4] https://developer.android.google.cn/studio/command-line/adb/.

assistants will have a wider range of application scenarios and more robust performance. To this end, we develop a closed-loop system for smart voice agents, powered by META-GUI [11].

Our contributions are as follows:

- We develop a web-client data collection platform and collect a dataset of dialogue traces, performing various tasks on 4 restaurant applications, with a wide coverage of the APP interface and functions.
- We build up a closed-loop system, that can understand both the user commands and screen contents to predict and perform a series of operations on the device.
- With user study, we validate the high availability, stability, and user-friendliness of our system for our target users.

2   System Architecture

The system architecture is shown in Fig. 1. It comprises two parts: the interaction Module and the Decision Module.

2.1   Interaction Module

The Interaction Module has three components: voice interaction, ASR module, and TTS module. The user would first wake up the assistant with special preset words and then raise the voice command. Then ASR (Automatic Speech Recognition) module can recognize the speeches and transfer them into texts, which will be further fed into the Decision Module. Finally, after receiving the system response from the Decision Module, TTS (Text To Speech) module will convert the texts into speeches and send them to the user. Furthermore, during the assistant's operations, the user can acquire the process feedback from the phone screen. In our project, we use the AISpeech Voice Wakeuper service[5] to wake up the assistant, and we also use APIs provided by AISPEECH to implement the ASR[6] module and TTS[7] module.

2.2   Decision Module

As described in [11], the Decision Module has three sub-modules: Natural Language Understanding (NLU), Action Executor, and Natural Language Generation (NLG). **CAM-GUI** omits the NLU module and directly inputs the user utterance to the Action Executor module.

The Action Executor module has a similar function to the Decision Making module in the Traditional TOD system, while the Action Executor module can directly operate the smartphone instead of calling backend APIs. Given the

---

[5] https://aispeech.com/core/wakeup.

[6] https://aispeech.com/core/asr.

[7] https://aispeech.com/core/tts.

user utterance, dialogue histories, the current screenshot, and the corresponding view hierarchy, the Action Executor module will predict the next action to be performed on the screen. And it will judge whether the operating process should be ended. If not, it will loop through the above process. After finishing the operation process, the NLG module will generate the system response according to the dialogues and the operating results (Table 1).

Table 1. Several example task goals. The content was translated from Chinese. The complete table can be seen in Appendix B.

Type	Task
dine-in order (DI)	You want to order a [meal_name] set meal at [restaurant_name] for [dine-in/take-away]
delivery order (DE)	You want to order [meal_name_1]...[meal_name_n] at [restaurant_name], and have it delivered to [delivery_address]
account setting(AS)	You want to turn [on/off] personalized recommendations in the privacy settings of [restaurant_name] app
account setting(AS)	You want to see which coupons are available for [restaurant_name], ask the assistant to tell the total number of coupons
account setting(AS)	You want to check the membership level of [restaurant_name], ask the assistant to tell the current level, number of points, and expiration date

Table 2. Distribution of various types of tasks for four applications. **DI** refers to *dine-in order*, **DL** refers to *delivery order*, **AS** refers to *account setting*, **TTr** refers to *Total Trace*, **TTu** refers to *Total Turn*, **TS** refers to *Total Screen*

Application	Package Name	#DI	#DL	#AS	#TTr	#TTu	#TS
KFC	com.yek.android.kfc.activitys	4	2	4	10	80	199
Lucky Coffee	com.lucky.luckyclient	0	0	4	4	28	65
McDonald's	com.mcdonalds.gma.cn	27	11	3	41	205	687
Starbucks	com.starbucks.cn	0	0	5	5	21	82
In total	–	31	13	16	60	334	1033

3   Dataset

To evaluate the performance of **CAM-GUI**, we manually collect 80 dialogues and their corresponding GUI operation traces, including various tasks on 4

Android applications. After cleaning, we get 60 dialogues in total, and the data statistics are shown in Table 2. In this section, we will introduce the task design, collection platform, collection process, and cleaning process respectively.

3.1   Task Design

To enhance the comprehensiveness of the obtained data for the APP interface, we design three types of tasks related to dine-in order (DI), delivery order(DE), and account setting(AS).

The former two tasks encompass a diverse range of ordering strategies that involve the selection of individual items and combo meals, along with additional user requirements. We construct multiple task templates for these two categories and the slot values are randomly selected with crawled menu information before annotating.

The last category of tasks involves a range of activities, such as user account login, setting delivery address, invoicing, as well as making queries about coupon expire-date and membership level. Task designers are chosen based on their extensive experience with ordering through peculiar applications and are required to thoroughly explore all relevant sections and check functional settings before designing the tasks.

3.2   Collection Platform

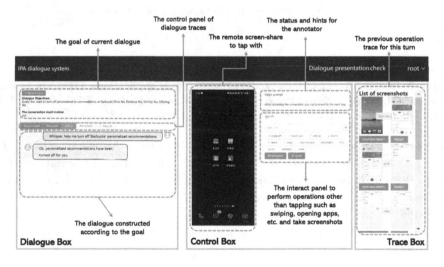

Fig. 2. The screenshot of data collection platform.

We build a web-based platform to collect data. The overview of the platform interface can be seen in Fig. 2. The annotation platform consists of three main parts.

Dialogue Box. The Dialogue Box contains the dialogue task and the annotators can edit the dialogue contents.

Control Box. After connecting with a real or virtual device, the screen will be shown in the Control Box, and the annotators can operate the phone on it. Control Box also contains some buttons to control the phone for convenience. We use the accessibility service to get screenshots and layouts, and use the ADB to interact with the phone, such as clicking, swiping, and other operations on the phone, all actions and corresponding parameters can be seen in Table 3.

Trace Box. The Trace Box can record and display the operation process.

Table 3. The actions in our dataset. There are 8 different actions with 4 different parameters.

Action	Description
Open(package=x)	Open application with package name of x
Click(item=x)	Click the item with index x on the screen
Swipe(direction=x)	Swipe screen towards direction x, which includes "up" and "down"
Input(text=x)	Input the text x to the smartphone
Enter( )	Press the "Enter" button on the keyboard
Clear( )	Clear the current input box
Back( )	Press the "back" button on the smartphone
End( )	Turn has been finished and it will go to the Response Generator module

3.3 Data Collection

We recruit 5 annotators to label the data, and they are all familiar with these 4 applications. Given the required tasks, they would first explore how to finish the task on the phone. After they are ready, they would act as users and agents alternatively to write the dialogues. When acting as the agent, they should also perform operations on the phone according to the dialogues. In total, we collect 80 dialogue traces.

3.4 Quality Check and Data Cleaning

For the annotated trace data, the evaluator will assess and provide feedback on aspects of task completion, dialog integrity, and data preservation.

As the interface layout information is automatically gathered through the accessibility service, we employ a script to streamline the layout and extract solely the data pertaining to visible leaf nodes. Due to the differences in operating systems and the inaccuracy of View Hierarchy, the target item may be missing in some traces. We manually fix these traces and discard the traces that can not be fixed up. After cleaning, we get 60 traces in total.

4   Experiment Demonstrations

Based on the dataset we have collected, we train an action model and a reply model to implement the Action Executor module and NLG module respectively. The architecture of the action model is *BERT+mm*, which is described in [11], and we use BART-Large-Chinese [10][8] as the architecture of the reply model. We show a completed test case in Appendix C and a demonstration video in Appendix E, from which we can see that our assistant can successfully finish the commands proposed by the user. However, there is a mistake during the process, i.e., the price of the order is actually 25 CNY, but the model response is 34.5 CNY, which is the price of the McCrispy Chicken combo. The reason is that we do not use any structure or image information for the reply model, which means it is hard for the model to align the foods and their corresponding prices. We claim that with a more powerful model, that has a strong ability to understand the GUI interfaces and contents, the performance will be much better.

5   User Study

In order to assess how well **CAM-GUI** performs, we carry out a study among users on our campus. Participants with varying levels of familiarity with mobile phone assistants are chosen at random for the study. After a brief demonstration by the developers, the participants are allowed to freely use **CAM-GUI** and then asked to rate its performance compared to other existing assistants on several different criteria listed below using a score from 1 to 5:

- **Conversation-related Performance:** conversation naturalness, grammatical correctness and completeness.
- **Voice-related Performance:** voice naturalness, voice emotion, recognition accuracy.
- **System-related Performance:** response time, system stability.
- **Ease of Use:** the difficulty of using the voice assistant to becoming familiar with it.
- **Functionality:** richness of functions, difficulty, and complexity of executing tasks.
- **Compatibility:** fluency of operation when using different apps, systems, and devices.

---

[8] https://huggingface.co/fnlp/bart-large-chinese.

- **Fault Tolerance:** the ability to automatically correct or accept user manual corrections or interruptions for user errors.
- **Overall Usability:** the overall feeling of the voice assistant or the degree of inclination to use it.

The results of these scores can be found in Table 5. To visualize the results, we draw a radar diagram, which is shown in Fig. 3. From the results, we can find that our assistant is stronger than other existing assistants in the aspects of Ease of use, Functionality, Compatibility, Conversation-related performance, and Voice-related performance, which shows the promising ability of **CAM-GUI**. However, the System-related performance is not good enough. The response time of each operation is much longer than other API-based assistants. That is because the system would wait for the phone screen to be stable before operating, otherwise, the agent may predict a wrong operation, and this is not necessary for others. In the future, we will work hard to reduce the system response time, and bring users a better experience.

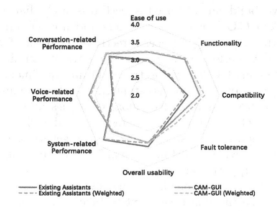

Fig. 3. The radar diagram of user study results.

6   Related Work

Research related to interface understanding and manipulation is gaining increasing attention. This requires knowing what is on the screen and performing what operations can achieve the final goal.

GUI Understanding. Some studies [2–4] have addressed this problem through computer vision techniques, such as optical character recognition (OCR) and target detection models to locate and analyze text and images on GUI screens and encode them into machine-recognizable information. In our system, we use both screenshots and layout trees as inputs to get complementary and complete information.

Command Understanding. There are some studies [8,9,14] using semantic parsing to extract target operations from natural language queries on websites.

Programming by Demonstration. Some studies [5,6] can understand user commands in dialogues, perform corresponding operations, and respond to users with text based on the screen results. They use Programming by Demonstration to learn specific operational flow by user's demonstration, performing some process-fixed tasks when given specific or similar instructions. PUMICE [7] enables users to define procedures and concepts by demonstrating.

Our goal is to build a robust agent that can handle various applications and tasks, with human instruction and correction in dialogue, even if the objective is unfamiliar. With the development of large language models, great potential has shown [12] to make use of their reasoning ability.

7   Conclusion and Future Work

In this paper, we build an advanced speech-to-speech dialogue assistant on mobile GUI (**CAM-GUI**), which can understand user commands and screen contents, and perform action sequences directly on GUI interface to finish user goals. To evaluate the performance, we collect a dataset containing 60 dialogues and their corresponding GUI operation traces. From experiment demonstrations and user studies, we show that our assistant can reach a promising result.

In future work, we plan to collect more dialogues and interaction traces, include a wider variety of application categories and task types, cover more close-to-life application scenarios, and improve the generalization capability of the model. The system will support more complex tasks involving multiple applications. It will also support user preference settings for memory and automatic filling, simplifying the process for high-frequency commands.

We hope to enable the smart assistant to take into account the strong reasoning ability of large language models, which can retain memories of past conversations, and we will try to use multi-modal Large Language Models (LLMs) for prediction. Additionally, we hope to invite more target users to experience the system and further improve its usability.

Acknowledgements. We thank Ershuai Fan and Pengxiang Wu, from AISpeech for API and SDK support. We thank teachers and students from Shanghai Jiao Tong University for participating in system test and data annotation. We also thank all the anonymous reviewers for their thoughtful comments. This work has been supported by the China NSFC Project (No. 62106142 and No. 62120106006), Shanghai Municipal Science and Technology Major Project (2021SHZDZX0102).

A   Ethics and Broader Impact Statement

Crowdsourcing. Data collection is done by the authors and no third-party crowdsourcers have participated.

Data Privacy. Data collection is based on a locally deployed platform and is performed using physical mobile phones. Each data annotator is provided with a phone specifically for data collection and is provided with a virtual phone number and email address to protect privacy. The use of provided information is not mandatory, however, annotators may also use their own devices and information for collection on a voluntary basis.

In order to obtain a richer data distribution, we do not restrict the device model, system, and application version (considering that there is a task to check for application updates) for data collection.

The content involving user privacy has been processed as necessary for disclosure in the paper and supplementary files.

Intended Use. The system is designed to help users efficiently use mobile applications to perform complex operations and difficult commands. Users send their commands and further instruction with either voice or text to the agent, which in turn comprehends, predicts, and executes the target operation. The system can subsequently be deployed on smart devices including GUIs such as mobile phones, tablets, and computers to significantly enhance the usability of smart voice assistants.

Potential Misuse. The use of voice or text commands to control the GUI interface requires the agent to get a high degree of operational freedom and access to control, and will most likely involve a large amount of user privacy information. Considering that decisions are often made on remote servers, a remote hijacking control or a service provider's failure to follow user privacy protocols may lead to user privacy leakage, data loss, or damage to safety.

Environmental Impact. There are a large number of elderly and people with disabilities around the world. We hope that this smart agent can improve their accessibility to the fast-evolving electronic devices, lower the threshold of using complex applications, reduce their fragmentation from the modern information society, and finally promote social equality.

B   Task Templates

We provide all task templates that we used when collecting data in Table 4.

C   An Example for Dialog Traces

A complete annotated example of a dialogue can be seen in Fig. 4.

Table 4. The complete templates for task goals.

Type	Task (Origin)	Task (Translation)
dine-in order (DI)	你想要在【商家名】点餐【菜品】并且选择【堂食/外卖】。	You want to order a [meal_name] set meal at [restaurant_name] for [dine-in/take-away].
dine-in order (DI)	你想要查看【商家名】有哪些可用的套餐及个数。	You want to see which coupons are available and the number of them for [restaurant_name].
delivery order (DE)	你想要在【商家名】点餐【菜品1...菜品N】并且送餐到【地点】。	You want to order [meal_name_1]...[meal_name_n] at [restaurant_name], and have it delivered to [delivery_address].
delivery order (DE)	你想要查看【商家名】账户中现有的优惠券的数量。	You want to check the number of coupon for [restaurant_name].
account setting(AS)	你想要在【商家名】的隐私设置中【开启/关闭】个性化内容推荐。	You want to turn [on/off] personalized recommendations in the privacy settings of [restaurant_name] app.
account setting(AS)	你想要查看在【商家名】的会员等级、积分和过期时间。	You want to check the membership level, number of points, and expiration date of [restaurant_name].
account setting(AS)	你想要使用【手机号】登录【商家名】。	You want to log in to [restaurant_name] with your phone number of phone_number.
account setting(AS)	你想要确认【商家名】的会员卡是否过期，如果过期了，就按照性价比（对比月卡和季卡）续费一张季卡，用【支付方式】支付，请手机助手告诉续费的价格；如果没过期，请手机助手告诉还有多久有效期。	You want to confirm whether the membership card of [restaurant_name] has expired. If it has expired, renew a quarter card according to the cost performance ratio (compare monthly card and quarter card) and pay with [payment_method]. Ask the mobile assistant to tell you the renewal price; If it is not expired, ask the mobile assistant to tell you how long it is still valid.
account setting(AS)	你想要给最近的一笔订单开发票，填写发票【抬头】，并设置接收邮箱为【邮箱】。	If you want to invoice the most recent order, fill in the invoice title [invoice] and set the receiving email to [email_address].
account setting(AS)	你想要保存送餐地址【地址】，并设置为默认地址。	You want to save the delivery address [address] and set it as the default address.
account setting(AS)	你想要检查【商家名】软件版本更新，并请手机助手告当前软件版本和最新软件版本，和更新包大小。	You want to check the [restaurant_name] software version update and ask the mobile assistant to tell the current software version and the latest software version, and the update package size.

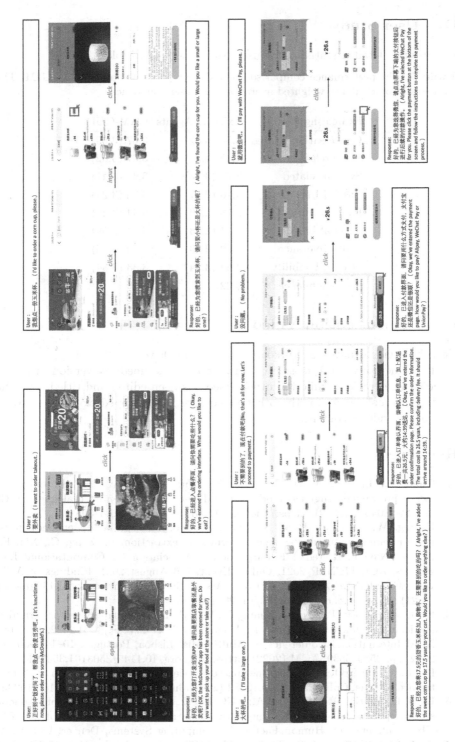

Fig. 4. An example for dialog traces. The figure has been rotated counterclockwise by 90°.

D   User Study

We provide detailed statistics for user study results in Table 5.

Table 5. The performance comparison of Existing Assistants and **CAM-GUI**. The numbers in brackets are weighted averages using user confidence.

Evaluating Factor	Existing Agents	CAM-GUI
Conversation-related	3.56 (3.52)	3.67 (3.69)
Voice-related	3.00 (2.97)	3.67 (3.62)
System-related	3.78 (3.76)	3.44 (3.41)
Ease of Use	3.00 (2.97)	3.22 (3.24)
Functionality	2.78 (2.83)	3.44 (3.48)
Compatibility	3.11 (3.03)	3.44 (3.59)
Fault Tolerance	2.89 (2.83)	3.00 (3.10)
Overall Usability	3.44 (3.34)	3.33 (3.34)
User confidence	3.22	–

E   Demonstration Video

Two short videos have been uploaded to Google Drive.

The video `original_demonstration.mp4` is the unedited version of a dialogue demonstration, while the other video has been edited and titled with the prefix `submission`. As for the other videos, we have skipped or fast-forwarded part of the video's content.

References

1. Chen, Z., Liu, Y., Chen, L., Zhu, S., Wu, M., Yu, K.: OPAL: ontology-aware pretrained language model for end-to-end task-oriented dialogue. CoRR abs/2209.04595 (2022). https://doi.org/10.48550/arXiv.2209.04595
2. Hwang, W., Yim, J., Park, S., Yang, S., Seo, M.: Spatial dependency parsing for semi-structured document information extraction. In: Zong, C., Xia, F., Li, W., Navigli, R. (eds.) Findings of the Association for Computational Linguistics: ACL/IJCNLP 2021, Online Event, 1–6 August 2021. Findings of ACL, vol. ACL/IJCNLP 2021, pp. 330–343. Association for Computational Linguistics (2021). https://doi.org/10.18653/v1/2021.findings-acl.28
3. Li, J., Xu, Y., Lv, T., Cui, L., Zhang, C., Wei, F.: DiT: self-supervised pre-training for document image transformer. In: Magalhães, J., et al. (eds.) MM 2022: The 30th ACM International Conference on Multimedia, Lisboa, Portugal, 10–14 October 2022, pp. 3530–3539. ACM (2022). https://doi.org/10.1145/3503161.3547911
4. Li, M., et al.: TrOCR: transformer-based optical character recognition with pretrained models. CoRR abs/2109.10282 (2021). https://arxiv.org/abs/2109.10282
5. Li, T.J., Azaria, A., Myers, B.A.: SUGILITE: creating multimodal smartphone automation by demonstration. In: Mark, G., et al. (eds.) Proceedings of the 2017 CHI Conference on Human Factors in Computing Systems, Denver, CO, USA, 06–11 May 2017, pp. 6038–6049. ACM (2017). https://doi.org/10.1145/3025453.3025483

6. Li, T.J., Chen, J., Xia, H., Mitchell, T.M., Myers, B.A.: Multi-modal repairs of conversational breakdowns in task-oriented dialogs. In: Iqbal, S.T., MacLean, K.E., Chevalier, F., Mueller, S. (eds.) UIST 2020: The 33rd Annual ACM Symposium on User Interface Software and Technology, Virtual Event, USA, 20–23 October 2020, pp. 1094–1107. ACM (2020). https://doi.org/10.1145/3379337.3415820

7. Li, T.J., Radensky, M., Jia, J., Singarajah, K., Mitchell, T.M., Myers, B.A.: PUMICE: a multi-modal agent that learns concepts and conditionals from natural language and demonstrations. In: Guimbretière, F., Bernstein, M.S., Reinecke, K. (eds.) Proceedings of the 32nd Annual ACM Symposium on User Interface Software and Technology, UIST 2019, New Orleans, LA, USA, 20–23 October 2019, pp. 577–589. ACM (2019). https://doi.org/10.1145/3332165.3347899

8. Mazumder, S., Riva, O.: FLIN: a flexible natural language interface for web navigation. In: Toutanova, K., et al. (eds.) Proceedings of the 2021 Conference of the North American Chapter of the Association for Computational Linguistics: Human Language Technologies, NAACL-HLT 2021, Online, 6–11 June 2021, pp. 2777–2788. Association for Computational Linguistics (2021). https://doi.org/10.18653/v1/2021.naacl-main.222

9. Pasupat, P., Jiang, T., Liu, E.Z., Guu, K., Liang, P.: Mapping natural language commands to web elements. In: Riloff, E., Chiang, D., Hockenmaier, J., Tsujii, J. (eds.) Proceedings of the 2018 Conference on Empirical Methods in Natural Language Processing, Brussels, Belgium, 31 October–4 November 2018, pp. 4970–4976. Association for Computational Linguistics (2018). https://aclanthology.org/D18-1540/

10. Shao, Y., et al.: CPT: a pre-trained unbalanced transformer for both Chinese language understanding and generation. arXiv preprint arXiv:2109.05729 (2021)

11. Sun, L., Chen, X., Chen, L., Dai, T., Zhu, Z., Yu, K.: META-GUI: towards multi-modal conversational agents on mobile GUI. In: Goldberg, Y., Kozareva, Z., Zhang, Y. (eds.) Proceedings of the 2022 Conference on Empirical Methods in Natural Language Processing, EMNLP 2022, Abu Dhabi, United Arab Emirates, 7–11 December 2022, pp. 6699–6712. Association for Computational Linguistics (2022). https://aclanthology.org/2022.emnlp-main.449

12. Wang, B., Li, G., Li, Y.: Enabling conversational interaction with mobile UI using large language models. CoRR abs/2209.08655 (2022). https://doi.org/10.48550/arXiv.2209.08655

13. Wu, C., Hoi, S.C.H., Socher, R., Xiong, C.: TOD-BERT: pre-trained natural language understanding for task-oriented dialogue. In: Webber, B., Cohn, T., He, Y., Liu, Y. (eds.) Proceedings of the 2020 Conference on Empirical Methods in Natural Language Processing, EMNLP 2020, Online, 16–20 November 2020, pp. 917–929. Association for Computational Linguistics (2020). https://doi.org/10.18653/v1/2020.emnlp-main.66

14. Xu, N., et al.: Grounding open-domain instructions to automate web support tasks. In: Toutanova, K., et al. (eds.) Proceedings of the 2021 Conference of the North American Chapter of the Association for Computational Linguistics: Human Language Technologies, NAACL-HLT 2021, Online, 6–11 June 2021, pp. 1022–1032. Association for Computational Linguistics (2021). https://doi.org/10.18653/v1/2021.naacl-main.80

A Pilot Study on the Prosodic Factors Influencing Voice Attractiveness of AI Speech

Yihui Wang, Haocheng Lu, and Gaowu Wang[✉]

Beijing Normal University, Beijing 100875, China
{202321080024,202111998172}@mail.bnu.edu.cn, wgw@bnu.edu.cn

Abstract. This study investigates the attractiveness of AI-synthesized voice and its influencing factors from the perspective of speech prosody. Firstly, a comparative analysis, including MOS (Mean Opinion Score) listening test, was conducted on the vocal and facial attractiveness levels of four most popular AI-synthesized voices with 20 listeners, and the best-performing ChatGPT (Chat Generative Pre-trained Transformer) was selected for subsequent analysis. Secondly, a segment of ChatGPT speech was re-synthesized using Praat and Python with different prosodic acoustic parameters, such as overall fundamental frequency, intonation, and duration, respectively. Then a MOS listening test was carried out to evaluate the voice attractiveness of re-synthesized segments in four dimensions: power, competence, warmth, and honesty. The statistical results, including ANOVA (Analysis of Variance) and HSD (Honestly Significant Difference), indicate that altering prosodic parameters, particularly an overly high fundamental frequency and an excessively high speech rate, would significantly reduce the attractiveness of AI-synthesized voice, while the opposite is not the same. Thirdly, to exclude semantic factors, a speech segment in Finnish, which is a completely unfamiliar language to listeners, was used for MOS listening test, and the results are consistent in different languages.

Keywords: Voice Attractiveness · Speech Synthesis · Artificial Intelligence

1 Introduction

Since 2018, OpenAI has introduced a series of language models known as GPT (Generative Pre-trained Transformer). On November 30, 2022, OpenAI launched the highly popular ChatGPT built upon GPT-3.5. This product, a chatbot driven by generative AI, focuses on conversation generation. In September 2023, ChatGPT introduced a voice feature, converting written text into spoken voice via Text-to-Speech (TTS) technology. This voice model, with its human-like quality and fluency, has a very high attractiveness. Although AI-synthesized voices lack an actual speaker, listeners imagine one based on the voice. Generative AIs like ChatGPT have a broad prospect in the field of intelligent assistants, and more attractive synthesized voices could lead to a more approachable and trustworthy user experience. Therefore, this paper focuses on AI-synthesized voices, attempting to research the following questions: What is the level of ChatGPT's voice attractiveness? How do prosodic acoustic parameters influence ChatGPT's voice attractiveness?

© The Author(s), under exclusive license to Springer Nature Singapore Pte Ltd. 2024
J. Jia et al. (Eds.): NCMMSC 2023, CCIS 2006, pp. 316–329, 2024.
https://doi.org/10.1007/978-981-97-0601-3_27

Voice attractiveness refers to the attractiveness a speaker's voice conveys to the listener during conversation. From a psychological perspective, Zuckerman & Driver [1] have demonstrated the existence of vocal attractiveness stereotypes, where people with more attractive voices are perceived to possess ideal personality traits. Collins [2] was the first to study male voice attractiveness, recording the voices of 34 male subjects and measuring four vocal parameters: power spectrum, harmonic frequencies, peak frequency, and formant frequencies. She found that men with voices in which there were closely spaced, low frequency harmonics were judged as being more attractive. Currently research on voice attractiveness mainly focuses on English speech, generally from two perspectives: the speaker's voice attractiveness factors, such as internal hormonal levels, manners of articulation, facial attractiveness; and the listener's perspective, comparing subjective evaluations and expectations of the speaker. Currently, voice attractiveness research often approaches from a psychological angle, incorporating experimental phonetics. For example, Zhou et al.'s [3] study on the physical and social factors affecting voice attractiveness compared the impacts of vocal and semantic variables in speech. Liu & Wang's [4] research summarizes recent domestic and international findings, introducing six acoustic parameters commonly used in voice attractiveness studies and physiological, psychological, and social factors influencing judgments of voice attractiveness.

The current mainstream method for evaluating synthesized speech is the Mean Opinion Score (MOS). This study draws on the experimental methods of Zuckerman & Driver [1], and refers to the acoustic parameters affecting voice attractiveness proposed by Liu & Wang's [4]. The following tasks are undertaken: Firstly, a comparison of the four most popular AI Mandarin synthesized male voicesis conducted, exploring whether there is consistency in artificial synthesized voice attractiveness preferences based on listeners' subjective perceptions. Subsequently, AI voice samples are selected for changes in fundamental frequency, intonation, and duration to determine which factors contribute more significantly to voice attractiveness.

For this study, Adobe Audition 2023 was used for recording, and voice processing was achieved through Praat software [5] and Python calling the Parselmouth library [6]. Data processing utilized Excel and SPSS tools.

2 AI Voice Attractiveness Comparative Evaluation

2.1 Attractiveness of Four AI Voices

This study uses self-recorded materials, with four AI male voices respectively from: S Voice Assistant, ChatGPT Voice, K Voice Assistant, and H Voice Assistant (referred to as S, C, K, and H, respectively). To eliminate the influence of semantics on the experimental results, the chosen text is the story "The North Wind and the Sun," composed of everyday Mandarin vocabulary, without underlined sentiment. The text consists of 124 characters, and the duration of the four generated speeches is about 30 s each. The voices were synthesized by AI product mobile applications, recorded internally, with a sampling rate of 48 kHz and a 32-bit float sampling format, processed and stored as WAV files using Adobe Audition 2023.

The experiment was conducted in a quiet indoor environment with 20 listeners (10 males and 10 females aged between 18 and 24). The four voices were played in a random order, and listeners used a 7-point Likert scale to evaluate the four AI male voices.

Zuckerman & Driver's [1] research, which integrated subjective evaluations with objective measurements, found that subjective rating scores could predict voice attractiveness well, with an explanatory power of 74% [7]. This paper will draw upon the experimental methods of Zuckerman, employing subjective evaluations and voice analysis techniques to investigate the reasons behind the consistency in preferences for AI voice attractiveness. This study's rating scale comprised four aspects: intelligibility (clarity of the message in the voice), naturalness (this speech sounds natural and coherent), voice quality preference (I like the speaker's voice), and friendliness (I would like to be friends with the speaker). Ratings ranged from "1 strongly disagree" to "7 strongly agree". The evaluation results are shown as Fig. 1.

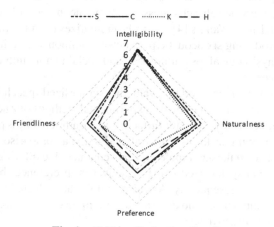

Fig. 1. AI Voice Evaluation Results

A reliability test was conducted using SPSS, yielding a Cronbach's alpha of 0.912, indicating high reliability of the scale. A validity test was performed using factor analysis, with the KMO values of all four dimensions exceeding 0.6 and the Bartlett's test p-values being less than 0.05, indicating good validity of the scale. The data showed significant differences across all four dimensions ($p < 0.01$). The variance calculation revealed the order of differences: intelligibility < naturalness < friendliness < preference ($0.007 < 0.125 < 0.264 < 0.693$), indicating that AI voices can generally convey text information naturally and coherently, but there is a significant variation in voice attractiveness. The ChatGPT voice performed the best, particularly in terms of voice quality preference and friendliness, making it more acceptable to the listeners compared to other AI voices. Post-experiment inquiries with the subjects revealed a general consensus that the ChatGPT voice sounded "very attractive," leading to higher evaluations and the perception that the expected speaker possesses more desirable qualities. Additionally, compared to other AI voices, ChatGPT's synthesized voice includes detailed simulation of real human conversational elements, contributing to its realism and attractiveness. For example,

ChatGPT's voice includes "breath sounds" and uneven speech rates with varying pauses between clauses, closely resembling real human speech patterns.

Research by Zuckerman [1] and Berry [8] has explored the relationship between vocal attractiveness and facial attractiveness. Although AI voices are trained using human speech and are highly realistic, there is no actual "speaker". In human-computer interaction scenarios, listeners create an image of the speaker in their expectations. To study the correlation between predicted facial attractiveness ("I assume the speaker has an attractive appearance") and voice attractiveness, another MOS survey was conducted, asking 20 listeners to evaluate the predicted facial attractiveness of the four voices using the Likert scale. The predicted facial attractiveness for S, C, K, and H voices is shown as Fig. 2.

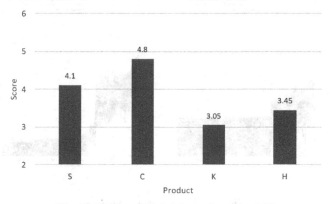

Fig. 2. Predicted Facial Attractiveness of AI

Attractive faces and voices, though different in sensory form, largely imply and convey overlapping information such as sexual attractiveness, health, age, or aesthetics, with facial attractiveness having a relative advantage [9]. The ChatGPT voice, which has higher voice attractiveness, also scored the highest in predicted facial attractiveness. This suggests that users' expected images of AI voices are consistent with their voice attractiveness for more attractive voices, people psychologically create a more ideal image, making them more willing to interact with such intelligent assistants.

2.2 Comparison of AI Voices Power Spectra

Collins' [2] research compared the information in the power spectra of 34 male voices and found a strong consistency between female listeners' predictions of male vocal features and physical characteristics. Previous research on individual speakers indicates that measurements of continuous speech stabilize after approximately 30 s [10]. The lengths of the four speech samples compared in this paper are around 30 s, making the power spectra relatively referential. Figure 3 presents the power spectra of four speech segments generated using Praat and Python. These power spectra display the distribution of energy in the voice signal across various frequencies. The horizontal axis represents frequency, while the vertical axis indicates the sound pressure level. As shown in Fig. 3, the ChatGPT voice exhibits a bandwidth of about 12 kHz, corresponding to an approximate sampling rate of 24 kHz. This indicates that ChatGPT's voice can encompass a richer range of frequency information, making it more akin to human conversation.

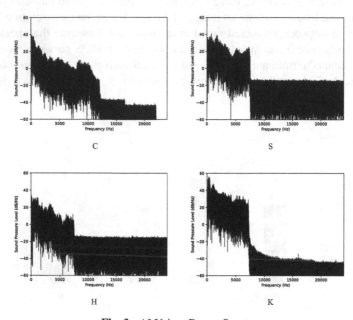

Fig. 3. AI Voices Power Spectra

3   How Acoustic Prosodic Parameters Affect AI Voice Attractiveness

To explore the extent to which acoustic prosodic parameters contribute to the attractiveness of the ChatGPT voice, a third experiment was conducted. This experiment involved altering the fundamental frequency, intonation, and duration of segment C, resulting in variants C_{i1}-C_{i4}, C_{j1}-C_{j2}, C_{k1}-C_{k4}. Twenty listeners were asked to subjectively rate these 11 voice segments across four dimensions, using a 7-point Likert scale. In the perception of attractiveness, physical attractiveness generalizes to the moral domain. Thus, a person with an attractive appearance is often stereotypically attributed with more ideal personality traits [11]. This halo effect also exists in voice attractiveness, where more attractive voices lead to the formation of more ideal personality stereotypes [1]. The dimensions for this study's scale were based on the four identified by Berry [8] in her research on voice attractiveness and stereotypes: power (I think the speaker is confident and influential), competence (I think the speaker has strong professional skills), warmth (I think the speaker is warm and approachable), and honesty (I think the speaker is honest and trustworthy).

3.1 Fundamental Frequency

In modifying the fundamental frequency, this study employed the concept of 'octaves' from acoustics to proportionally adjust the overall fundamental frequency. The original voice C's fundamental frequency was lowered by one octave (fundamental frequency * 1/2), denoted as C_{i1}; lowered by half an octave (fundamental frequency * $2^{-1/2}$), denoted as C_{i2}; raised by half an octave (fundamental frequency * $2^{1/2}$), denoted as C_{i3}; and raised by one octave (fundamental frequency * 2), denoted as C_{i4} (Fig. 4).

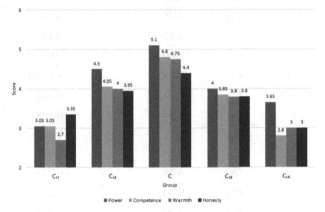

Fig. 4. Scores of Voice Attractiveness (C_{i1}-C_{i4}, F_0 varied)

ANOVA (Analysis of Variance) was used to test the differences in the quantitative matrices of C, C_{i1}-C_{i4}. The F-values and p-values are shown as Table 1.

Table 1. Results of One-way ANOVA (F_0 varied)

	Power	Competence	Warmth	Honesty
F	7.132	7.736	7.027	3.889
p	<0.001	<0.001	<0.001	0.006
Sig.	*	*	*	*

Changes in fundamental frequency showed significant differences in the perception of attractiveness across all four dimensions, with all p-values less than 0.05. Based on these findings, further analysis was conducted using Tukey's HSD (Honestly Significant Difference) test for the four dimensions. The paired differences and significance are shown as Table 2.

Table 2. Results of HSD Post Hoc Multiple Comparisons (F_0 varied)

Factor			Mean Difference	p	Sig.
Power	C	C_{i1}	2.05	<0.001	*
		C_{i2}	0.60	0.602	
		C_{i3}	1.10	0.070	
		C_{i4}	1.45	0.007	*
Competence	C	C_{i1}	1.75	<0.001	*
		C_{i2}	0.75	0.360	
		C_{i3}	0.95	0.147	
		C_{i4}	2.00	<0.001	*
Warmth	C	C_{i1}	2.05	<0.001	*
		C_{i2}	0.75	0.428	
		C_{i3}	0.95	0.198	
		C_{i4}	1.75	0.001	*
Honesty	C	C_{i1}	1.05	0.061	
		C_{i2}	0.45	0.775	
		C_{i3}	0.60	0.537	
		C_{i4}	1.40	0.004	*

Both C_{i1} and C_{i4} showed significant differences in the perception of attractiveness compared to the original voice. As illustrated in Fig. 5, C_{i4} exhibited the greatest degree of difference, indicating that a substantial increase of the fundamental frequency led listeners to perceive the speaker as less attractive. This corroborates the findings of Re et al. [12] who studied the attractiveness of male voices in English. Generally, male voices with lower fundamental frequencies are more attractive, but this low frequency has a limit (above 96 Hz) excessively low frequencies are associated with illness.

3.2 Intonation Variability

Apart from the overall proportional changes in fundamental frequency, the relative pitch variations within the same segment of speech also impact the voice effect, mainly manifesting in the degree of intonation variability. To alter the relative pitch for varying degrees of intonation change, the original text was first marked for stress, with 14 words using distinct stress. In Praat, the process "to pitch(ac)-to manipulation-select-multiply pitch frequency" was executed. To maintain natural auditory coherence as much as possible, the highest and lowest pitch points of the 14 stressed words, along with 3–5 adjacent points, were manually selected. The pitch difference was then expanded (highest point * 1.2, lowest point * 0.5), denoted as C_{j1}; and contracted (highest point * 0.5, lowest point * 1.2), denoted as C_{j2}.

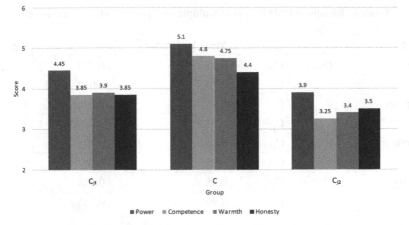

Fig. 5. Scores of Voice Attractiveness (Intonation varied)

Differences in the quantitative matrices of C, C_{j1}, and C_{j2} were tested, with the resulting F-values and p-values shown as Table 3.

Table 3. Results of One-way ANOVA (Intonation varied)

	Power	Competence	Warmth	Honesty
F	3.934	8.759	5.089	2.386
p	0.025	<0.001	0.009	0.101
Sig.	*	*	*	

P-values in 'power', 'competence', and 'warmth' were less than 0.05, indicating that the relative changes in fundamental frequency within stressed words are related to voice attractiveness. The dimension of 'competence' had a p-value far less than 0.05, showing the most significant difference. An HSD test was conducted. These results are shown as Table 4.

Table 4. Results of HSD Post Hoc Multiple Comparisons (Intonation varied)

Factor			Mean Difference	p	Sig.
Power	C	C_{j1}	0.65	0.290	
		C_{j2}	1.20	0.019	*
Competence	C	C_{j1}	0.95	0.036	*
		C_{j2}	1.55	<0.001	*
Warmth	C	C_{j1}	0.85	0.125	
		C_{j2}	1.35	0.007	*
Honesty	C	C_{j1}	0.55	0.388	
		C_{j2}	0.90	0.086	

Compared to the original voice, reducing the relative fundamental frequency changes in stressed words negatively impacts all three dimensions of perception. Specifically, a relatively gentle and unvarying intonation is perceived as less competent, less warm, and less trustworthy. In essence, a monotone intonation significantly diminishes the attractiveness of the voice. This finding aligns with Rodero's [13] research, which indicated that in speaking contexts, participants found moderate pitch variation and moderate intensity of gestures to be the most effective and attractive strategies.

3.3 Duration

Variations were made in the duration of the speech: faster (duration * 0.8), slightly faster (duration * 0.9), slightly slower (duration * 1.1), and slower (duration * 1.2), respectively denoted as C_{k1}, C_{k2}, C_{k3}, C_{k4}. The results are shown as Fig. 6.

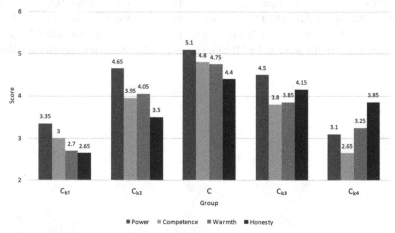

Fig. 6. Scores of Voice Attractiveness (C_{k1}-C_{k4}, Duration varied)

Differences in the quantitative matrices of C, C_{k1}-C_{k4} were tested, with the resulting F-values and p-values shown as Table 5.

Table 5. Results of One-way ANOVA (Duration varied)

	Power	Competence	Warmth	Honesty
F	6.834	7.543	7.039	5.824
p	<0.001	<0.001	<0.001	<0.001
Sig.	*	*	*	*

All p-values were far less than 0.05, indicating significant differences. An HSD analysis was conducted. The results are shown as Table 6.

Table 6. Results of HSD Post Hoc Multiple Comparisons (Duration varied)

Factor			Mean Difference	p	Sig.
Power	C	C_{k1}	1.75	0.003	*
		C_{k2}	0.45	0.873	
		C_{k3}	0.60	0.706	
		C_{k4}	2.00	<0.001	*
Competence	C	C_{k1}	1.80	0.001	*
		C_{k2}	0.85	0.297	
		C_{k3}	1.00	0.155	
		C_{k4}	2.15	<0.001	*
Warmth	C	C_{k1}	2.05	<0.001	*
		C_{k2}	0.70	0.452	
		C_{k3}	0.90	0.205	
		C_{k4}	1.50	0.005	*
Honesty	C	C_{k1}	1.75	<0.001	*
		C_{k2}	0.90	0.169	
		C_{k3}	0.25	0.971	
		C_{k4}	0.55	0.643	

Compared to the original speech, a faster speaking rate has a significantly negative impact on evaluations across all four dimensions. A slower speaking rate, except for honesty, also affects the other dimensions. Speaking rate is an important prosodic parameter in conveying emotions in speech. A slow speaking rate often implies hesitancy and lack of certainty in the speaker, while a fast speaking rate may give an impression of tension and urgency, appearing less friendly and approachable.

4 The Impact of Finnish Language Prosodic Feature Variations on AI Voice Attractiveness

To further validate the impact of changes in fundamental frequency and duration on the attractiveness of ChatGPT's voice, Finnish language text was used to generate speech samples, eliminating the influence of semantic factors. The experimental text consisted of 16 words, with the original synthesized voice lasting 10 s, recorded using Adobe Audition 2023 at a sampling rate of 48 kHz, a 32-bit float sampling format, and stored as a WAV file, denoted as voice E.

1984 George Orwellilta on dystooppinen romaani, joka tutkii totalitaarista hallintoa ja valvonnan sekä sensuurin vaikutuksia yhteiskuntaan. *('1984' by George Orwell is a dystopian novel that examines totalitarian governance and the effects of surveillance and censorship on society.)*

A subjective evaluation experiment was conducted again, using a 5-point Likert scale. Twenty listeners were asked to rate the 'overall attractiveness' of the voice, none of whom were Finnish speakers.

4.1 Fundamental Frequency

Variations in Fundamental Frequency: In an experiment to determine the attractiveness of female voices, Borkowska & Pawlowski [14] extracted the fundamental frequencies of five vowels from 58 Polish female university students, categorizing them into low, medium, high, and very high. They then altered 10 representative samples by either decreasing or increasing the frequency by 20 Hz, which were then rated for attractiveness by 144 young men. To find the optimal fundamental frequency threshold, we made smaller changes to the fundamental frequency of voice E, adjusting it in 5 Hz increments, including -25 Hz, -20 Hz, -15 Hz, -10 Hz, -5 Hz, $+5$ Hz, $+10$ Hz, $+15$ Hz, $+20$ Hz, $+25$ Hz, respectively denoted as E_{i1}-E_{i10}. Interviews were conducted with the subjects, who indicated that without access to semantic information, it was challenging to predict if a speaker is 'confident, competent, warm, or honest' based on voice alone. Consequently, we made some adjustments, opting to use a single-dimensional 'preference' scale for subjective evaluations. The results are shown as Fig. 7.

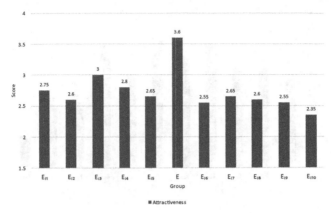

Fig. 7. Scores of Voice Attractiveness (E_{i1}-E_{i10}, F_0 varied)

The results of the ANOVA analysis showed an F-statistic value of 2.063 with a p-value of 0.029. Since $p < 0.05$, this indicates a statistically significant difference in the ratings given by listeners to the original voice E and the modified voices E_{i1} to E_{i10}. This means that even after excluding semantic factors, minor changes in fundamental frequency in an unfamiliar language can affect the level of voice attractiveness.

Further HSD tests were conducted. The results showed that the modified voice E_{i10} had a statistically significant difference in ratings compared to the original voice E, indicating that a slightly higher fundamental frequency negatively impacts voice attractiveness, whereas a slight decrease in fundamental frequency has little effect. This confirms the general trend observed across different languages, where male voices with lower fundamental frequencies tend to be more attractive. A slightly higher fundamental frequency might reduce the gender-specific traits of the voice (such as the correlation between a deep male voice and dominance), thereby lowering the listener's sensitivity.

4.2 Duration

Duration Variations: Changes in speech speed were made at *0.75, *0.8, *0.85, *0.9, *0.95, *1.05, *1.1, *1.15, *1.2, *1.25, respectively denoted as E_{k1}-E_{k10}.

The results of the ANOVA analysis showed an F-statistic value of 4.651 with a p-value of 5.44E−06. With a p-value far less than 0.05, there is a significant difference in the ratings between the original voice E and the modified voices E_{i1} to E_{i10}. This indicates that changes in duration also significantly affect the level of voice attractiveness. The results are shown as Fig. 8.

Further HSD tests were conducted. The results showed that the modified voices E_{k8}, E_{k9}, and E_{k10} had statistically significant differences in ratings compared to the original voice E, indicating that a slightly faster speech rate negatively impacts voice attractiveness, while a slower speed has little effect. This finding differs from the study by Ferdenzi et al. [15], who used Praat to manipulate the duration of different French voice samples (vowels, words, sentences) and found that lengthening the duration reduced voice attractiveness (especially in vowel samples), while shortening the duration slightly

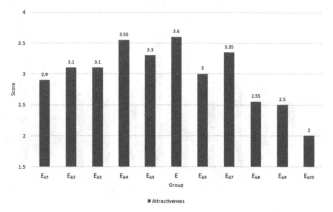

Fig. 8. Scores of Voice Attractiveness (E_{k1}-E_{k10}, Duration varied)

could enhance attractiveness. This discrepancy may be due to the inherent differences between French and Finnish languages. French typically has a faster speech rate, so shorter durations align with native speakers' linguistic habits. In contrast, Finnish as an unfamiliar language in this study is perceived with increased difficulty at faster speech rates.

5   Conclusion

The conclusions of this paper can be summarized in three key points: First, the ChatGPT voice exhibits high voice attractiveness, primarily characterized by its clear, natural fluency, and also garners positive recognition in terms of vocal preference, friendliness, and the predicted attractiveness of the face associated with the voice. Second, alterations in prosodic parameters significantly influence the perception of voice attractiveness. Notably, excessively high variations in fundamental frequency and overly rapid speech rate have the most significant negative impact; a monotonic intonation also tends to decrease voice attractiveness. Third, in Finnish, changes in fundamental frequency and duration reveal consistent findings: slightly higher fundamental frequency and a faster speech rate adversely affect voice attractiveness. Unlike human voices, which are limited by the speaker's vocal conditions and therefore have random and uncontrollable attractiveness, AI voices can be continuously optimized through parameteric adjustments, synthesizing more popular voices. Therefore, this paper's exploration of the acoustic factors affecting AI voice attractiveness can provide insights for optimizing the attractiveness of AI voices.

Acknowledgments. This research was supported by the National Natural Science Foundation of China (11974054) and the Phonetics and Artificial Intelligence Course Development Project of Beijing Normal University (20-01-27).

References

1. Zuckerman, M., Driver, R.E.: What sounds beautiful is good: the vocal attractiveness stereotype. J. Nonverbal Behav. **13**, 67–82 (1989). https://doi.org/10.1007/BF00990791
2. Collins, S.A.: Men's voices and women's choices. Anim. Behav. **60**, 773–780 (2000)
3. Zhou, A., Zhang, R., Ma, X., Hou, L.: Vocal attractiveness exploration: evidence from physical and social attribution. Psychol. Explor. **34**, 333–338 (2014). (in Chinese)
4. Liu, Y., Wang, Y.: Acoustic parameters and factors affecting voice attractiveness. Chin. J. Phon. **2**, 24–32 (2022). (in Chinese)
5. Boersma, P., Weenink, D.: Praat: doing phonetics by computer. http://www.praat.org/. Accessed 15 Dec 2023
6. Jadoul, Y., Thompson, B., de Boer, B.: Introducing parselmouth: a Python interface to Praat. J. Phon. **71**, 1–15 (2018)
7. Xu, M.: The preference for natural voice: effects and reasons (2013). (in Chinese)
8. Berry, D.S.: Vocal types and stereotypes: joint effects of vocal attractiveness and vocal maturity on person perception. J. Nonverbal Behav. **16**, 41–54 (1992). https://doi.org/10.1007/BF00986878
9. Liu, M., Sommer, W., Yue, S., Li, W.: Dominance of face over voice in human attractiveness judgments: ERP evidence. Psychophysiology **60**, e14358 (2023)
10. Li, K.-P., Hughes, G.W., House, A.S.: Correlation characteristics and dimensionality of speech spectra. J. Acoust. Soc. Am. **46**, 1019–1025 (1969)
11. Dion, K., Berscheid, E., Walster, E.: What is beautiful is good. J. Pers. Soc. Psychol. **24**, 285–290 (1972)
12. Re, D.E., O'Connor, J.J., Bennett, P.J., Feinberg, D.R.: Preferences for very low and very high voice pitch in humans. PLoS ONE **7**, e32719 (2012)
13. Rodero, E.: Effectiveness, attractiveness, and emotional response to voice pitch and hand gestures in public speaking. Front. Commun. **7**, 869084 (2022)
14. Borkowska, B., Pawlowski, B.: Female voice frequency in the context of dominance and attractiveness perception. Anim. Behav. **82**, 55–59 (2011)
15. Ferdenzi, C., Patel, S., Mehu-Blantar, I., Khidasheli, M., Sander, D., Delplanque, S.: Voice attractiveness: influence of stimulus duration and type. Behav. Res. Methods **45**, 405–413 (2012). https://doi.org/10.3758/s13428-012-0275-0

The DKU-MSXF Diarization System for the VoxCeleb Speaker Recognition Challenge 2023

Ming Cheng[1], Weiqing Wang[1], Xiaoyi Qin[1], Yuke Lin[1], Ning Jiang[2], Guoqing Zhao[2], and Ming Li[1(✉)]

[1] Data Science Research Center, Duke Kunshan University, Kunshan, China
ming.li369@dukekunshan.edu.cn
[2] Mashang Consumer Finance Co., Ltd., Chongqing, China

Abstract. This paper describes the DKU-MSXF submission to track 4 of the VoxCeleb Speaker Recognition Challenge 2023 (VoxSRC-23). Our system pipeline contains voice activity detection, clustering-based diarization, overlapped speech detection, and target-speaker voice activity detection, where each procedure has a fused output from 3 sub-models. Finally, we fuse different clustering-based and TSVAD-based diarization systems using DOVER-Lap and achieve the 4.30% diarization error rate (DER), which ranks first place on track 4 of the challenge leaderboard.

Keywords: Speaker diarization · Target-speaker voice activity detection · VoxSRC-23

1 Introduction

Speaker diarization is the task of breaking up multi-party conversational audio into speaker-homogeneous segments, which aims to solve the problem of "Who spoke when". In this paper, we focus on the speaker diarization task and present the details of our submitted system to track 4 of the VoxCeleb Speaker Recognition Challenge 2023.

Figure 1 depicts the framework of our developed system. First, the voice activity detection (VAD) module removes non-speech regions from the input audio. The remaining is split into multiple short segments, followed by speaker embedding extraction. Then, the initial clustering-based diarization results can be obtained by agglomerative hierarchical clustering (AHC) with overlapped speech detection (OSD) as the post-processing. We replace speaker embedding models trained under different conditions to repeat the above process three times and obtain the fused clustering-based results by Dover-Lap [14]. Next, target-speaker voice activity detection (TSVAD) models with different speaker embedding models are adopted to refine the clustering-based results. In the end, clustering-based and TSVAD-based results are fused again to obtain the final prediction.

J. Jia et al. (Eds.): NCMMSC 2023, CCIS 2006, pp. 330–337, 2024.
https://doi.org/10.1007/978-981-97-0601-3_28

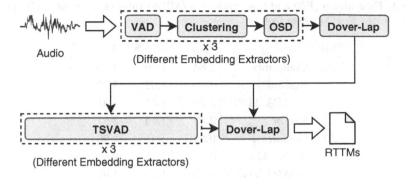

Fig. 1. Framework of The Developed System.

In general, the framework integrates the advantages of clustering-based and TSVAD-based methods, which is similar to our previous submissions in VoxSRC-21 [18] and VoxSRC-22 [20]. The main differences are improved speaker embedding models and Seq2Seq-TSVAD models from our recent works [1,2].

2   Dataset Description

According to challenge rules, any data except the test set is allowed in this task. The datasets used to train each model in our system are described as follows.

- Voice activity detection (VAD) and overlapped speech detection (OSD): Vox-Celeb 1&2 [4,11] for data simulation and VoxConverse [3] for adaptation and validation.
- Speaker embedding: VoxCeleb 1&2 [4,11] and VoxBlink-Clean [10] for training and evaluation.
- Clustering-based diarization: VoxConverse [3] for hyper-parameter tuning.
- TSVAD-based diarization: VoxCeleb 1&2 [4,11] for online data simulation and VoxConverse [3] for adaptation and validation.
- Data augmentation: MUSAN [15] and RIRs [9] corpora.

3   Model Configuration

This section describes each model in our submitted system. If not specified, the input acoustic features of all models are 80-dim log Mel-filterbank energies with a frame length of 25 ms and a frameshift of 10 ms. MUSAN [15] and RIRs [9] are applied as the data augmentation.

3.1   VAD and OSD

As the implementations of voice activity detection (VAD) and overlapped speech detection (OSD) tasks are very similar, we utilize the same neural networks to train these two tasks. Adopted model architectures are described as follows.

Table 1. False alarm (FA) and miss detection (MISS) rates of different VAD and OSD models on the VoxConverse test set.

Task	Model	FA (%)	MI (%)	Total (%)
VAD	Conformer	2.84	1.09	3.93
	ResNet34	3.20	1.02	4.22
	ECAPA-TDNN	2.70	1.51	4.21
	Fusion	2.83	1.14	3.97
OSD	Conformer	0.59	1.41	2.00
	ResNet34	0.54	1.51	2.05
	ECAPA-TDNN	0.52	1.45	1.97
	Fusion	0.44	1.45	1.89

Conformer [7] is the first backbone network. All encoder layers share the same settings: 256-dim attentions with 4 heads and 1024-dim feed-forward layers with a dropout rate of 0.1. The kernel size of convolutions in Conformer blocks is 15. Finally, a linear layer with sigmoid activation is adopted to transform the dimension of Conformer outputs into one, representing the frame-level posterior probability of VAD or OSD.

ResNet34 [8] is the second backbone network, where the widths (number of channels) of the residual blocks are $\{64, 128, 256, 512\}$. At the end of convolutional blocks, the spatial average pooling extracts frame-level features from the convolutional outputs. Finally, a linear layer with sigmoid activation predicts the frame-level posterior probability of VAD or OSD.

ECAPA-TDNN [6] is the third backbone network. The number of filters in the convolutional layers is set to 1024. The scale dimension in the Res2Block is set to 8. The dimension of the bottleneck in the SE-Block and the attention module is set to 128. Finally, a linear layer with sigmoid activation predicts the frame-level posterior probability of VAD or OSD.

VAD or OSD models with different network architectures are fused by averaging their predictions at the score level. Table 1 shows the VAD and OSD models' performance on the VoxConverse test set, respectively. We adopt the fused results as the final predictions in this part.

3.2 Speaker Embedding

To ensemble learning, we train three speaker embedding models with diverse network architectures and training data. The first one is SimAM-ResNet34 [13] with statistics pooling (SP) [16]. The second one is ResNet101 [8] with attentive statistics pooling (ASP) [12]. The third one is SimAM-ResNet100 with ASP. The first two models are trained on the VoxCeleb2 [4] dataset. For the last model, we additionally mix the VoxBlink-Clean [10] dataset into the training.

All speaker embedding models utilize the same back-end part. After the pooling layer projects the variable-length input audio to the fixed-length vector,

Table 2. Equal error rates (EERs) of different speaker embedding models on the Vox-O trial.

#	Model	Training Data	EER (%)
Spk1	SimAM-ResNet34+SP	Vox2	0.81
Spk2	ResNet101+ASP	Vox2	0.49
Spk3	SimAM-ResNet100+ASP	Vox2+VoxBlink	0.44

Table 3. Thresholds (THRs) of different AHC-based diarization systems.

#	Segment THR	Stop THR	Speaker THR
Ahc1	0.54	0.60	0.20
Ahc2	0.62	0.62	0.20
Ahc3	0.66	0.68	0.30

a 256-dim fully connected layer is adopted as the speaker embedding layer, and the ArcFace (s = 32, m = 0.2) [5] is used as a classifier. The detailed configuration of model training is the same as [13]. The performance of different trained speaker embedding models is shown in Table 2.

3.3   Clustering-Based Diarization

We adopt agglomerative hierarchical clustering (AHC) to implement the clustering-based diarization system, which is the same as we used in previous years [18,20]. First, speaker embeddings are extracted from the uniformly segmented speech with a length of 1.28 s and a shift of 0.32 s, pairwisely measured by cosine similarity. Two consecutive segments are merged into a longer segment if their similarity exceeds a segment threshold. Then, we perform a plain AHC on the similarity matrix with a relatively high stop threshold to obtain clusters with high confidence. These clusters are split into "long clusters" and "short clusters" by the total duration in each cluster, and the central embedding of each cluster is the mean of all speaker embeddings within the cluster. Finally, each short cluster is assigned to the closest long cluster by the similarity between their central embeddings. If a short cluster is too different from all long clusters, the similarity between them is lower than a speaker threshold, and then we treat it as a new speaker. As post-processing based on the OSD model, overlapped speech regions are assigned to the two closest speaker labels.

Based on speaker embedding models in Table 2, we develop three AHC-based diarization systems following the above approach. All the hyper-parameters are directly tuned on the VoxConverse test set by grid search. The duration for classifying long and short clusters is 6 s for all models. The other thresholds are shown in Table 3.

3.4   TSVAD-Based Diarization

Given speaker profiles (e.g., x-vector [16]), the TSVAD method can estimate each speaker's frame-level voice activities and perform robustly even in complex acoustic environments. We adopt the Seq2Seq-TSVAD [2] consisting of the front-end extractor with segmental pooling [19], Conformer [7] encoder, and proposed speaker-wise decoder. The front-end extractor is initialized by the pre-trained speaker embedding model, the same as the one for extracting speaker profiles.

Each model training starts from BCE loss & Adam optimizer with a learning rate of $1e-4$ and a linear warm-up of 2,000 iterations. The whole process is described as follows.

- First, the model with a frozen front-end extractor is trained on simulated data until back-end convergence.
- Second, all model parameters are unfrozen to train on both simulated (VoxCeleb 1&2) and real data (VoxConverse dev set) at the ratio of 0.8/0.2. Also, the learning rate decreases to $1e-5$ in the second half of this phase.
- Third, all data simulation and augmentation are removed, inspired by the large margin finetuning (LMFT) [17] in speaker verification. Meanwhile, to use much training data as possible, we mix the first 186 samples of the VoxConverse test set into finetuning and leave the last 46 samples as validation, namely the VoxConverse test46 set.

During inference, a clustering-based diarization is required first to extract speaker profiles from each speaker's speech segments. Then, each test audio is cut into fixed-length chunks with a stride of 1 s and fed into the TSVAD model with extracted speaker profiles. Chunked predictions are stitched by averaging the overlapped predicted regions, which can also be viewed as a score-level fusion.

Using speaker embedding models in Table 2 as different profile extractors, we develop three TSVAD-based diarization systems. The first one equipped with SimAM-ResNet34 is trained and inferred under audio chunks of 64 s. The last two equipped with ResNet101 and SimAM-ResNet100 are trained and inferred under audio chunks of 16 s, limited by higher GPU memory costs of the deeper networks. The other configurations share the same settings: 512-dim attentions with 8 heads, convolutions with a kernel size of 15, and 1024-dim feed-forward layers with a dropout rate of 0.1. The speaker capacity and output VAD resolution are set to 30 and 0.08 s, respectively.

4   Experimental Results

Table 4 illustrates the performance of our developed diarization systems on the VoxConverse test set, VoxConverse test46 set, and VoxSRC-23 challenge test set. As the TSVAD models utilize part of the VoxConverse test set in training, their performance is only evaluated on the other two datasets.

Systems #1–3 represent the AHC-based diarization with different speaker embedding models, obtaining the DERs of 5.51%, 5.32%, and 5.36% on the

Table 4. Diarization error rates (DERs) of different systems on the VoxConverse test set, VoxConverse test46 set, and VoxSRC-23 challenge test set.

#	Method	DER (%)		
		VoxConverse Test	VoxConverse Test46	VoxSRC-23 Challenge Test
1	Ahc1	4.83	4.14	5.51
2	Ahc2	4.49	3.92	5.32
3	Ahc3	4.55	3.92	5.36
4	Dover-Lap (#1–3)	–	3.81	5.19
5	+ TSVAD with Spk1	–	2.85	4.49
6	+ TSVAD with Spk2	–	2.93	4.57
7	+ TSVAD with Spk3	–	2.91	4.53
8	Dover-Lap (#4–7)	–	2.73	4.30

challenge test set. Then, we adopt the Dover-Lap method to fuse them and obtain system #4 to achieve a 5.19% DER on the challenge test set.

Systems #5–7 represent the TSVAD-based diarization with different speaker embedding models. Using the fused clustering-based results to extract speaker profiles, these three models have very close DERs varying from 4.49% to 4.57% on the challenge test set. Finally, systems #4–7 are fused by the Dover-Lap again to obtain system #8, achieving the best 4.30% DER on the challenge test set.

5   Conclusions

This paper describes our system development for track 4 of the VoxSRC-23. This year, we mainly focus on improving the speaker embedding, AHC-based diarization, and TSVAD-based diarization models. To achieve the best performance, we also train diverse sub-models for each part of the whole framework. The finally fused method shows significant improvement, obtaining the DERs of 2.73% on the VoxConverse test46 set and 4.30% on the VoxSRC-23 challenge test set, respectively.

Acknowledgments. This research is funded in part by the National Natural Science Foundation of China (62171207) and Science and Technology Program of Suzhou City (SYC2022051). Many thanks for the computational resource provided by the Advanced Computing East China Sub-Center.

References

1. Cheng, M., Wang, H., Wang, Z., Fu, Q., Li, M.: The WHU-Alibaba audio-visual speaker diarization system for the MISP 2022 challenge. In: Proceedings of the ICASSP, pp. 1–2 (2023). https://doi.org/10.1109/ICASSP49357.2023.10095802

2. Cheng, M., Wang, W., Zhang, Y., Qin, X., Li, M.: Target-speaker voice activity detection via sequence-to-sequence prediction. In: Proceedings of the ICASSP, pp. 1–5 (2023). https://doi.org/10.1109/ICASSP49357.2023.10094752

3. Chung, J.S., Huh, J., Nagrani, A., Afouras, T., Zisserman, A.: Spot the conversation: speaker diarisation in the wild. In: Proceedings of the Interspeech, pp. 299–303 (2020). https://doi.org/10.21437/Interspeech. 2020–2337

4. Chung, J.S., Nagrani, A., Zisserman, A.: VoxCeleb2: deep speaker recognition. In: Proceedings of the Interspeech, pp. 1086–1090 (2018). https://doi.org/10.21437/Interspeech. 2018–1929

5. Deng, J., Guo, J., Xue, N., Zafeiriou, S.: ArcFace: additive angular margin loss for deep face recognition. In: Proceedings of the CVPR, pp. 4685–4694 (2019). https://doi.org/10.1109/CVPR.2019.00482

6. Desplanques, B., Thienpondt, J., Demuynck, K.: ECAPA-TDNN: emphasized Channel Attention, Propagation and Aggregation in TDNN Based Speaker Verification. In: Proceedings of the Interspeech, pp. 3830–3834 (2020). https://doi.org/10.21437/Interspeech. 2020–2650

7. Gulati, A., et al.: Conformer: convolution-augmented transformer for speech recognition. In: Proceedings of the Interspeech, pp. 5036–5040 (2020). https://doi.org/10.21437/Interspeech.2020-3015

8. He, K., Zhang, X., Ren, S., Sun, J.: Deep residual learning for image recognition. In: Proceedings of the CVPR, pp. 770–778 (2016)

9. Ko, T., Peddinti, V., Povey, D., Seltzer, M.L., Khudanpur, S.: A study on data augmentation of reverberant speech for robust speech recognition. In: Proceedings of the ICASSP, pp. 5220–5224 (2017). https://doi.org/10.1109/ICASSP.2017.7953152

10. Lin, Y., Qin, X., Cheng, M., Li, M.: VoxBlink: X-large speaker verification dataset on camera (2023)

11. Nagrani, A., Chung, J.S., Zisserman, A.: VoxCeleb: a large-scale speaker identification dataset. In: Proceedings of the Interspeech, pp. 2616–2620 (2017). https://doi.org/10.21437/Interspeech.2017-950

12. Okabe, K., Koshinaka, T., Shinoda, K.: Attentive statistics pooling for deep speaker embedding. In: Proceedings of the Interspeech, pp. 2252–2256 (2018). https://doi.org/10.21437/Interspeech.2018-993

13. Qin, X., Li, N., Weng, C., Su, D., Li, M.: Simple attention module based speaker verification with iterative noisy label detection. In: Proceedings of the ICASSP, pp. 6722–6726 (2022). https://doi.org/10.1109/ICASSP43922.2022.9746294

14. Raj, D., et al.: DOVER-lap: a method for combining overlap-aware diarization outputs. In: Proceedings of the SLT, pp. 881–888 (2021). https://doi.org/10.1109/SLT48900.2021.9383490

15. Snyder, D., Chen, G., Povey, D.: MUSAN: a music, speech, and noise corpus (2015)

16. Snyder, D., Garcia-Romero, D., Sell, G., Povey, D., Khudanpur, S.: X-vectors: robust DNN embeddings for speaker recognition. In: Proceedings of the ICASSP, pp. 5329–5333 (2018). https://doi.org/10.1109/ICASSP.2018.8461375

17. Thienpondt, J., Desplanques, B., Demuynck, K.: The Idlab voxsrc-20 submission: large margin fine-tuning and quality-aware score calibration in DNN based speaker verification. In: Proceedings of the ICASSP, pp. 5814–5818 (2021). https://doi.org/10.1109/ICASSP39728.2021.9414600

18. Wang, W., et al.: The DKU-DukeECE-Lenovo system for the diarization task of the 2021 VoxCeleb speaker recognition challenge (2021)

19. Wang, W., Lin, Q., Cai, D., Li, M.: Similarity measurement of segment-level speaker embeddings in speaker diarization. IEEE/ACM Trans. Audio Speech Lang. Process. **30**, 2645–2658 (2022). https://doi.org/10.1109/TASLP.2022.3196178
20. Wang, W., Qin, X., Cheng, M., Zhang, Y., Wang, K., Li, M.: The DKU-DukeECE diarization system for the VoxCeleb speaker recognition challenge 2022 (2022)

Chinese EFL Learners' Auditory and Visual Perception of English Statement and Question Intonations: The Effect of Lexical Stress

Qiunan Xu[ID] and Ping Tang[(⊠)][ID]

School of Foreign Studies, Nanjing University of Science and Technology, Nanjing, China
{qiunan,ping.tang}@njust.edu.cn

Abstract. English is a stress-timed language, where native speakers rely on acoustic cues such as pitch to perceive word stress and intonation. In contrast, Chinese learners of English as a Foreign Language (EFL) primarily use pitch as a perceptual cue and are influenced by the interaction between word stress and intonation. However, the specific impact of stress patterns (iambic or trochee) and English proficiency (high or low) on their perception of English intonation remains unclear. Furthermore, the extent to which visual cues affect Chinese EFL learners' perception of intonation requires further investigation. This study investigated the effects of auditory and visual cues on the perception of English intonation among Chinese EFL learners. Thirty Chinese college students participated in this study. The stimuli were presented in two modalities: audio-only and audio-visual, encompassing both statement and question intonations. Participants rated the degree of question intonation on a scale ranging from 1 to 7, with higher scores indicating a stronger perception of intonation. The results indicate that sentence-final stress patterns have an impact on the perception of intonation, specifically in questions, with a relatively smaller effect observed for trochee. Furthermore, high-proficient Chinese EFL learners demonstrated an advantage in perceiving statement intonation. However, visual cues did not significantly enhance the perception of English intonation, likely due to limited exposure to natural visual cues. These findings contribute to our understanding of how word stress and English proficiency affect intonation perception, providing insights for language instruction and curriculum development.

Keywords: Stress-intonation interplay · Chinese EFL learners · Audio-visual cues

1 Introduction

Intonation employs suprasegmental phonetic features to convey pragmatic meanings at the sentence level [1]. Pitch patterns are critical for distinguishing intonation types, such as question and statement intonation [2]. In English, it is widely accepted that sentences expressing completion or assertion are characterized by a low or falling pitch, while those indicating inquiry or questioning exhibit a high or rising pitch [3]. English is a stress-based language, and previous research suggests a close relationship between English

J. Jia et al. (Eds.): NCMMSC 2023, CCIS 2006, pp. 338–345, 2024.
https://doi.org/10.1007/978-981-97-0601-3_29

lexical stress and pitch [4]. Fry [5] proposed that a higher pitch tends to be perceived as stressed by listeners. In practice, intonation and stress work together to convey meaning, sharing the same pitch cues. Therefore, it is necessary to explore how listeners perceive intonation within this context.

Second language acquisition (SLA) has predominantly focused on the segmental level, neglecting suprasegmental aspects. However, suprasegmental features, particularly word stress, significantly impact L2 acquisition [6, 7]. Mandarin and English possess distinct prosodic systems, requiring Chinese EFL learners to reacquire new acoustic correlates. Previous research indicates that L2 learners encounter difficulties in perceiving and producing non-native contrasts, including unfamiliar phonetic features [8, 9]. The study revealed that stress-intonation interactions did not affect word stress perception for native English speakers. In contrast, Chinese EFL learners exhibited significantly lower stress identification rates under interplay. This suggests their heightened susceptibility to stress-intonation interactions. Interestingly, English proficiency did not influence Chinese EFL learners' perception of word stress under different stress-pitch patterns. Nonetheless, a research gap exists in investigating intonation perception under stress-intonation interplay and its relationship with English proficiency, emphasizing the need for further exploration.

In addition, speech multimodality aids intonation perception [10]. There is a relationship between prosody and visible movements of the neck, head, and mouth [11, 12]. Studies have shown that stressed syllables exhibit pronounced facial features. Srinivasan and Massaro [13] found that both auditory and visual cues are reliable for conveying English intonation to native speakers, although visual cues have a weaker influence compared to auditory cues. However, the effects of visual cues on Chinese EFL learners' English intonation perception remain unclear. Furthermore, research on English intonation perception lacks investigations into stress-intonation interactions, indicating the need for further exploration in this area.

Therefore, the current study aimed to explore Chinese EFL learners' perceptions of English intonation based on auditory and visual cues. Specifically, we focused on three questions: First, how do sentence-final word stress patterns (trochee, iambic) affect the perception of English intonation (statement, question) when only auditory cues are available? Second, how do differences in English proficiency (low, high) affect intonation perception? Third, how do additional visual cues affect the perception of English intonation?

2 Methods

2.1 Participant

Thirty Chinese EFL learners (18 female; 12 male) aged from 20 to 23 at Nanjing University of Science and Technology participated in this experiment. All participants were born and raised in Mandarin-speaking families. They have learnt English as a foreign language for at least eleven years in the formal instruction context and were divided into two groups according to their English proficiency: English major group (15 students; i.e., high proficiency group) and non-English major group (15 students; i.e., low proficiency group). The English major students had received extensive training in phonetics and devoted considerable effort to pronunciation learning, in contrast to their non-English major counterparts. No participants reported any hearing or speech problems.

2.2 Stimuli

The experiment investigates the interplay between two stress patterns and two intonation conditions. Sixty disyllabic words, comprising 30 iambic and 30 trochee stress pattern words, were selected from the word entry of CET-4 to mitigate the potential influence of word frequency. The experimental design employs four-syllable-bearing sentences, specifically the phrase "She just said X". After conducting the prediction validation test, a total of 120 target sentences (30 syllables × 2 stress patterns × 2 intonations) were constructed.

The stimuli in this study were produced by a female native English speaker who followed written instructions to generate sentences with a statement or question intonation. The speaker underwent the production process without any coaching or feedback.

2.3 Procedure

In this study, a 7-point Likert scale was employed, with responses ranging from 1 (none question intonation) to 7 (strong question intonation). Utilizing this scale facilitated the assessment of question intonation in sentences, not only introducing complexity to the experimental design but also enabling the discernment of nuanced variations in the impact of distinct stress positions on intonation across a spectrum of values.

The experiment was conducted using Psychopy and was structured into two blocks: an audio-only modality and an audio-visual modality. In the audio-only block, participants simply listened to the audio stimuli, whereas in the audio-visual block, they watched the video screen and performed the task. After each stimulus, a 7-point Likert scale was displayed on the screen, prompting participants to rate the degree of question intonation by clicking the mouse. The audio-only block consistently preceded the audio-visual block in the experimental sequence. To mitigate order effects, the trial sequences within each block were randomized for every participant. Each block consisted of 120 trials, with a 3-min intermission between the two blocks. Prior to commencing the main experiment, participants underwent a practice session designed to familiarize them with the experimental procedures. This practice session consisted of 10 trials, evenly distributed across the two modalities (5 trials per modality).

2.4 Data Analysis

All data analyses were performed in R. Linear mixed-effect models were employed to capture the interplay between stress patterns, intonations, modalities, and English proficiency levels. Specifically, the "ordinal" package was utilized for model construction. Subsequent to model fitting, Tukey-HSD post hoc comparisons were performed using the "emmeans" package [14]. To visually represent the results, the "ggplot2" package [15] was employed for generating graphics.

3 Results

3.1 Effects of Stress Position

Figure 1 illustrated participants' mean rating of the question intonation degree in different stress patterns and intonations. A linear mixed-effect model was performed on the participants' scoring data, with two fixed effects "Stress" (2 levels: trochee and iambic) and "Intonation" (2 levels: statement and question), and two random factors "Participants" (30 participants) and "Words" (30 words). The model reported a significant interaction of "Stress × Intonation" ($p = 0.018$), indicating differences in scoring outcomes between statement and question intonation across two stress patterns.

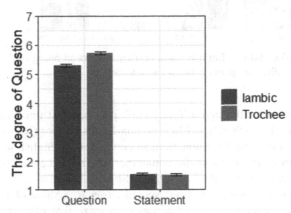

Fig. 1. Participants' mean rating of the English question intonation degree in different stress patterns and intonations

A Turkey-HSD post-hoc test was performed for this interaction to compare the score difference between iambic and trochee in intonation contrast. Our findings reveal that, in questions, scores are significantly higher when sentences conclude with a trochee rather than an iambic pattern (contrast: Iambic-Trochee, $\beta = -0.413$, SE $= 0.128$, z $= -3.124$, $p < 0.01$). In contrast, no significant differences in scores based on stress patterns were observed in the statements. These results emphasize the influence of stress patterns on question intonation perception, suggesting that it is more perceptible when the questions end with a trochee.

3.2 Effects of English Proficiency

Figure 2 displayed the mean scores for question intonation perception among English major and non-English major students in both iambic and trochee cases. A linear mixed-effect model was performed, with three fixed effects "Stress" (2 levels: trochee and iambic), "Intonation" (2 levels: statement and question) and "Major" (2 levels: English major and non-English major), and two random factors "Participants" (30 participants) and "Words" (30 words). The statistical analysis revealed a significant interaction denoted as "Intonation × Major" ($p < 0.05$), indicating that intonation perception is influenced by English proficiency.

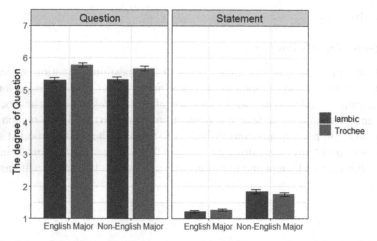

Fig. 2. English major and non-English major participants' mean rating of the English question intonation degree in different stress patterns and intonations

Further examination through a Tukey-HSD post-hoc test unveiled no discernible difference between English major (EM) and non-English major (Non-EM) participants in their perception of question intonation. In contrast, non-English major participants scored higher than their English major counterparts in the perception of statement intonation (contrast: EM-(Non-EM), $\beta = -0.524$, SE = 0.178, z = -3.050, p < 0.01). In other words, English major participants scored closer to a value of 1 in their perception of statement intonation. These findings emphasize that participants with higher English proficiency perceive statement intonation more accurately than those with lower proficiency.

3.3 Effects of Visual Cues

Figure 3 investigated visual cues across stress patterns, intonations, and English proficiency levels. A linear mixed-effect model was conducted on the participants' scoring data, with four fixed effects "Stress" (2 levels: trochee and iambic), "Intonation" (2 levels: statement and question), "Major" (2 levels: English Major and Non-English major), and "Modality" (AO and AV), and two random factors "Participants" (30 participants) and "Words" (30 words). Notably, the analysis did not reveal any interaction between "Modality" and other factors. This finding implies that the presence of visual cues had no discernible impact on the perception of intonation. In other words, participants consistently demonstrated the ability to perceive and differentiate intonation, irrespective of whether they had visual cues or relied solely on auditory information.

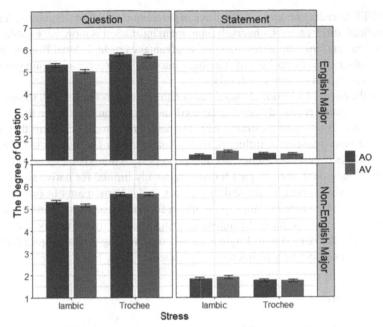

Fig. 3. English major and non-English major participants' mean rating of the English question intonation degree in different stress patterns, intonations and modalities

4 Discussion

The current study first explored how sentence-final stress patterns affect the perception of English intonation when only auditory cues are available. The results demonstrate a significant impact of stress patterns on the perception of English intonation, aligning with previous research. Chinese EFL learners, owing to their extensive experience with tonal languages, tend to rely on pitch as a prominent cue. However, English relies on pitch cues to convey both lexical stress and intonation, where stress patterns can occasionally impede intonation perception. Notably, the influence of stress patterns was observed exclusively in questions, with no discernible impact on statements. But the reasons underlying this phenomenon will require further investigation in the future. Furthermore, Chinese EFL learners exhibited better perception of final stress when it ended with trochee compared to when it ended with iambic.

We then explored the impact of English proficiency differences on intonation perception. Our results revealed Chinese EFL learners with higher English proficiency outperformed their counterparts with lower English proficiency in the statement intonation perception. However, there was no significant variance in intonation perception across different English proficiency levels when it came to questions. This discrepancy could be attributed to the fact that the high-level English learners, who were primarily Chinese English-major students, had limited exposure to natural English input. Moreover, as question intonation is intrinsically more challenging for Chinese EFL learners compared to statement intonation, it is plausible that both high and low proficiency

Chinese EFL learners encounter difficulties in perceiving question intonation, leading to non-significant differences. Conversely, statement intonation is more accessible for Chinese EFL learners, thereby accentuating the advantages of high-level English learners, leading to observable disparities in statement intonation perception among participants with varying English proficiency levels.

As for the effect of additional visual cues on the perception of English intonation, the current investigation revealed that visual cues do not significantly enhance the perception of English intonation among Chinese EFL learners. Prior studies confirm that auditory cues exert a more substantial influence than visual cues for English native speakers in distinguishing between statements and questions [13]. While visual cues do contribute to intonation judgment, their impact remains relatively limited for native speakers. This observation lends support to the findings of our study. Participants in our experiment acquired English as a foreign language within a formal instructional context, affording them limited exposure to natural English language usage. Consequently, they may not have been well-acquainted with English visual cues, making it challenging for them to utilize these cues effectively in intonation judgment.

5 Conclusion

Our study investigated the English intonation perception among Chinese EFL learners and the role of auditory and visual cues. Our findings revealed that word stress patterns influence intonation perception, with particular significance in the case of questions, where trochee enhances accuracy. Variations in participants' English proficiency levels were observed in the perception of statements. Notably, due to limited exposure to natural English input, visual cues do not significantly contribute to intonation perception among Chinese EFL learners. All this has implications for language instruction, emphasizing the significance of word stress patterns, customized teaching approaches, and the need for authentic language input to enhance intonation perception among Chinese EFL learners.

Acknowledgments. This research is funded by The National Social Science Fund of China (20CYY012).

References

1. Ladd, D.R.: Intonation phonology. Cambridge University Press, Cambridge (2008)
2. Mcroberts, G.W., Studdert-Kenndy, M., Shankweiler, D.P.: The role of fundamental frequency in signaling linguistic stress and affect: evidence for a dissociation. Percept. Psychophys. **57**(2), 59–74 (1995)
3. Liu, F., Xu, Y.: Parallel encoding of focus and interrogative meaning in mandarin intonation. Phonetica **62**(2), 70–87 (2005)
4. Vogel, I.: The acquisition of prosodic phonology: challenges for the L2 learner. In: Structure, Acquisition, and Change of Grammars: Phonological and Syntactic Aspects, pp. 32–50 (2000)
5. Fry, D.B.: Duration and intensity as physical correlates of linguistic stress. J. Acoust. Soc. Am. **26**(1), 138 (1954)

6. Juffs, A.: Tone, syllable structure and interlanguage phonology: Chinese learner's stress errors. Int. Rev. Appl. Linguist. **28**(2), 99–118 (1990)
7. Dupoux, E., Peperkamp, S., Sebastián-Gallés, N.: A robust method to study stress "deafness." J. Acoust. Soc. Am. **110**(3), 1606–1618 (2001)
8. Zhang, Y., Nissen, S.L., Francis, A.L.: Acoustic characteristics of English lexical stress produced by native Mandarin speakers. J. Acoust. Soc. Am. **123**(6), 4498–4513 (2008)
9. Tyler, M.D., Best, C.T., Faber, A., Levitt, A.G.: Perceptual assimilation and discrimination of non-native vowel contrasts. Phonetica **71**(1), 4–21 (2014)
10. McGurk, H., MacDonald, J.: Hearing lips and seeing voices. Nature **264**(5588), 746–748 (1976)
11. Chen, T.H., Massaro, D.W.: Mandarin speech perception by ear and eye follows a universal principle. Psychophys **66**, 820–836 (2004)
12. Chen, T.H., Massaro, D.W.: Seeing pitch: visual information for lexical tones of Mandarin-Chinese. J. Acoust. Soc. Am. **123**(4), 2356–2366 (2008)
13. Srinivasan, R.J., Massaro, D.W.: Perceiving prosody from the face and voice: distinguishing statements from echoic questions in English. Lang. Speech **46**, 1–22 (2003)
14. Lenth, R.V.: emmeans: Estimated Marginal Means, aka Least-Squares Means. R package version 1.6.0 (2021). https://CRAN.R-project.org/package=emmeans
15. Wickham, H.: ggplot2-Elegant Graphics for Data Analysis. Springer, Cham (2016). https://doi.org/10.1007/978-3-319-24277-4

An Improved System for Partially Fake Audio Detection Using Pre-trained Model

Jianqian Zhang, Hanyue Liu, Mengyuan Deng, Jing Wang$^{(\boxtimes)}$, Yi Sun, Liang Xu, and Jiahao Li

School of Information and Electronics, Beijing Institute of Technology, Beijing, China
{3120210828,3120220704,wangjing,xuliang}@bit.edu.cn,
m13032870917@163.com, ljhaaa20010601@163.com

Abstract. The technology of speech synthesis and conversion has made good progress with the development of deep learning. However, such technology can also do harm to information security and may be applied for illegal uses. Therefore, researchers have conducted a lot of research on the task of speech deep forgery detection recently. Corresponding to this, the Audio Deep Synthesis Detection Challenge 2023 (ADD 2023) is held. In this paper, we propose a fake audio detecting system using a pre-trained model, focusing on partially fake audio detection tasks. We have presented our models to the ADD 2023 challenge. In the final competition, our system got a score of 0.4855 in the manipulation region location track.

Keywords: deep-learning · audio deepfake detection · manipulation · fake region location

1 Introduction

With the development of deep learning, speech synthesis and speech conversion have made great progress. [6,7,16,17] Automated voice technology interaction, like voiceprint recognition and intelligent voice, facilitates people's lives and meets the needs of the public in pursuit of a better life. However, as the generation model can generate more realistic speech, more people can not effectively distinguish the differences between true speech and the manipulated sample. Also, if such technology is abused, it will also pose a great threat to information security and may cause many illegal results. Under this background, how to realize the effective recognition of synthetic speech has become an important problem that has to be faced with the wide application of speech technology.

The most original speech forgery method is to invite audience to judge the speech source directly, and then calculate the mean opinion score(MOS). This method is effective, and is widely used. However, it has the shortcomings of costing a lot of effort. [11] proposed to use frame-level features based on cochlear

This work was supported by National Nature Science Foundation of China (Grant No. 62071039) and in part by Beijing Natural Science Foundation (Grant No. L223033).

filter cepstrum coefficients and instantaneous frequency changes, and then used Gaussian mixture model for discrimination. Gomez-Alanis et al. [4] introduced a network named LC-GRNN to extract frame-level features and learning temporal correlation are realized simultaneously. Jung et al. [5] adopted an end-to-end deep neural network to replace the process of manual extraction of acoustic features, effectively completing the detection task without the assistance of professional knowledge. Similarly, some different blocks, like the fusion of SincNet and VGG, introducing attention layers to the network [9,20] and so on. These researches show that the integrated approach is beneficial to improve the robustness of the speech authentication model [3]. Besides, some more efforts have been made for manipulation detection tasks recently [2,19]. As for now, many works have been done for judging if there is manipulation in the whole speech, but not for detecting the specific location of the fake part. This work is often treated as the sub-task [18], but it is also necessary. Besides, although most systems are able to perform well on the training set, it has insufficient generalization ability to detect attacks that are not seen during training.

In this paper, we introduce a system based on RawNet [15] to solve the task of locating the manipulated regions in partially fake audio. In our proposed system, we use data augmentation methods and adopt the pre-trained model of wav2vec2.0 [1] system to better the result of our system. Both whole-level and frame-level possibilities are given for judgment. Besides, we introduce Temporal Contrastival Loss to better the effect of detection are applied to improve the performance of our system.

2   Proposed System Structure

The main structure of our model could be divided into several parts. Firstly, in addition to the Mel-spectrogram, the pre-trained model has been added to enrich the feature information. Secondly, augmentation methods are used when training. The main network is based on the RawNet, and we add new a loss function besides the cross entropy loss to improve the performance of the location detection work. Also, post-processing is applied following the output of the networks. The main structure of our system is provided in Fig. 1.

2.1   Pre-trained Model

Pre-training is a commonly used method in deep learning algorithms. Its main function is to improve model performance in the case of few labeled data sets. Among several pre-trained models, we choose to use several wav2vec2.0 models to extract features, and concatenate them with the Mel-spectrogram, in order to enrich the information of input speech signals [13]. The pre-trained models are used as the speech representation in building the model, and can supplement some deep information of samples, so as to make the system more effective when facing the manipulation not encountered during the training process.

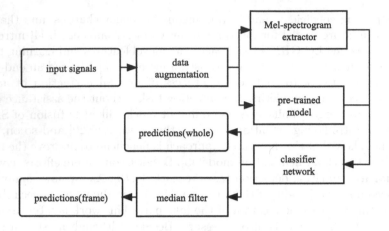

Fig. 1. The main structure of the system

2.2   Feature Extractor

In our system, the feature extractor is an important block to get the main information of the input speech signals. Mel-spectrogram has been used as a kind of feature of speech signals for subsequent acoustic analysis and learning. Besides, in some other detecting systems, SincNet has been applied to be the feature extractor [12]. This network is a kind of novel CNN architecture. We compared the performance among some features and finally chose to use the Mel-spectrogram as the basic input feature feeding to the network.

2.3   Data Augmentation

In the audio deepfake detection task, we need a large amount of data including many kinds of manipulation methods. As a result, the data augmentation becomes necessary. So, in our system, we use SpecAugment [10], and only apply frequency masking and do not use the time mask because it may confuse the location boundary. For every epoch we randomly set a mask to the input features so that the limited data could be made more use of in the training process to improve the generalization. Also, to make the system able to distinguish the more kinds of manipulation, we added noise from the MUSAN [14] dataset. Adding noise to the samples can also work a lot, and we mix them with the random SNR ranging from -10 to 0 db so that the model can be able to distinguish the more kinds of manipulation

2.4   Networks

We build the structure based on residual blocks, and add a fully connected linear layer called FC-attention as a self-attention layer to help better find out further information. The input speech signals are sent to the extractor, and after

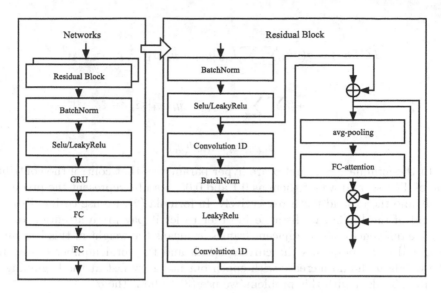

Fig. 2. The main structure of the network

applying data augmentation, they are concatenated together and pass through the networks with the shape of (B, C, T), where B stands for the batch size process, C means the number of feature bins for each time segment and T stands for the time sequences. Then, there are a series of blocks, each containing two 1D convolution layers with normalization and activation function, and it repeats six times, as shown in Fig. 2.

In the residual block, to gather more information from the features, we apply a structure of the self-attention layer, which contains an average pooling for reshaping and a fully connected layer. We apply the kernel size as $C \times 3$ to the convolution layer. After that, the further features with the number of bins set to 128 are fed to a GRU followed by two fully connected linear layers to get the possibility of the predictions. One for judging the whole sentence if there is a fake part and the other is for frame-level judgment. Thus, the model's output will directly give out the prediction for both the whole speech and its frames.

2.5   Temporal Contrastival Loss

Besides the cross entropy loss of the whole samples and frame-level, to make the system more sensitive to the boundary judgment of abnormal events, we introduced Temporal Contrastival Loss (TCLoss), which comes from an effort of the sound event detection task [8]. This loss is describe in Eq. (1) (2):

$$TCLoss = \sigma(-\alpha \sum_{i=1}^{N} \sum_{t=2}^{T'} l_{>0} |y_{i,t} - y_{i,t-1}| |\hat{z}_{i,t} - \hat{z}_{i,t-1}|^2$$

$$+\beta \sum_{i=1}^{N} \sum_{t=2}^{T'} l_{=0} |y_{i,t} - y_{i,t-1}| |\hat{z}_{i,t} - \hat{z}_{i,t-1}|^2) \tag{1}$$

$$\hat{z}_{i,t} = z_{i,t}/mod(z_i) \tag{2}$$

In the formula (1), α and β are hyper parameters that control the contribution of TCloss, and we set them as 0.1 and 0.05. z and y represent the predicted result and the ground truth respectively. In formula (2), we use cosine distance instead of the Euclidean distance in order to let it pay more attention to the relative differences of the adjacent frames. σ means the weight of this loss function, and e, E represents the current epoch and the total number of epochs. The weight σ should increase with epoch but not grow fast at the beginning of training. To deal with this problem, we use (3) to limit the σ.

$$\sigma = \begin{cases} 0, & 1 \le e \le E/2 \\ 0.01 \times (e - E/2), & E/2 < e \le E \end{cases} \tag{3}$$

When meeting the fake events, the label of the boundary has a mutation value. So we apply this loss to increase the sensitivity at the edge of the event, so that the system will perform better after training with this loss function.

2.6   Post-processing

In our experiments, we observe the result given by the networks and found that the locations it judged were sometimes abnormal, especially at the boundary of fake events. So we set a post-processing block using the median filter to deal with the possibility of the classifier's result for the frame-level periods. With this block as a filter to process the outputs, we can get the final results which have less possibility to be misjudged.

3   Experiments

3.1   Dataset

The dataset for the ADD 2023 challenge contains the training set, development set, and test set. All the speech signals are at a sample rate of 16000 Hz. There are 53,093 samples in the training set and 17,823 in the development set, with labels both for the whole sample and for frame-level segments. The test set contains 50,000 samples.

3.2   Measurement

The task requires not only to judge whether there is a fake part in the speech but also to find the location of the manipulation if exists. Therefore, multiple metrics factor into the final score. Sentence accuracy measures the ability of the model to correctly distinguish between genuine and fake audio, and is defined as Eq. (4):

$$A = \frac{TP + TN}{TP + TN + FP + FN} \tag{4}$$

In addition, we use "Precision(P)", "Recall(R)", and "F_1" for the segment to measure the ability of the model to correctly identify fake areas from fake audio. They are defined, respectively, as (5):

$$P = \frac{TP}{TP + FP}, \qquad R = \frac{TP}{TP + FN}, \qquad F_1 = \frac{2 \cdot P \cdot R}{P + R} \tag{5}$$

where $TP, TN\ FP$, and FN denote the numbers of true positive, true negative, false positive, and false negative samples respectively. And the final score is the weighted sum of "Sentence Accuracy" and "Segment F1" as (6):

$$Score = 0.3 \times A + 0.7 \times F_1 \tag{6}$$

3.3   Experiments and Results

When constructing the system, we use two RTX3090 for the training work. We choose the Adam optimizer algorithm for the network parameters. The learning rate is set to 0.0001 with the weight decay to be 0.0001, a total of 100 epochs to train.

Table 1. The scores of different systems. w/o or w/ mean with or without.

feature	GRU layer type	TCLoss	pre-trained model		
			w/o pre-trained model	base-960h	XLR-S-300m
Sinconv	GRU	w/o TCLoss	0.2847	0.3511	0.3835
		w/ TCLoss	0.3286	0.4196	0.4348
	BIGRU	w/o TCLoss	0.3877	0.4193	0.4190
		w/ TCLoss	0.4025	0.3798	0.4571
Mel	GRU	w/o TCLoss	0.3218	0.4522	0.4481
		w/ TCLoss	0.3603	0.4491	**0.4855**
	BIGRU	w/o TCLoss	0.2855	0.3235	0.4695
		w/ TCLoss	0.3441	0.4347	0.4792

We used the Mel-spectrogram and Sinconv as input features and utilized the base-960h and XLR-S-300m model as pre-trained model. Besides, we tried to change the GRU structure and decided to use a three-layer GRU with more nodes

Table 2. The scores of systems using Mel-spectrogram with different FFT size.

feature	FFT size	GRU layer type	
		GRU	BiGRU
Mel	256	0.2396	0.2230
	512	0.2621	0.2855
	1024	**0.3603**	**0.3441**

Table 3. The scores of systems using Mel-spectrogram with or without (w/o) data augmentation.

feature	Augmentation	GRU layer type	
		GRU	BiGRU
Mel	w/o augmentation	0.3603	0.3441
	Spec	**0.4522**	0.4347
	MUSAN	0.4492	0.3235
	Spec + MUSAN	0.4481	**0.4695**

instead of a two-layer BiGRU. Eventually, an 11-point median filter is following the output to smooth the possibility. Some results with different parameters or structures are displayed in Table 1. Also, in Table 2, a comparison between the size of FFT is shown and the best result among them appears when the FFT size comes to 1024. Similarly, we show the scores of different augmentation methods in Table 3, which proves the effect of these additional methods. Among lots of models we have tried and submitted to the challenge, the score comes to 48.55 as our best result in the final submission.

4   Conclusion

In this paper, we proposed a system for audio deepfake detection task. We apply data augmentation methods, pre-trained model wav2vec, changes on the base model, the post smooth filter, and a new loss function to improve the system performance. Finally, we equip the system with the discriminate capacity between real and partially fake audio, with the frame-level judgments to locate the fake audio. At last, we get 0.4855 as the final score in the ADD 2023 challenge.

References

1. Baevski, A., Zhou, Y., Mohamed, A., Auli, M.: wav2vec 2.0: a framework for self-supervised learning of speech representations. In: Advances in Neural Information Processing Systems, vol. 33, pp. 12449–12460 (2020)
2. Chen, T., Kumar, A., Nagarsheth, P., Sivaraman, G., Khoury, E.: Generalization of audio deepfake detection. In: Odyssey, pp. 132–137 (2020)

3. Chettri, B., Stoller, D., et al.: Ensemble models for spoofing detection in automatic speaker verification. In: Proceedings of Interspeech 2019, pp. 1018–1022 (2019)
4. Gomez-Alanis, A., Peinado, A.M., Gonzalez, J.A., Gomez, A.M.: A light convolutional GRU-RNN deep feature extractor for ASV spoofing detection (2019)
5. Jung, J.W., Shim, H.J., Heo, H.S., Yu, H.J.: Replay attack detection with complementary high-resolution information using end-to-end DNN for the ASVspoof 2019 challenge. arXiv preprint arXiv:1904.10134 (2019)
6. Kameoka, H., Kaneko, T., Tanaka, K., Hojo, N.: StarGAN-VC: non-parallel many-to-many voice conversion using star generative adversarial networks. In: 2018 IEEE Spoken Language Technology Workshop (SLT), pp. 266–273. IEEE (2018)
7. Kaneko, T., Kameoka, H.: CycleGAN-VC: non-parallel voice conversion using cycle-consistent adversarial networks. In: 2018 26th European Signal Processing Conference (EUSIPCO), pp. 2100–2104. IEEE (2018)
8. Kothinti, S., Elhilali, M.: Temporal contrastive-loss for audio event detection. In: ICASSP 2022-2022 IEEE International Conference on Acoustics, Speech and Signal Processing (ICASSP), pp. 326–330. IEEE (2022)
9. Monteiro, J., Alam, J., Falk, T.H.: End-to-end detection of attacks to automatic speaker recognizers with time-attentive light convolutional neural networks. In: 2019 IEEE 29th International Workshop on Machine Learning for Signal Processing (MLSP), pp. 1–6. IEEE (2019)
10. Park, D.S., et al.: SpecAugment: a simple data augmentation method for automatic speech recognition. In: Interspeech 2019. ISCA (2019)
11. Patel, T.B., Patil, H.A.: Combining evidences from MEL cepstral, cochlear filter cepstral and instantaneous frequency features for detection of natural vs. spoofed speech. In: Sixteenth Annual Conference of the International Speech Communication Association (2015)
12. Ravanelli, M., Bengio, Y.: Speaker recognition from raw waveform with SincNet. In: 2018 IEEE Spoken Language Technology Workshop (SLT), pp. 1021–1028. IEEE (2018)
13. Schneider, S., Baevski, A., Collobert, R., Auli, M.: wav2vec: unsupervised pre-training for speech recognition. In: Proceedings of Interspeech 2019, pp. 3465–3469 (2019)
14. Snyder, D., Chen, G., Povey, D.: MUSAN: a music, speech, and noise corpus. arXiv preprint arXiv:1510.08484 (2015)
15. Tak, H., Patino, J., Todisco, M., Nautsch, A., Evans, N., Larcher, A.: End-to-end anti-spoofing with RawNet2. In: IEEE International Conference on Acoustics, Speech and Signal Processing (ICASSP), pp. 6369–6373 (2021)
16. Wang, Y., et al.: Tacotron: towards end-to-end speech synthesis. In: Proceedings of Interspeech 2017, pp. 4006–4010 (2017)
17. Wang, Y., et al.: Style tokens: unsupervised style modeling, control and transfer in end-to-end speech synthesis. In: Dy, J., Krause, A. (eds.) Proceedings of the 35th International Conference on Machine Learning. Proceedings of Machine Learning Research, vol. 80, pp. 5180–5189. PMLR (2018)
18. Wu, H., et al.: Partially fake audio detection by self-attention-based fake span discovery. In: ICASSP 2022-2022 IEEE International Conference on Acoustics, Speech and Signal Processing (ICASSP), pp. 9236–9240. IEEE (2022)
19. Yi, J., et al.: Half-truth: a partially fake audio detection dataset. arXiv preprint arXiv:2104.03617 (2021)
20. Zeinali, H., Stafylakis, T.: Athanasopoulou: detecting spoofing attacks using VGG and SincNet: BUT-Omilia submission to ASVspoof 2019 challenge. In: Proceedings of Interspeech 2019, pp. 1073–1077 (2019)

Leveraging Synthetic Speech for CIF-Based Customized Keyword Spotting

Shuiyun Liu, Ao Zhang, Kaixun Huang, and Lei Xie[✉]

Audio, Speech and Language Processing Group (ASLP@NPU),
School of Computer Science, Northwestern Polytechnical University, Xi'an, China
lxie@nwpu.edu.cn

Abstract. Customized keyword spotting aims to detect user-defined keywords from continuous speech, providing flexibility and personalization. Previous research mainly relied on similarity calculations between keyword text and acoustic features. However, due to the gap between the two modalities, it is challenging to obtain alignment information and model their correlation. In our paper, we propose a novel method to address these issues. Firstly, we introduce a text-to-speech (TTS) module to generate the audio of keywords, effectively addressing the cross-modal challenge of text-based customized keyword spotting. Furthermore, we employ the Continuous Integrate-and-Fire (CIF) mechanism for boundary prediction to get token-level acoustic representations of keywords thus solving the keyword and speech alignment problem. Our experimental results on the Aishell-1 dataset demonstrate the effectiveness of our proposed method. It significantly outperforms both the baseline method and the Dynamic Sequence Partitioning (DSP) method in terms of keyword spotting accuracy. Compared with the DSP method, our model can achieve a significant improvement in the relative wake-up rate of 72.7% when the false accept rate is fixed at 0.02. And our model represents a 64% improvement over the baseline model.

Keywords: Keyword spotting · Speech synthesis · Continuous Integrate-and-Fire

1 Introduction

Keyword Spotting (KWS) refers to the detection of pre-defined keywords within audio segments. This technology finds widespread application in various edge devices, exemplified by Apple's "Hey Siri" and Xiaomi's "Xiao Ai" mobile assistant. Although this type of KWS system, which spots specific keywords, can achieve high accuracy, it requires a large amount of training data that includes the specific keywords, and keywords cannot be changed after training. This necessitates users to remember specific keywords, which is not user-friendly. Therefore, to deal with such limitations, customized KWS, which allows users to customize keywords, has gained popularity in recent years.

J. Jia et al. (Eds.): NCMMSC 2023, CCIS 2006, pp. 354–365, 2024.
https://doi.org/10.1007/978-981-97-0601-3_31

In customized KWS tasks, there have been numerous studies. Some previous studies utilize query-by-example (QbyE) methods, which rely solely on audio signals as input [17,26]. In QbyE methods, reference keyword speech is enrolled and compared with new input speech queries [3,7,8]. However, these methods require users to record the speech of keywords during customization, and their performance is affected by the quality of the reference keyword speech. In contrast, customizing keywords based on text simplifies the process of setting keywords and performs stable across different speakers. Common text-based customized KWS methods typically rely on ASR acoustic models to transform speech data into posteriorgrams, followed by post-processing techniques like HMMs or various search and decoding approaches to detect keywords [1,11,29]. Extensive experiments have shown that such a two-stage approach increases the likelihood of false alarms [13,23], and its training using ASR loss did not completely consistent with the KWS objectives.

Hence, it is essential to design an end-to-end KWS system. Subsequent research studies [10,21,28] begin to detect keywords directly by correlation between keyword text and input speech. To address the issue of modal mismatch, Wei, B et al. [28], Shin et al. [21] utilize cross-attention mechanism to model the agreement between keyword text and speech and thus perform keyword detection. However, Nishu, K. et al. point out that the text encoder does not adequately model the textual information to solve the cross-modal problem and the method of attention lacks focus on alignment, resulting in suboptimal performance. They propose two solutions, one is an audio-compliant text encoder and the other is the dynamic sequence partitioning (DSP) method. The audio-compliant text encoder [15] transforms provided phonemes into vectors and acquires text embeddings through vector concatenation, alleviating the issues of cross-modal. The DSP [16] method has introduced a dynamic sequence partitioning algorithm to align audio with text, allowing for the discovery of optimal audio sequence boundaries, which results in obtaining token-level acoustic representations. Thereby calculating the distance between text representation and this acoustic representation to detect target keywords. However, the construction of the text encoder is cumbersome, and it lacks generalizability. Additionally, the DSP method exhibits poor computational robustness.

To address these challenges, we propose a new customized KWS model designed to mitigate these issues. Our model utilizes a Text-to-Speech (TTS) module to generate corresponding audio based on keyword text. The real audio and synthesized audio are then passed through an acoustic encoder to obtain acoustic embeddings. Subsequently, we employ the Continuous Integrate-and-Fire (CIF) mechanism [5] to derive token-level acoustic representations and calculate the distances between these representations.

2   Related Work

2.1   Text-to-Speech

In recent years, there have been substantial advancements in the training of ASR and KWS models using synthetic speech data and the use of synthetic data for data augmentation has been widely adopted [14,19,24,27]. In the domain adaptation domain, given the lack of paired text-audio data in the target domain, how to achieve domain adaptation using only textual data gains much attention. One of the challenges lies in the modal differences between text and speech. The use of Text-to-Speech (TTS) synthesized data is an important method to overcome these modal differences [22].

Taking inspiration from this, we employ synthetic data to address the cross-modal challenges in the customized KWS task. By extracting keyword embeddings from synthetic keyword audio, we bridge the gap between the text and audio modalities.

2.2   Continuous Integrate-and-Fire

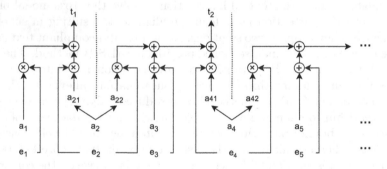

Fig. 1. Illustration of the calculation of CIF on an encoder outputs embedding $E = \{e_1, e_2, e_3, e_4, e_5, ...\}$ with predicted weights $a = \{0.3, 0.8, 0.5, 0.6, 0.2, ...\}$. The integrated embedding $t_1 = 0.3 * e_1 + 0.7 * e_2$, $t_2 = 0.1 * e_2 + 0.5 * e_3 + 0.4 * e_4$.

In order to focus on alignment, DSP calculates token-level acoustic representations and computes similarity based on these token-level representations. Here, we adopt the CIF model to obtain token-level acoustic representations. CIF [5], is an alignment mechanism applied between the encoder and decoder, providing word boundaries. It has already found successful applications in ASR models. The encoder takes acoustic features $X = \{x_1, x_2, \ldots, x_t, \ldots, x_T\}$ and produces encoder outputs $E = \{e_1, e_2, \ldots, e_u, \ldots, e_U\}$. Figure 1 illustrates the CIF implementation process, the CIF module receives encoder outputs E and predicts frame-level generated weights $a = \{a_1, a_2, \ldots, a_u, \ldots, a_U\}$. Subsequently,

CIF accumulates these weights along the time axis and during the accumulation process, a threshold of $\beta = 1$ serves as an activation indicator. When the cumulative value reaches this threshold, it indicates the acoustic boundary corresponding to a specific target label. At boundary points, the weight a_u is partitioned into two segments: one for immediate token weight accumulation until the threshold is met, and the other reserved for accumulating weights for subsequent tokens. Then, CIF integrates the relevant encoded outputs in a weighted sum to obtain the acoustic embedding $T = \{t_1, t_2, \ldots, t_i, \ldots, t_L\}$. The decoder receives the acoustic embedding T and generates the final sequence $S = \{s_1, s_2, \ldots, s_i, \ldots, s_L\}$. In CIF, the frame-level weights effectively align the acoustic representation with output tokens and obtain token boundaries upon reaching the threshold. This capability empowers us to explore the feasibility of utilizing CIF for natural alignment finding useful information in KWS.

In the context of KWS tasks, our emphasis is on capturing the information pertaining to keywords. This aligns with the idea of the DSP method, which derives word boundaries and obtains token-level acoustic embeddings by aligning text and audio sequences. Through the integration of the CIF mechanism, we improve our ability to predict alignment between text and audio sequences of varying lengths.

3   Method

In this section, we describe our proposed method, which consists of two components, a TTS generator and a CIF-based end-to-end customized keyword detector, as shown in Fig. 2. It primarily consists of four modules: a TTS generator, a shared encoder, a pattern extractor, and a verifier.

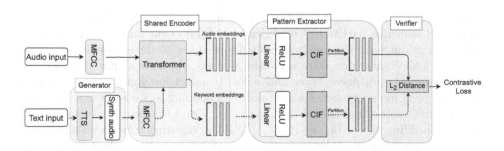

Fig. 2. Overview of our proposed model

3.1   TTS Generator

The TTS generator consists of three main components: the front-end processing, the acoustic model, and the vocoder. During the front-end processing, the provided transcripts are converted into phonemes, and Mel-spectrograms are extracted from real audio data. The acoustic model, inspired by DelightfulTTS 2 [12], incorporates changes into the FastSpeech model [18]. DelightfulTTS 2 has replaced the original transformer blocks with conformer blocks, allowing the model to better focus on local information, thereby enhancing the quality of synthesized Mel-spectrograms. Furthermore, we introduce speaker information modeling by utilizing a speaker lookup table, enabling multi-speaker speech synthesis. To reconstruct high-quality speech waveforms from the Mel-spectrograms generated by the acoustic model, we employed the HiFi-GAN [9] as the vocoder. HiFi-GAN is a state-of-the-art neural vocoder based on generative adversarial networks (GANs) to generate high-quality speech. During training, we randomly use speaker embeddings that vary to improve the robustness of our model.

3.2 Shared Encoder

The inclusion of synthesized audio addresses the previous issue of text and audio modality mismatch, allowing us to simplify our network architecture as both audio types can share a common encoder. We utilize a transformer [25] architecture as the shared encoder, incorporating self-attention layers to capture both global and local audio context. For the encoder network, we use 8 transformer self-attention blocks.

More specifically, we train the shared encoder using the CTC loss [6]. When dealing with input audio denoted as $\hat{a}$, we represent the output embeddings of the shared encoder as $\hat{a} = \{\hat{a}_1, \hat{a}_2, \dots, \hat{a}_m\}$. Similarly, for synthesized audio $\hat{s}$, we obtain the embedding output as $\hat{s} = \{\hat{s}_1, \hat{s}_2, \dots, \hat{s}_n\}$.

3.3 Pattern Extractor

The pattern extractor consists of three components: linear projection, CIF alignment and sequence partition. To project real audio embeddings and synthetic audio embeddings into a common embedding space of dimensionality d, both types of audio embeddings are separately passed through a simple projection layer composed of a linear layer and a ReLU activation function. For a given audio embedding $\hat{a}$ and synthesized audio embedding $\hat{s}$, we represent the projected audio embedding as $a = \{a_1, a_2, \dots, a_m\}$ and the projected synthesized audio embedding as $s = \{s_1, s_2, \dots, s_n\}$. CIF accomplishes token-level alignment and generates token-level embeddings. For a given projected audio embedding a and synthesized audio embedding s, we denote the CIF-aligned audio embedding and the CIF-aligned synthesized audio embedding as $h_a = \{h_{a1}, h_{a2}, \dots, h_{al}\}$ and $h_s = \{h_{s1}, h_{s2}, \dots, h_{sl}\}$ respectively. Since the CIF acoustic sequence T aligns rigorously with the text sequence during training, we employ the Mean Absolute Error (MAE) to guide the learning. Ultimately, through sequence partitioning, we isolate the subsequence containing the keyword information of primary interest. This subsequence is crucial for accurate keyword detection.

3.4   Verifier

The verifier is responsible for discerning whether the audio input and the synthesized audio input correspond to the same keyword (positive sample) or differ in their keywords (negative sample). This determination relies on the calculation of the L_2 distance $z(H_a, H_s)$ between the two embeddings, which is expressed by the following formula:

$$z(H_a, H_s) = \|\mathbf{H_a} - \mathbf{H_s}\|_2 \tag{1}$$

To achieve the goal of minimizing the L_2 distance between audio embeddings and synthesized audio embeddings in positive sample pairs and maximizing this distance in negative sample pairs, we employ contrastive learning loss as our training objective.

3.5   Training Process

In this subsection, we present the complete workflow of our entire model. Firstly, we acquire a trained TTS generator. Using this TTS generator, we can synthesize audio from multi-speakers. Then, both the real audio and synthesized audio are fed into the shared encoder to obtain acoustic embeddings. For extractor training, we adopt the keyword generation method proposed by Wei, B et al. [28]. This method allows us to create both positive and negative samples for training. Specifically, we randomly select consecutive sub-samples from the speech transcripts as keywords to generate positive samples, and retain the start and end labels. Simultaneously, we randomly select word combinations from the lexicon that are absent in the speech transcripts and do not exceed the transcript's word count to generate negative samples. The quantities of positive and negative samples remain consistent. Taking the real audio processing flow as an example, the pattern extractor takes audio embeddings, as well as the start and end labels as input. For positive samples, the audio embeddings are transformed into token-level acoustic embeddings through the CIF module and are partitioned based on the start and end labels. For negative samples, as there is no relevant label information, we randomly select a segment not exceeding the token length for partitioning. The processing flow for synthetic audio is the same. Finally, we calculate the similarity between the two embeddings to determine whether they match.

4   Experiment

4.1   Dataset

We train a multi-speaker voice synthesis system using the Aishell-3 dataset [20] and use the training Aishell-1 dataset [2] for training the KWS model.

For TTS inference process, we utilize transcripts from the Aishell-1 dataset and synthesized audio with nearly a hundred distinct vocal characteristics. This greatly imporves the diversity of voice characteristics, simulates real-world scenarios, and improves robustness. In downstream tasks, we randomly select audio samples with different vocal characteristics and fed them into the shared encoder.

4.2   Model Details

The input audio signal is represented using 40-dimensional Mel-Frequency Cepstrum Coefficient (MFCC). These features are computed with a 25ms window
and a 10ms frame shift.

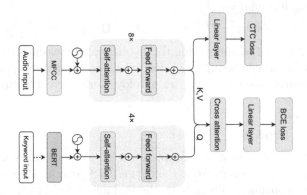

Fig. 3. Vanilla E2E KWS [28]

Baseline Training. We also follow the approach of Bo Wei et al. [28], using the
Vanilla E2E KWS model as our baseline model, as shown in Fig. 3. This model
employs a cross-attention mechanism, where speech embeddings are extracted by
the acoustic model and fed to the attention layer as K and V matrices. Keyword
embeddings are extracted by the keyword encoder and fed to the attention layer
as the Q matrix. The first vector from the attention output serves as the classification vector, which, after passing through a linear layer, yields the probability
of keyword presence in the speech utterance.

DSP Method Training. We replicate the DSP method, and for the text
encoder, we utilized the BERT model [4] to extract textual information. Subsequently, the textual information has been fed into a transformer architecture to
obtain text embeddings.

4.3   Evaluation

In the inference phase, we evaluate the model using the same keyword generation
method as during training. We assess performance based on Receiver Operating
Characteristic (ROC) curves, which are constructed by scanning thresholds over
all possible confidence values and plotting the False Reject (FR) rate against the
False Accept (FA) rate.

Similarly, we also evaluate all models based on wws score, which is the sum
of the False Alarm Rate (FAR) and the False Reject Rate (FRR). This metric

mitigates the possibility of overly optimistic evaluations stemming from highly imbalanced class distributions, with FAR and FRR defined as follows

$$FAR = \frac{FP}{FP + TN} \tag{2}$$

$$FRR = \frac{FN}{FN + TP} \tag{3}$$

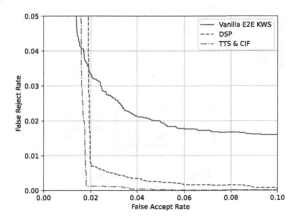

Fig. 4. ROC curves of three KWS methods on Aishell-1 dataset

4.4 Results

Table 1. Performance of our experimental models

Model Architectures	FAR ↓	FRR ↓	SCORE ↓
Vanilla E2E KWS	0.021	0.032	0.053
DSP	0.020	0.007	0.027
TTS & CIF [ours]	**0.018**	**0.001**	**0.019**
Only TTS	0.017	0.021	0.038
Only CIF	0.014	0.009	0.023

Comparison with Baseline. Figure 4 shows the ROC curves comparing the performance of all proposed models on the Aishell-1 dataset. It can be observed that the model combining the TTS generator and CIF soft alignment achieves better results, with a reduction in the Area Under the Curve (AUC). Compared to the DSP method, our proposed method further reduces both the False Reject Rate (FRR) and False Accept Rate (FAR). Our method improves performance when FAR is higher than 0.02. From Table 1, it is evident that our model achieved a score of 0.019, representing a 64% improvement over the baseline model. Furthermore, when compared to the DSP method, our approach improves FRR performance.

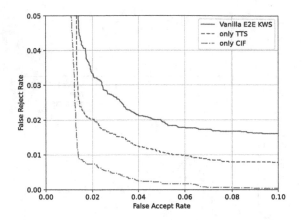

Fig. 5. Ablation result

Ablation Study. To investigate the impact of different components in our proposed model on KWS performance, we conduct ablation experiments, as illustrated in Fig. 5. It is evident that both the sole use of TTS-synthesized audio and the introduction of the CIF alignment mechanism contribute to significant performance improvements in the model. In the *only CIF* experiment, our implementation details closely resemble those of the DSP method. For the text encoder, we utilize BERT to extract textual information. Table 1 also showcases the results of our ablation experiments. While our approach shows a marginal decrease in FAR performance compared to the DSP method, there is a substantial enhancement in FRR performance. In the aggregate, we achieve a score of 0.23, reflecting a significant 14% improvement over the DSP results.

5 Conclusions

In this paper, we propose a novel end-to-end customized keyword recognition model that unites TTS generation and soft monotonic alignment. This includes addressing the mismatch between text and audio embeddings by using TTS-synthesized audio. Additionally, we introduce the CIF mechanism, which accurately predicts word boundaries to obtain token-level acoustic embeddings. The experiments demonstrate that our results achieve the state-of-the-art performance. We also set up additional experiments to explore the effectiveness of the two components.

References

1. Bluche, T., Primet, M., Gisselbrecht, T.: Small-footprint open-vocabulary keyword spotting with quantized LSTM networks. CoRR abs/2002.10851 (2020)
2. Bu, H., Du, J., Na, X., Wu, B., Zheng, H.: AISHELL-1: an open-source mandarin speech corpus and a speech recognition baseline. In: 20th Conference of the Oriental Chapter of the International Coordinating Committee on Speech Databases and Speech I/O Systems and Assessment, O-COCOSDA 2017, Seoul, South Korea, 1–3 November 2017, pp. 1–5. IEEE (2017)
3. Chen, G., Parada, C., Sainath, T.N.: Query-by-example keyword spotting using long short-term memory networks. In: 2015 IEEE International Conference on Acoustics, Speech and Signal Processing, ICASSP 2015, South Brisbane, Queensland, Australia, 19–24 April 2015, pp. 5236–5240. IEEE (2015)
4. Devlin, J., Chang, M., Lee, K., Toutanova, K.: BERT: pre-training of deep bidirectional transformers for language understanding. In: Burstein, J., Doran, C., Solorio, T. (eds.) Proceedings of the 2019 Conference of the North American Chapter of the Association for Computational Linguistics: Human Language Technologies, NAACL-HLT 2019, Minneapolis, MN, USA, 2–7 June 2019, Volume 1 (Long and Short Papers), pp. 4171–4186. Association for Computational Linguistics (2019)
5. Dong, L., Xu, B.: CIF: continuous integrate-and-fire for end-to-end speech recognition. In: 2020 IEEE International Conference on Acoustics, Speech and Signal Processing, ICASSP 2020, Barcelona, Spain, 4–8 May 2020, pp. 6079–6083. IEEE (2020)
6. Graves, A., Fernández, S., Gomez, F.J., Schmidhuber, J.: Connectionist temporal classification: labelling unsegmented sequence data with recurrent neural networks. In: Cohen, W.W., Moore, A.W. (eds.) Machine Learning, Proceedings of the Twenty-Third International Conference (ICML 2006), Pittsburgh, Pennsylvania, USA, 25–29 June 2006. ACM International Conference Proceeding Series, vol. 148, pp. 369–376. ACM (2006)
7. Hou, J., Xie, L., Fu, Z.: Investigating neural network based query-by-example keyword spotting approach for personalized wake-up word detection in mandarin Chinese. In: 10th International Symposium on Chinese Spoken Language Processing, ISCSLP 2016, Tianjin, China, 17–20 October 2016, pp. 1–5. IEEE (2016)
8. Kim, B., Lee, M., Lee, J., Kim, Y., Hwang, K.: Query-by-example on-device keyword spotting. In: IEEE Automatic Speech Recognition and Understanding Workshop, ASRU 2019, Singapore, 14–18 December 2019, pp. 532–538. IEEE (2019)
9. Kong, J., Kim, J., Bae, J.: HiFi-GAN: generative adversarial networks for efficient and high fidelity speech synthesis. In: Larochelle, H., Ranzato, M., Hadsell, R., Balcan, M., Lin, H. (eds.) Advances in Neural Information Processing Systems 33: Annual Conference on Neural Information Processing Systems 2020, NeurIPS 2020, 6–12 December 2020, virtual (2020)
10. Lee, Y., Cho, N.: Phonmatchnet: phoneme-guided zero-shot keyword spotting for user-defined keywords, pp. 3964–3968 (2023)
11. Lengerich, C.T., Hannun, A.Y.: An end-to-end architecture for keyword spotting and voice activity detection (2016)
12. Liu, Y., Xue, R., He, L., Tan, X., Zhao, S.: Delightfultts 2: end-to-end speech synthesis with adversarial vector-quantized auto-encoders. In: Ko, H., Hansen, J.H.L. (eds.) Interspeech 2022, 23rd Annual Conference of the International Speech Communication Association, Incheon, Korea, 18–22 September 2022, pp. 1581–1585. ISCA (2022)

13. Liu, Z., Li, T., Zhang, P.: RNN-T based open-vocabulary keyword spotting in mandarin with multi-level detection. In: IEEE International Conference on Acoustics, Speech and Signal Processing, ICASSP 2021, Toronto, ON, Canada, 6–11 June 2021, pp. 5649–5653. IEEE (2021)

14. Mimura, M., Ueno, S., Inaguma, H., Sakai, S., Kawahara, T.: Leveraging sequence-to-sequence speech synthesis for enhancing acoustic-to-word speech recognition. In: 2018 IEEE Spoken Language Technology Workshop, SLT 2018, Athens, Greece, 18–21 December 2018, pp. 477–484. IEEE (2018)

15. Nishu, K., Cho, M., Dixon, P., Naik, D.: Flexible keyword spotting based on homogeneous audio-text embedding. CoRR abs/2308.06472 (2023)

16. Nishu, K., Cho, M., Naik, D.: Matching latent encoding for audio-text based keyword spotting, pp. 1613–1617 (2023)

17. R, K., Kurmi, V.K., Namboodiri, V.P., Jawahar, C.V.: Generalized keyword spotting using ASR embeddings. In: Ko, H., Hansen, J.H.L. (eds.) Interspeech 2022, 23rd Annual Conference of the International Speech Communication Association, Incheon, Korea, 18–22 September 2022, pp. 126–130. ISCA (2022)

18. Ren, Y., Hu, C., Tan, X., Qin, T., Zhao, S., Zhao, Z., Liu, T.: Fastspeech 2: fast and high-quality end-to-end text to speech. In: 9th International Conference on Learning Representations, ICLR 2021, Virtual Event, Austria, 3–7 May 2021. OpenReview.net (2021)

19. Rosenberg, A., et al.: Speech recognition with augmented synthesized speech. In: IEEE Automatic Speech Recognition and Understanding Workshop, ASRU 2019, Singapore, 14–18 December 2019, pp. 996–1002. IEEE (2019)

20. Shi, Y., Bu, H., Xu, X., Zhang, S., Li, M.: AISHELL-3: a multi-speaker mandarin TTS corpus and the baselines. CoRR abs/2010.11567 (2020)

21. Shin, H., Han, H., Kim, D., Chung, S., Kang, H.: Learning audio-text agreement for open-vocabulary keyword spotting. In: Ko, H., Hansen, J.H.L. (eds.) Interspeech 2022, 23rd Annual Conference of the International Speech Communication Association, Incheon, Korea, 18–22 September 2022, pp. 1871–1875. ISCA (2022)

22. Sim, K.C., et al.: Personalization of end-to-end speech recognition on mobile devices for named entities. In: IEEE Automatic Speech Recognition and Understanding Workshop, ASRU 2019, Singapore, 14–18 December 2019, pp. 23–30. IEEE (2019)

23. Tian, Y., Yao, H., Cai, M., Liu, Y., Ma, Z.: Improving RNN transducer modeling for small-footprint keyword spotting. In: IEEE International Conference on Acoustics, Speech and Signal Processing, ICASSP 2021, Toronto, ON, Canada, 6–11 June 2021, pp. 5624–5628. IEEE (2021)

24. Ueno, S., Mimura, M., Sakai, S., Kawahara, T.: Multi-speaker sequence-to-sequence speech synthesis for data augmentation in acoustic-to-word speech recognition. In: IEEE International Conference on Acoustics, Speech and Signal Processing, ICASSP 2019, Brighton, United Kingdom, 12–17 May 2019, pp. 6161–6165. IEEE (2019)

25. Vaswani, A., et al.: Attention is all you need. In: Guyon, I., et al. (eds.) Advances in Neural Information Processing Systems 30: Annual Conference on Neural Information Processing Systems 2017, 4–9 December 2017, Long Beach, CA, USA, pp. 5998–6008 (2017)

26. Vygon, R., Mikhaylovskiy, N.: Learning efficient representations for keyword spotting with triplet loss. In: Karpov, A., Potapova, R. (eds.) SPECOM 2021. LNCS, vol. 12997, pp. 773–785. Springer, Cham (2021). https://doi.org/10.1007/978-3-030-87802-3_69

27. Wang, H., Jia, Y., Zhao, Z., Wang, X., Wang, J., Li, M.: Generating TTS based adversarial samples for training wake-up word detection systems against confusing words. In: Zheng, T.F. (ed.) Odyssey 2022: The Speaker and Language Recognition Workshop, 28 June–1 July 2022, Beijing, China, pp. 402–406. ISCA (2022)
28. Wei, B., et al.: End-to-end transformer-based open-vocabulary keyword spotting with location-guided local attention. In: Hermansky, H., Cernocký, H., Burget, L., Lamel, L., Scharenborg, O., Motlícek, P. (eds.) Interspeech 2021, 22nd Annual Conference of the International Speech Communication Association, Brno, Czechia, 30 August–3 September 2021, pp. 361–365. ISCA (2021)
29. Zhuang, Y., Chang, X., Qian, Y., Yu, K.: Unrestricted vocabulary keyword spotting using LSTM-CTC. In: Morgan, N. (ed.) Interspeech 2016, 17th Annual Conference of the International Speech Communication Association, San Francisco, CA, USA, 8–12 September 2016, pp. 938–942. ISCA (2016)

Author Index

J. Jia et al. (Eds.): NCMMSC 2023, CCIS 2006, pp. 367–368, 2024.
https://doi.org/10.1007/978-981-97-0601-3

Printed in the United States
by Baker & Taylor Publisher Services